The Earthscan reader in

Tropical forestry

Edited by
Simon Rietbergen

EARTHSCAN
Earthscan Publications Ltd, London

First published in the UK in 1993 by
Earthscan Publications Limited
120 Pentonville Road, London N1 9JN

A catalogue record for this book is available from the British Library

ISBN: 1 85383 127 1

Typeset by Castlefield Press, Kettering, Northants.
Printed in England by Clays Ltd, St Ives plc

Earthscan Publications Limited is an editorially independent subsidiary of Kogan Page Limited and publishes in association with the International Institute for Environment and Development and the World Wide Fund for Nature.

Contents

Contents

List of illustrations

Tables

Figures

Preface

Tropical forests are at the centre of everyone's attention, and so is tropical forestry, the science and practice concerned with managing these forests. Tropical deforestation regularly makes newspaper headlines and no self-respecting magazine has passed the chance to run lavishly illustrated features on the subject. All this attention is a recent phenomenon, however: for most of the time since the start of professional tropical forestry by the German forester Brandis in the teak forests of South Asia in the 1860s tropical forests have been a preoccupation of a handful of people working in remote areas. This relative isolation has left the forestry profession ill-prepared for the contemporary debates concerning tropical forests, on issues such as development aid, timber trade, biological diversity conservation and, more recently, climate change.

If the first controversies regarding tropical forests raged between scientists and practitioners, and were therefore inaccessible to the public at large, tropical forests have since entered the political arena, and discussions have become more public and publicized, if not more transparent. Many participants in the debate have political and/or pecuniary axes to grind, and do so with great skill. The resulting contradictions are a confusing mixture of good and bad science and ideology.

The readings selected for this book, in covering a wide range of scientific disciplines and opinions concerning tropical forests, are intended to make the debate more transparent for interested lay people. The readings are divided into six sections: policy and management; land use and land degradation; international trade; forestry development aid; indigenous peoples; and biological diversity conservation. These represent the main areas of controversy.

The overview and introduction attempt to pull together the main conclusions from the selected readings and other key sources, and to reduce the tropical forests debate to more understandable proportions. They also contain sections highlighting issues on which no overview papers were included in the readings: such as the Tropical Forestry Action Plan and the relation between tropical forests and climate change. The readings in this volume cannot do justice to the vast literature on tropical forests. Guidance for further reading is contained in the introduction and additional references are provided at the end of each article.

Acknowledgements

The publishers would like to thank the authors and copyright holders of the following chapters for permission to reprint them in this volume.

Ch 1 from *No Timber Without Trees: sustainability in the tropical forest*, copyright 1989 International Tropical Timber Organization; ch 2 from *Public Policies and the Misuse of Forest Resources*, Cambridge University Press 1988, copyright 1988 World Resources Institute; ch 3 from *Conservation in Africa: people, policies and practice*, copyright 1987 Cambridge University Press; ch 4 by kind permission of the authors; ch 5 from *Land Degradation and Society*, copyright 1987 Methuen Press; ch 6 from *Alternatives to deforestation: steps towards sustainable use of the Amazon rain forest*, copyright 1990 Columbia University Press; ch 7 by kind permission of the author; ch 8 copyright 1991 UK Forestry Commission; ch 9 from *A reappraisal of forestry development in developing countries*, copyright 1983 Nijhoff/Junk Publishers; ch 10 from *Beyond the Woodfuel Crisis: people, land and trees in Africa*, copyright 1988 Gerald Leach and Robin Mearns; ch 11 from *Tribal peoples and development issues*, John Bodley (ed), copyright 1988 Mayfield Publishing Company; ch 12 from Impact of Science on Society 30, No. 3, copyright 1980 UNESCO; ch 13 by kind permission of Nels Johnson.

I would like to thank Jennifer Rietbergen-McCracken for editorial help, inspiration and discipline. Thanks are also due to Raymond Noronha for taking time to comment on a draft version of the introduction; to Jonathan Sinclair Wilson and Jo O'Driscoll, both at Earthscan Publications, for the original idea and limitless patience respectively; and to the ex-colleagues at the International Institute for Environment and Development and colleagues at the World Bank, and many other people who have wittingly or unwittingly contributed to this book by generously sharing their ideas.

Overview

Tropical forests are diverse environments, both biologically and culturally. Rural populations in many developing countries depend on tropical forest resources for part of their livelihoods, and the poorer people among them do so in particular. Superimposed on these local uses are national and international demands for forest produce and services such as watershed protection, timber and non-timber products, biological diversity and carbon sequestration. The many different uses of tropical forests are not always fully compatible: choices have to be made, and different interests reconciled for forest management to be sustainable. Rational decisions regarding forest land use require reliable facts and figures on the impact of human interventions on tropical forest ecosystems. Presently available data need to be treated with caution: the uncertainty caused by the complexity of tropical forest environments is compounded by poor science and conflicting perceptions and interests.

FORESTS: NATURE AND CULTURE

What are tropical forests? The word 'tropical' refers to the area between the tropics of Cancer and Capricorn.[1] A forest, to most present-day Europeans and North Americans, is an area where trees form a more or less continuous canopy. In the Food and Agricultural Organization's (FAO's) tropical forest assessment, however, forests were defined as vegetation types where trees cover more than 10 per cent of the land area. This definition includes vegetation types that most would not call forests, but rather parklands or open woodlands. Worldwide, two-fifths of all tropical forests consist of such open woodlands; in the African tropics two-thirds of the forests are open woodlands.[2] Many of these woodlands are intermittently farmed and fallowed (as are a considerable proportion of the closed forests) by the people living in or around them, and grazed and browsed by the animals they possess. This state of affairs has important implications for the way we look at tropical forests, and the terms we use to describe them.

First of all, what do we call forests that are farmed or grazed, or both? In present-day western Europe, with its clear boundaries between farm and forest, terms no longer exist for such temporary forest fields or pasture; in this introduction they will be referred to as swiddens and woodland pasture respectively.[3] Secondly, many tropical forests are cultural rather than natural entities: even if their tree cover consists only of indigenous species that have grown spontaneously (unlike European forests, which consist mainly of planted trees, whether exotic or indigenous), they are strongly influenced by the farming and herding practices of the local inhabitants, past and present. A useful distinction often made is that between primary and secondary forests: primary forests consist of virgin forest essentially unmodified by human activity or only slightly modified by hunting and gathering practices of indigenous peoples; secondary forests have been subjected to farming or timber harvesting, but again still support a forest cover of solely or mainly indigenous species. The terms primary and secondary forest are simplifications: in reality they grade almost imperceptibly into one another.[4]

The media often equates tropical forests with tropical rainforests. But there are many other kinds of tropical forest for example seasonal tropical forests, tropical dry forests, woodlands, riverine forests, mangroves, freshwater swamp forests and cloud forests.[5] Some of these, such as seasonal tropical forests which have been settled intensively because of their more hospitable climate and fertile soils, are more threatened than tropical rainforests; others, such as the slow-growing cloud forests in tropical mountainous areas, are more fragile.[6] Distinctions between these types of forest are based on broad differences in climate (rainfall, temperature) and soil, but even within each forest type there is enormous variation. A term such as tropical rainforest is only a first approximation, a broad category covering many distinct forest types according to more subtle differences in climate, soil, constituent species etc. Furthermore, forest ecosystems consist not only of vegetation, but also of other plant and animal life. Large animals such as elephants influence the forest vegetation in obvious ways, but even small animals play a crucial role (eg specialized pollinators such as bats and insects) and are therefore an integral part of forest ecosystems.

An additional complication is that both the natural and the social factors shaping tropical forests are not static, but changing over time. The impact of societal changes on tropical forests is now almost daily fare in the media, but naturally occurring changes get much less attention.[7] However, many areas now covered in dense forests supported drier vegetation types (grasslands and open woodlands) during the last ice age, and were only relatively recently recolonized by moist forest vegetation.[8] In conclusion, the 'tropical forest' does not exist. What looks from the air to be an endless and uniform green carpet is in reality a complex patchwork of different forest ecosystems overlaid with varying types and intensities of human use, past and present.

Tropical forests and society

Interactions between forests and society in the tropics are fundamentally different from those in the developed world.[9] Even if forests in the latter provide timber, conserve soil and water, and constitute a favoured environment for open air recreation, they are a direct source of daily subsistence for only a small minority of the population, such as workers in forest industries and forest-related tourism. In the tropics a much larger proportion of people live off the land, and use a variety of forest and tree products on a daily basis, such as woodfuels, timber for construction and household implements, foods of vegetable and animal origin, browse and fodder, and medicine. In many cases their farming systems benefit from soil and water conservation services provided by forests in their vicinity. Indeed, their material and spiritual culture is provided and inspired by forests and trees to a large extent.[10] Superimposed on these local uses are demands for produce and services valued at national and international level, including watershed protection, industrial timber, traded non-wood products such as rattans and bushmeat, potentially useful genetic material, and carbon sequestration to mitigate global warming. This variety of forest products and services often benefits different groups of people, who may live close to, or far removed from, the forest, or who even belong to as yet unborn generations. As a consequence tropical forests, much more than their sub-tropical, temperate and boreal counterparts, have been increasingly viewed as part of an international environmental context.[11] So-called 'dread claims', dire predictions about global change made by green activists, were integral to the ascension of the environmental movement. Two out of the three components of this global change – the greenhouse effect and the loss of biodiversity – have been strongly linked to the degradation and disappearance of tropical forests.[12]

Forest management

In the broadest sense, forest management can be defined as taking a firm decision about the future of any area of forest, applying it, and monitoring the application.[13] In turn, forestry can be defined as the science and practice involved with the management of forests. Management can be for a single use, if its objective is incompatible with any other use – as is the case with strict ecosystem preservation – or multiple use, where a compromise is struck between a number of objectives.[14] Originally, management was mainly in the form of localized common property arrangements; with the emergence of nation states, management authority has been vested increasingly in public agencies at various levels of government.[15] In the developed world, professional foresters manage large areas of forests for both protection and production purposes. But in the tropics, where forest resources are extensive and professional foresters a scarce commodity, local people whose livelihoods depend on the forests are often the main frontline managers – even if this is rarely recognized officially.

In an ideal world, the fact that a variety of people value forests for different products and services could give rise to a peaceful co-existence of different management regimes. In the real world, however, especially where forest resources are scarce and many peoples' interests not adequately represented, there are those who win and those who lose: interventions in forest ecosystems which benefit some people may harm others. Forest management, therefore, has not only involved technical operations, but also political negotiations to reconcile divergent interests. The existence of important externalities provides a strong case for the involvement of some form of public authority in the management of forest resources.[16] So public agencies have been given the responsibility for establishing forest management regimes that yield optimal results to society as a whole while respecting individual resource use rights as much as possible. But achieving a 'balance of use' is not an easy task, even for the most well-intentioned authorities. Economically rational and socially fair decisions about forest management require not only political commitment, but also knowledge: about the various parties involved and their resource use and needs; the trade-offs and compatibilities between their interests; and the technical options available for managing forests for a variety of objectives. There is a need for fora where local, national and international interests can be represented and management agreements negotiated, to make sure that they are acceptable to most if not all parties affected. And, last but not least, decisions need to be applied, and implementation and monitoring capacity established in the public agencies concerned.

Sustainability and forestry

> If the seasons of husbandry be not interfered with, the grain will be more than can be eaten. If close nets are not allowed to enter the pools and ponds, the fishes and turtles will be more than can be consumed. If the axes and bills enter the hills and forests only at the proper time, the wood will be more than can be used. When the grain and fish and turtles are more than can be eaten, and there is more wood than can be used, this enables the people to nourish their living and mourn for their dead, without any feelings against any. (Mencius, China, 4th Century BC)[17]

As the above quote shows, the principle of sustainability has been known for quite some time. The concept of sustainable development was firmly put on the international agenda by the Brundtland Commission publication *Our Common Future* in 1987, but can be traced as far back as 1980, to the 'World Conservation Strategy: Living Resource Conservation for Sustainable Development', published by IUCN.[18] Sustainable development is defined by the Brundtland Commission as development that meets the needs of the present without compromising the ability of future generations to meet their needs. Implicit in this definition of sustainable development are three dimensions: social justice, environmental sustainability, and economic efficiency.

The preoccupation of the forestry profession with sustainability dates back much further: in Germany mathematical calculus was applied to determine sustainable timber yields more than 200 years ago.[19] But the concept of sustainability used by the forestry profession is much narrower in scope than the Brundtland one, referring mainly to the continued production of timber. Other forest benefits (soil and water conservation, flora and fauna preservation, recreation) are supposed to accrue automatically, in the wake of sustained timber yields. This paradigm effectively masks incompatibilities between management objectives and conflicts between forest users, and thus transfers forest policy choices out of the political into the technical arena. But with increased mechanization on the one hand and growing importance of nature conservation on the other, conflicts between timber production and other forest management objectives have multiplied, society's awareness of these conflicts has mushroomed, and the credibility of the forestry profession has suffered as a consequence.[20] In many countries, the term sustainability has served as an ideological decoy for governments wishing to appropriate forest resources and extinguish local people's customary rights to use them: the latter's forest management systems were labelled 'unsustainable' and in need of replacement by 'rational' practices.[21] The European experience with legislation reserving forests from the local people was mirrored in similar laws promulgated in African and Asian colonies, some of which survive in developing countries to this day. The use of the term sustainability in this context is far from an innocent statement for the common good, reflecting eternal human values such as altruism, equity and restraint, but rather thinly veiled politics.

Sustainability, therefore, is a term which should not be used loosely, but defined for each specific forest area and management objective: what product or service should be provided in perpetuity and what condition should the forest be maintained in? At the level of the forest stand, for the management objective of sustainable timber production, a simple definition should suffice: nothing is done to irreversibly reduce the potential of a forest site to produce marketable timber.[22] Similarly simple definitions can be given for protection and nature conservation objectives at that level. But for larger areas eg a nation's forest resources as a whole, a definition of what is sustainable management is a much more relative matter. For example, a country that has set aside a sizable share of its forests for watershed protection and nature conservation purposes can afford to manage its timber production forests more intensively without risking species extinction than another country which has opened most of its forests to logging. The latter country has to adhere to much more conservative management practices in its production forests, if it wants to achieve a similar degree of biodiversity conservation.

Facts and figures: uncertainty and complexity

Maybe unavoidably in an increasingly technocratic society, much of the debate about tropical forests is between scientists, or between people who base their arguments on facts and figures established by scientists. Facts and figures seem to be an indispensable ingredient in debates about tropical forests: dozens of football pitches deforested in an hour, thousands of species lost in a year, billions of tonnes of oil-equivalent worth of woodfuel burnt in a decade. But many of these facts and figures are highly suspect. Broadly speaking there are two reasons for this: first the limited relevance and poor quality of – as well as inherent limitations to – the science that has been brought to bear; and second the conflicting perceptions and interests of the different observers and parties involved.

Surprisingly little scientific research relating directly to the management of tropical moist forests has been done since the compilation of manuals on silviculture and management in Africa and South East Asia, from the late 1950s until the early 1970s. Since then there have been many descriptive, small-scale studies especially of untouched ecosystems, but large-scale analytical studies of, for instance, the impact of various silvicultural practices on tropical forest ecosystems have been few and far between.[23] Thorough analyses of the impact of human intervention on other tropical forest ecosystems seem to be similarly scarce.[24]

Tropical forest management has not only suffered from a lack of research, but also from poor quality science: vague or conflicting definitions; deficient measurements; faulty interpretation of data; poor establishment of causality; and the extrapolation of small-scale findings to universal phenomena. Some of these shortcomings will be illustrated by means of an examination of the science that has been applied to tropical deforestation. The wide disparity in published data concerning tropical deforestation is sufficient illustration in itself of the dangers involved in taking scientific data at face value, without asking how they were arrived at. But there are many other tropical forest-related fields with notoriously unreliable facts and figures: such as land degradation (Blaikie and Brookfield, this volume); woodfuels (Leach and Mearns, this volume); climate change (pp. 31–34, this volume) and biological diversity (Johnson, this volume).

In order to make scientifically valid statements about a phenomenon, one has to define it first. But deforestation is a slippery word: its definition has not been agreed upon and it is used by different people to mean different things. The consequences of such diverse activities as fuelwood and industrial timber harvesting; shifting cultivation; outright forest conversion to annual and tree crops or pasture; and grazing or burning in existing forest have all been termed deforestation, although the impact of these activities varies widely, and many of them leave a substantial part of the original forest vegetation intact.[25] Obviously, lumping such diverse impacts under the one heading of deforestation gives rise to meaningless figures.

A definition of deforestation: what's in a name?

In this overview, deforestation has been defined as the permanent conversion of forest land to non-forest use.[26] Obviously, according to such a definition, selective timber felling or browsing would not qualify as deforestation, and would have to be called something else. But would felling a closed forest, burning the debris and establishing a farm field equal deforestation? It depends: yes, if the farm field is permanent, or abandoned but turned into grassland or scrub by regular fires; no, if the farm field is left fallow and allowed to regenerate to (secondary) forest cover, as is the case in farming systems based on rotational bush-fallowing.[27] The word 'permanent' is all-important in this respect; it implies land use subsequent to the felling of a forest area has been taken into account. Why such an equivocal definition? Take the simplified case of a family practising a bush-fallow system on 20 hectares of secondary forest, on a 20-year rotation. Every year, they slash and burn one hectare and plant it with crops, leaving last year's field to regenerate to forest. After 20 years, if subsequent land use was not taken into account, one would accuse the family of having deforested 20 hectares of forest, whereas in reality the forest cover on their 20 hectares has not significantly changed. Secondly, such a definition helps to draw attention to the fact that subsequent land use is the crucial factor in determining the environmental impact of deforestation. The beautiful and highly productive irrigated rice fields of South East Asia were once covered in lowland rain and monsoon forest; they are now certainly less biologically diverse, but have supported large populations for millennia without undergoing major soil degradation processes.

Measuring deforestation

Most deforestation estimates are based on the interpretation of remote sensing imagery – satellite imagery, aerial photographs – and secondary data. However, there are some inherent limitations in remote sensing technology. Large expanses of closed forests and large-scale clearings are easy enough to distinguish, but small-scale patchworks of farms, fallows and intact forest as are found in shifting cultivation areas and open woodland areas are very hard to interpret and quantify. These inherent limitations are compounded by human errors. The World Resources Institute's (WRI's) estimate of deforestation in Brazil in 1987 (80,000 square kilometres), more than twice as high as that of Whitmore and Sayer (35,000 square kilometres), was based on faulty interpretation of satellite imagery.[28] Secondary sources are often of an even more dubious nature. In some studies deforestation is calculated as the estimated number of farm families in a forest area times the amount of land an average family is likely to clear each year. Needless to say that in the absence of good census data and measurements of forest clearing activities of a representative sample population such figures are highly unreliable. Many such calculations, however, obtain a ring of truth through repeated quoting by increasingly respectable sources.

Much energy has been devoted to improving tropical deforestation statistics. The latest FAO estimates indicate that tropical deforestation has increased from 11 million hectares per annum in the 1970s to about 17 million hectares per annum in the 1980s. (For a critical review of various published tropical deforestation estimates, see Alan Grainger, *Controlling Tropical Deforestation*, Earthscan, London 1993). Although better statistics on deforestation have their uses, it would seem obvious that knowledge of the causes of deforestation is a more important precondition for successful remedial action than is knowing whether, say, 10 or 20 million hectares of tropical forests are being lost each year.

It is often difficult to establish the causes of deforestation.[29] Many researchers attribute deforestation and other environmental degradation processes to one single underlying cause. Favourite scapegoats include population growth, shifting cultivation, logging, poverty and indebtedness. Such simplistic 'single cause' explanations are totally at odds with a reality in which many factors are interdependent. Rapidly growing South East Asia, stagnating sub-Saharan Africa and heavily indebted Latin America are united by their common failure to achieve sustainable natural forest management.[30] Through the 1970s and 1980s, population growth rates have decreased in Brazil and increased in Africa, but both continents have witnessed accelerating deforestation. And to quote Warford and Pearce: 'Poverty does not necessarily in and of itself lead to environmental degradation. That depends on the options available to the poor and on their responses to outside stimuli and pressures'.[31]

In reality, deforestation is rarely a simple process. More often than not, it results from a complex chain of events, involving a number of different agents and causes in each locality and point in time. Palo has proposed a model of system causality to describe the highly complex processes involved, and distinguish between agents, driving forces and accelerating forces of deforestation and forest degradation (see Figure I.1).[32] Prominent agents are land clearing for agriculture and cattle ranching, infrastructure development, and timber and fuelwood harvesting; driving forces are growing and migrating human populations and their demands, international demands, technology and accessibility; and accelerating forces are poor government policies and market failures. Instead of attributing deforestation to one single cause, the model introduces the notion of 'system causality' to describe the way in which these driving and accelerating forces mutually reinforce each other, and how the consequences of deforestation sometimes intensify the forces that caused it in the first place. The value of such an approach is obvious: the description and analysis of deforestation processes in different localities and at different points in time will enable us to find out where the 'pressure points' are and how to intervene to prevent deforestation where it is undesirable.

Different scientific disciplines focus on different aspects of changes in land use. Hence, the gradual and ongoing colonization of central African tropical moist forests by farmers over the centuries could be termed forest ecosystem degradation by an ecologist. A geographer, however, could perceive the very same process as the transformation of wilderness by humans into a cultural landscape, in this case a complex patchwork of homesteads, temporary farm fields and permanent gardens, forest fallows in various stages of regeneration and intact forest. Although these observations seem to be describing different events they are not; they are describing the same events and they are both equally true. Their divergence is explained by the difference in the angle of observation.[33] Such differences in perceptions are of course magnified in the case of different groups of tropical forest land users, who have direct interests at stake. Even if facts are agreed between scientists, conflicting values or interests (of various forest land users) can lead to different interpretations of these facts. The absence of certainty stemming from unreliable facts and figures is compounded by the uncertainty generated by conflicting perceptions and interests.[34] This places us firmly in the realm of politics.

Economics and ethics

> Where there are threats of serious or irreversible damage, lack of full scientific certainty should not be used as a reason for postponing measures to prevent environmental degradation. (Bergen Declaration, 1990)

There have been many hopeful statements recently about the future of tropical

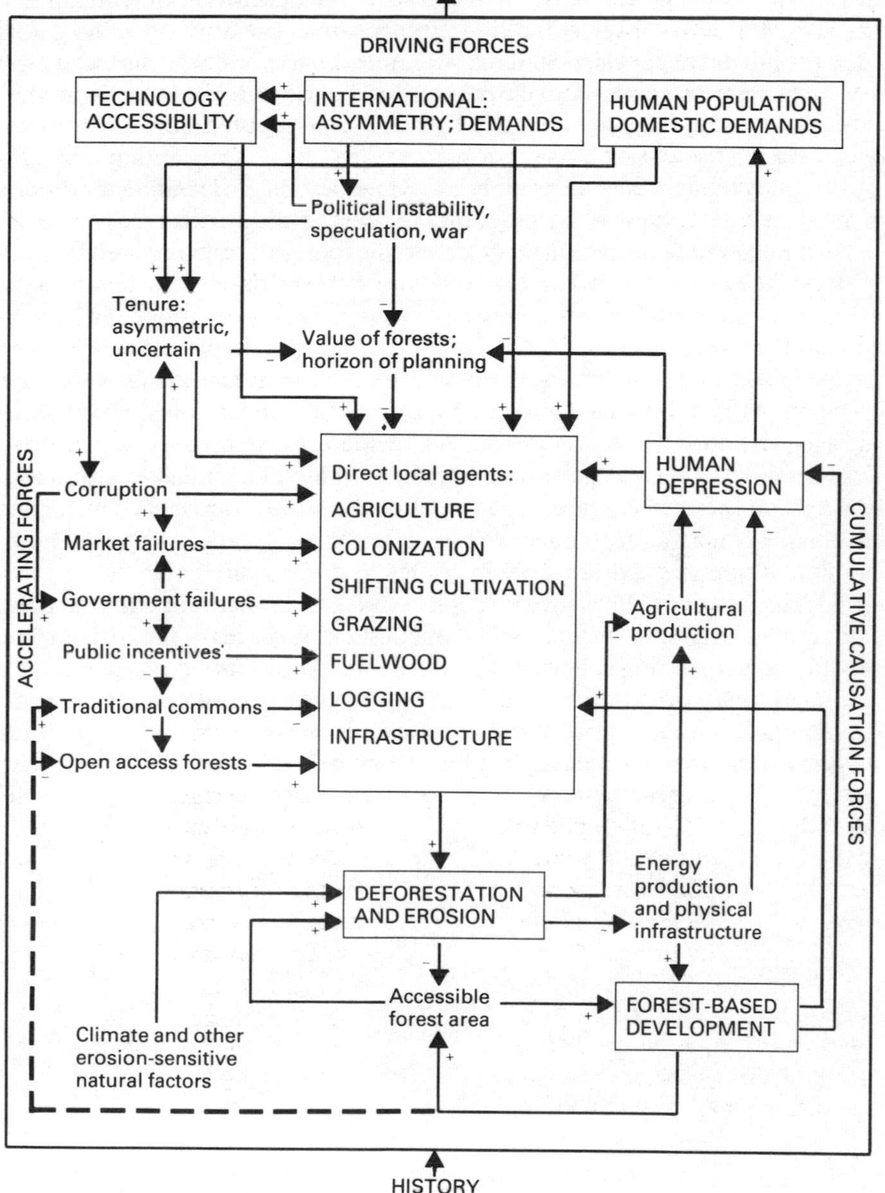

Figure I.1: System causality of deforestation and regressive development process in the Third World (modified from Palo 1987; plus sign = positive and minus sign = negative correlation)

forest conservation, suggesting that conservation is a necessity for economic reasons rather than a matter of deep green ethics. Some of this is due to advances in economic valuation methods for tropical forest products and services, which are diverse and often difficult to express in monetary terms. One oft-quoted paper calculated that the annual harvest of fruit and latex from one hectare of rainforest near Iquitos in Peru was worth about $700 (giving a Net Present Value of more than $6000), whereas a one-off timber felling would yield only $1000, and cause irrepairable damage.[35] Another reason for optimism is the increasing attempts by theorists to integrate natural resource scarcity and environmental degradation into economic calculus.[36]

Natural resource accounting is an economic tool with particular relevance to tropical forests. It is based on the view of natural resources as a capital stock which governments and private resource owners need to maintain if they want to continue to derive income from it.[37] Economists have also argued for applying low discount (interest) rates to forestry investments. Low discount rates would improve the economic viability of natural tropical moist forest management for timber production in comparison to its main alternative: plantations.[38] The concept of total economic value (TEV) makes a provision not only for direct and indirect use values of forests (eg tourism and watershed protection), but also for option and existence values. Option value relates to possible future use; existence value relates to the intrinsic value of the forest unrelated to its use.[39] Existence values could be substantial and might dominate use values.[40] Economic valuation techniques are useful tools for governments wanting to have a yardstick to decide which forests are worth more than others, and should receive priority attention in conservation programmes. But overall, the optimism about the purely economic viability of tropical forest conservation seems unfounded, for three reasons:

- Most of the methods referred to above have some fudge factor allowing conservation to look economically attractive: low discount rates, physical rather than monetary resource accounts, and economic values based on indirect measurement not real money.[41]
- It is uncertain to what extent economic planners and policy makers in developing countries and multi- and bilateral donor organizations can be convinced to use these environmentally more benign economic tools.
- The rigorous application of environmental economics may well prevent implementation of large-scale, destructive development projects in the short term, but it is much less clear how it would preserve much tropical forests in the long term.

Without doubt, the 10 acre watershed forest on a small Caribbean island that protects tourist-infested reefs against siltation would survive.[42] The last remnants of the tropical moist forests on the Atlantic coast of Brazil, full of endemic species and irreplaceable genes, would also be likely candidates for survival. But for large areas of lowland tropical moist forest which are relatively poor in biodiversity and not essential for soil and water conservation purposes (such as many forests in the central basin of Zaire), environmental economics is unlikely to save the day. The long-term cost of preventing ongoing colonization by farmers in such areas may well outweigh the benefits of preservation by far. Forests in such areas may persist on economic grounds, but this would be in a modified form, as forests managed for timber and non-timber production purposes. On the other hand, they may be preserved intact because governments choose to do so, for reasons of political expediency.

Nature conservation started out as an ethical pastime; its involvement with economics is of a far more recent date.[43] Most protected areas have their origin in the ethical and aesthetical sensibilities of individual human beings, often members of social elites. Tropical forest conservation, however, has a strong tradition of false morality. Historical examples include colonial forest policy statements, masking their intention to usurp the traditional land use rights of local people with numerous references to the need for conservation. A present-day equivalent is the reductionism of some environmental campaigners, meting out judgements of good and bad (green and non-green) with little if any regard for reality. Kenya has been praised for its support of an ivory ban, and Zimbabwe has been vilified for opposing it, but the issue at stake is one of comparative advantage rather than morals. In Kenya, the worth of ivory is in the tourist trade, whereas in the case of Zimbabwe, with much more dispersed wildlife resources, controlled hunting makes sense.[44]

Nevertheless, tropical forests and their conservation do provoke important questions of both social and environmental ethics. Is it justified to reserve forests from poor people, restricting their livelihoods for the greatest good of the greatest number of people?[45] Given that irreversible and harmful environmental change may come about before we can actually measure it, how conservative should we be in using forests and other natural resources? How much restraint should we show in the face of alarming statistics? Although improved economic methods and risk analysis can provide some guidance on the latter two questions, they are essentially ethical in nature. If people can agree on the answers to some of these questions, then there are ways to proceed. Governments could agree to adopt the precautionary principle (see quote from the 1990 Bergen Declaration above), and to shift part of the burden of proof about the long-term negative impact of tropical deforestation away from the conservers to the developers (precedents for this exist for pesticides and the pharmaceutical industry). But judging by the opposition to the idea of a forest convention at the Earth Summit in Rio in 1992, there seems to be little hope for such a consensus on environmental ethics at the global level.[46] The main blockage point is an issue that is both moral and financial; it is the question about the extent to which wealth of developed countries is based on exploitive relationships with developing countries (past and present), and whether or not developed countries have a moral obligation to re-transfer a much larger proportion of their wealth to developing countries. Developed and developing countries' views on this issue are, unsurprisingly, diametrically opposed. Whereas UNCED chairman Maurice Strong had called for $70 billion worth of environmental aid, the so-called 'Green increment', only $5 billion was committed. Whether we like it or not, tropical forests have become bargaining chips in the political negotiations in which developing countries ask industrialized nations for concessions in the form of reduction of the debt burden, improved access to export markets and accelerated technology transfer, in return for environmental conservation.

In summary, tropical forests turn out to be political and ethical minefields, and full of poor science as well. Any desire for simplicity and for clear distinctions between wrong and right has to make way for the acceptance of an uncertain and complex reality, shaped to a considerable extent by conflicting perceptions and interests. Put differently, there are many wrong ways of managing a given forest, but that does not mean that there is only one way of managing it that is right. For each management objective, a number of technical options are possible, each resulting in different winners and losers among the forest users concerned.

The introduction following attempts to shed more light on tropical forests and their users. It introduces the 13 articles selected for this reader under seven headings: policy and management; land use and land degradation; international trade; forestry development aid; indigenous peoples; biological diversity conservation; and climate change.[47] Many of these subject areas are closely interrelated; they are treated separately here for convenience, while cross-references are included to point out the linkages between them.

Introduction to reader articles

POLICY AND MANAGEMENT

Poore, in his introduction to *No Timber Without Trees: sustainability in the tropical rainforest*, describes how a socially and environmentally committed government could go about managing its forest resources: the 'art of the possible'. The piece argues that a balance between the development and conservation of forest lands can be found if forest management is conceived with its socio-economic and environmental contexts in mind. Although it was written with timber production in tropical rainforests in mind, the systematic approach to forest land allocation to various purposes it advocates (protection and conservation as well as production) can be applied equally well to other forest types and management objectives.

But how good a record do governments have for managing tropical forests? Politically speaking, natural forests and their inhabitants have always belonged to the physical and social margins of society.[48] Because of their sparse human populations and difficult access, forests and other wilderness areas have always proved difficult to control for the central authority laying claim to them. Governments, whether serving modern nation states or ancient kingdoms, have fought pitched battles to assert their sovereignty over forest areas. Natural forests have invested people dwelling in their vicinity with the power of evasion from time immemorial, and in some cases continue to do so until this day. In Java around 1500 BP (Before Present), the rulers of so-called hydraulic kingdoms derived their power from the control of irrigation water in their territories. Subjects of these kingdoms would flee into the upland forests to become shifting cultivators if exploitation of their labour became too extreme. As the rulers depended on having enough hands to work the ricefields for generating the surpluses that sustained their courts, they had to relax the terms of trade if too many people took to the forested hills.[49] From a government's point of view, forest dwellers pose two main problems. First of all, it is very hard to tax them: how does one assess taxes on extremely diverse livelihoods, that is, if one can find the people in the first place. Second, forest dwellers' ways of using the land may be in conflict with the national interest, or with what a government perceives to be the national interest.

How have governments dealt with this problem? Historically, governments endeavouring to increase control over remote forest areas and the people living there have resorted to two general policies, which have been pursued separately or in tandem: getting people out of the forests by reserving the forests from them;[50] or getting rid of the forests by promoting their colonization, often with non-native people. Significantly, Third World governments, whether socialist or capitalist, did not break with the traditions of their colonial predecessors in this respect. Thus, the at least partly politically-inspired resettlement of Javanese people in forest lands in the unruly

Lampung Province in South Sumatra in early 20th century Dutch East Indies has its equivalent, although on a much larger scale, in the so-called transmigration schemes in present-day Indonesia, under which millions of Javanese people have been resettled in the forested Outer Islands Sumatra, Kalimantan, Sulawesi and Irian Jaya. Similarly, the aggressive villagization programme pursued by Tanzania in the 1970s is different only in name from the Belgian, British, French and Portuguese campaigns to resettle the inhabitants of the vast African hinterlands along newly established roads in colonial times. The rationale which governments have commonly given for such policies is two-fold: to enhance socio-economic development of forest regions and their inhabitants in order to promote 'national integration', and to replace the forest dwellers' allegedly wasteful exploitation of natural resources by 'rational', sustainable land management practices.[51] The story is much the same in those Third World countries where such colonial precedent has been absent. The colonization schemes in the Amazon, a region largely neglected when it was under Portuguese and Spanish rule, are a case in point. Sadly, these large-scale colonization and resettlement programmes have rarely achieved their officially stated objectives. By concentrating people on fragile soils without the appropriate land use technologies, such schemes have caused untold damage to the environment and to indigenous peoples' livelihoods, and have wasted enormous amounts of money.[52]

Repetto and Gillis, in an excerpt from *Public Policies and the Misuse of Forest Resources*, analyze the ways in which governments have contributed to forest degradation problems, rather than being part of the solution. They argue that forestry and, more importantly, non-forestry policies which have led to economic waste of publicly owned forest resources have also undermined conservation efforts, regional development efforts and the achievement of other socio-economic goals. Generous tax treatment, heavily subsidized credit, direct government subsidies and pricing policies undervaluing timber and other forest resources have promoted agricultural, mining, dam and road programmes involving the clearing of forests, and stimulated investment in forest exploitation. Nationalization of forests has destroyed common property management regimes and transformed forests into open access resources, with the resulting free-for-all inviting reckless exploitation by outsiders, and forcing local people to steal from forests which they once owned and managed. *En passant*, Repetto and Gillis also quash the myth that developing country governments have a monopoly on bad forest policies, by demonstrating the similarity of the latter with policies pursued by the US government, including large-scale clearfelling of old-growth forest and below-cost timber sales.

David Anderson, in an excerpt from *Conservation in Africa: people, policies and practice*, provides a well-documented case history of the political conflicts involved in the management of Lembus Forest, Kenya, 1904–63: between the locals wanting to continue to utilize the forest for grazing; a commercial company wanting to exploit timber; and the colonial Forest Department seeking to conserve the forest and control its exploitation, in order to maintain financial returns while sustaining the resource. He concludes that colonial forest departments, much as their present-day counterparts, have not always been successful in controlling private commercial timber extraction and ensuring a proper flow of revenues to treasury and the forest department.

Shepherd, Shanks and Hobley, in their paper *Management of Tropical and Subtropical Dry Forests*, map out the challenge for government forest managers: to weigh against each other the dictates of national and international policy; technical feasibility; economic rationale; the needs and aspirations of local people; and to come up with concrete action plans. While Poore as well as Repetto and Gillis stress improving government policies from the top down, through better economic analysis, Shepherd *et al* emphasize the human factor and the importance of learning by doing. For foresters to be effective in the field, they need to conceive of technical solutions in terms of social strategies; therefore they need legislation and skills promoting the involvement of local users in the management of forest resources. Shepherd *et al* highlight the experience of forest departments in India and Nepal, where important changes in forest policy and law have resulted from extensive consultations and joint management activities with local people. Their paper also contends that sub-tropical dry forests are at least as important as moist forests: they retain moisture and carbon where it matters, support larger numbers of people and domestic animals per hectare, and provide both for export (gum, animal hides) and import substitution (manure and leaflitter instead of fertilizer).

Land use and land degradation

> It is crying for the moon to want to stop the world from turning, to freeze the landscape as it is. (Jack Westoby)[53]

Wherever humans have lived, they have influenced the landscape around them. Their impact on tropical seasonal forests in the Old World dates back at least as long as the domestication of fire, possibly as long as several hundreds of thousands of years ago.[54] It is uncertain, however, which changes in the area and character of forests can be ascribed to human influence and which to natural causes. Tropical forests have not been static since man's emergence two million years ago, but rather contracted and expanded as the climate grew drier or wetter during the ice ages and the warmer periods in between. Many areas now under tropical moist forest were not forested during major glaciation periods, when a considerable share of the world's moisture was locked up in ice caps covering large areas of Europe, Asia and North America. In these drier, cooler periods, climate also affected forests indirectly through its influence on human activity: lower sea levels created land bridges allowing colonization of new areas and prolonged dry periods facilitated the use of fires for hunting, rejuvenating pastures and clearing land, and limited the build-up of insect pests and human diseases. Even today, drier forest types such as monsoon forests and savannah woodlands are relatively more influenced by man than tropical moist forests.[55] The history of human settlement is an important determinant of the extent of natural forest cover today; it is no coincidence that Latin America, not settled until around 15,000 BP, has more than half of the world's remaining tropical moist forests.

The extent of human impact in any given area depends on the land use practices employed, which in turn depend to a large extent on population density. Hunter-gatherers influence the populations of the plants and animals they use and make small clearings to build camps along their seasonal migration routes, but occur at such low population densities (less than 4 persons per square kilometres) that their overall impact is very limited.[56] Nomadic and transhumant pastoralists occur at similarly low densities, but they have influenced large areas through grazing of herds and the use of fire. Agriculturalists living at higher population densities modify the landscapes in which they live in more easily discernible ways. Swidden cultivators turn considerable areas of forest into fallow vegetation, as they shift their fields from year to year, and their villages over longer periods of time – even if the part of their territory cultivated annually is limited.[57] At still higher population densities, farmers growing their crops in permanent fields create an almost wholly manmade landscape where natural vegetation, if any exists, is confined to relics which have often been influenced by some form of exploitation. In some areas, this process occurred quite a long time ago; the fertile lowlands of South East Asia, for instance, lost most of their forest cover more than 2000 years ago.[58] The process of agricultural intensification has taken two pathways: one leading to the creation of permanent fields of annual crops and mixed home gardens, and the other to replacement of fallow vegetation by permanent tree (cash) crop gardens.[59] Finally, the advent of the colonial era brought with it the large-scale clearing of forests, either directly for the establishment of plantations of annual crops such as sugar, tobacco, cotton, and perennial crops such as coffee, tea, rubber, teak, or indirectly, through the forest clearing activities of the farmers who had been displaced by such plantations.

This is of course a very schematic description: the neat terms used to categorize groups of people practising diverse livelihoods in tropical forest areas do not really do justice to the variation encountered on the ground. Many hunter-gatherers also practise some cultivation, planting or conserving useful species (fruit trees etc) near their camps;[60] many swidden cultivators also practise hunting, gathering and fishing, especially in the agricultural off-season, and have at least some permanent fields or gardens. Furthermore, there is often considerable specialization within communities, on the basis of gender, age and personal ability; and there are often extensive exchange relationships between people practising different livelihoods.[61] Rural livelihoods are thus best conceived of as being somewhere along a continuum between low-intensity (eg hunting and gathering) and high-intensity (permanent farming) land use, with many possible combinations in between. Over the whole of human history, the tendency has been away from the initial 100 per cent dependency on hunting, gathering and fishing, towards more and more intensive land use. But no land use systems are per definition better, or more sustainable, than others. People evolve land use technologies in response to environmental and social circumstances, and on the basis of their technical ability and cultural preferences, all of which change over time.

Wherever there are claims about land degradation processes in the tropics and their disastrous consequences, there are claims to the contrary, and the latter are not just the work of maverick scientists. What causes these statements of doom, and their antitheses? Most claims about land degradation are made by outsiders, who are at a disadvantage in making accurate observations for a number of reasons. They often do not have sufficient knowledge about local ecosystems and the project characteristics of the ecosystems in their areas of origin on to local ones. But what to them seems like an all-out disaster may in fact be a frequently occurring event.[62] Many observers get a snapshot perspective of the areas they study, and do not realize the longer-term processes involved, and the historical reasons for the phenomena they are witnessing.[63] They are also likely to overlook how local populations have adapted to changes in their environment over time, and may make much of a problem that no longer exists in the eyes of the local people.[64] Furthermore, outside observers may interpret data to suit their institutional or personal interests, whether or not this is intentional.[65] Robert Chambers has listed a whole series of biases which outside observers of tropical areas are subject to: eg roadside bias because they never stray far from tarmacked roads, elite bias because they are more likely to talk with village leaders than with the poor, and seasonal bias because they only make field visits in the dry season, when the roads are passable.[66]

But even if outsiders' observations are correct, there is still another problem. As Piers Blaikie and Harold Brookfield note in the introduction to their book *Land Degradation and Society*, land degradation processes such as deforestation are not only physical phenomena; they are also socially defined and there are often competing perceptions in their assessment. Blaikie and Brookfield propose to tackle the resulting uncertainty by considering land degradation in terms of altered benefits and costs accruing to people now and in the future. This would provide outside change agents (governments, aid projects) with a basis for negotiation with local land users. In order to establish costs and benefits of land degradation, it is often helpful to adopt the perspectives of the various land users concerned, and look for rationality within those perspectives rather than assuming irrationality up-front. These perspectives may include some unpleasant surprises, if it turns out that land users have little interest in long-term sustainability.[67]

The harvesting of woodfuels (ie fuelwood and charcoal) has often been mentioned as a cause of deforestation and land degradation. And yet, most developing country government and donor interventions have dealt with woodfuel issues as if they were simply energy problems, without taking into account their forestry, land use and macro-economic policy contexts. Gerald Leach and Robin Mearns, in their introduction to *Beyond the Woodfuel Crisis*, explain how such interventions were doomed to fail as a consequence. They also point out that the picture of a generalized woodfuel crisis causing large scale environmental damage in Africa (and developing country drylands in general) is a false one. The physical aspects of scarcity have been overestimated, and are often much less important than contextual issues such as labour shortages, land endowments, social constraints on access to wood resources, and cultural practices. Many of these issues are complex and dynamic, and undergo rapid adaptive change, which outsiders may easily miss. Moreover, although woodfuel harvesting may cause considerable forest degradation in areas of concentrated demand such as around cities and near rural industries, most woodfuels are consumed by rural households and derive from trees on agricultural land rather than forests. Leach and Mearns' critical examination of received wisdom yields a message of hope rather than desperation for sub-Saharan Africa: '. . . in many places hard work and creative innovations by people, governments and aid agencies are doing remarkable things to put the land into good shape, restore soils and a healthy cover of vegetation, and generally enhance livelihoods on a sustainable basis.'[68]

On a similarly positive note, Anthony Anderson examines the alternatives to deforestation in an excerpt from his book *Alternatives to Deforestation: steps towards sustainable use of the Amazon rainforest*. Deforestation in the Amazon, which has been accelerating until recently, has resulted mainly from socio-economic and political processes that originate outside the region itself. As Anderson asserts, these processes are frequently complex and elude simple analysis and simple solutions. One of the reasons for optimism, however, is that deforestation is not caused primarily by inexorable phenomena such as population growth, but rather by bad government policies which can be adjusted in the short term.[69] A second reason for qualified optimism is that there are alternatives to deforestation, such as agro-forestry, landscape recovery and natural forest management. However, these alternatives are not without problems. First, they are at an early stage of development in the Amazon, and there are numerous technical problems that need to be resolved. Second, investment to counter land degradation is rarely profitable in marginal areas.[70] Deforestation is seen as a global problem, and as a consequence there is a great clamour for global solutions. But, as Anderson demonstrates, solutions are of a decidedly more local nature: what works in one area is not necessarily applicable to the next.[71] The prevailing uncertainty provides a strong case for trying many different managers and technical options, rather than going all out down a road that may not be the right one.

International Trade

The widest traded and most controversial tropical forest product is timber, the bulk of which derives from natural tropical moist forests.[72] Harvesting of timber from natural tropical moist forest has been mainly 'mining', with little re-investment of private profits or government revenue in the future productivity of the forest resource. Indeed, buoyant markets for some tropical timber species have led to their economic extinction in the wild.[73] Most timber harvesting in tropical moist forest, however, is selective felling, which leaves the majority of the trees standing. Therefore, the main impact of logging is not its direct effect on the forest vegetation, but rather the access it provides to pioneer farmers and hunters, especially in Africa and South East Asia.[74] The International Tropical Timber Agreement (ITTA), which was signed in 1984, differs from other commodity agreements in that it has an environmental as well as a commercial mandate, through its commitment to the sustainable use of tropical forests and their genetic resources. The International Tropical Timber Organization (ITTO) administering the ITTA includes 22 producer and 24 consumer country members, and accounts for over 95 per cent of the international trade in tropical timber. Although the ITTO has adopted guidelines for sustainable forest management, and has set a target date for all internationally traded tropical timber to derive from sustainable sources by the year 2000, it seems to have had little impact in improving forest management on the ground to date.[75] And as far as the trade is concerned, the ITTA does not seem to have improved access of developing country markets to processed tropical timber products.[76]

Wood has for a long time been one of the key resources human civilizations depended upon to flourish. Worldwide it was the main source of fuel until the advent of coal in the eighteenth century, and billions of people in developing countries continue to rely on it to this day. It has been an integral part of mankind's material culture, and was irreplaceable in the North for the construction of strategically important items such as boats and bridges until the Industrial Revolution made it possible to produce large quantities of steel.[77] With the progressive exhaustion of European supplies of wood fit for boat building and the discontinuation of supplies from the North American continent after the American Revolution, the teak forests of South Asia became indispensable to the mercantile, colonial, and military expansion of the British empire. Today, all this has changed. As Arnold points out in his paper *Long Term Trends in Global Demand for and Supply of Industrial Wood*, tropical timber is no longer a strategic resource for the developed countries. In 1979, the share of tropical timber in the international trade peaked at 25 per cent of the total volume traded. This is now down to 10 per cent, and will decrease further due to the increasing depletion of easily accessible tropical moist forest; rapidly growing domestic markets in the main tropical timber producer countries; and increasing competition from growing temperate resources in the main importing countries (USA, Europe, Japan) and new exporting countries (Chile, New Zealand), which rapidly evolving wood technologies will allow to be processed into high quality products.

A recent study commissioned by the International Tropical Timber Organization found that only one eighth of one percent of productive tropical moist forest in ITTO member countries (India excepted) was being managed for sustainable timber production. The study summarized the conditions necessary for such management to succeed as follows:

> government resolve to set aside a forest estate for the production of timber and to manage it sustainably; a sound political case for the selection of a permanent forest estate as part of a national land use policy; guaranteed security for the forest estate, once chosen; an assured and stable market for forest produce; adequate information for the selection of the forest estate and for planning and controlling its management; a flexible predictive system for planning and control based on reliable information about growth and yield; the resources and conditions needed for control; and the will needed by all concerned to accomplish effective control.[78]

and concluded that these conditions were far from being fulfilled in most of ITTO's member countries.[79]

Much attention has been paid in the developed country media to the impact of the international tropical timber trade on tropical forests, as it seems to be one of the areas where consumers can exert direct influence on tropical forest destruction. Many NGO campaigns and government policy initiatives in Europe, the US and Australia have focused on restricting or banning the imports of tropical logs and timber products into the developed countries.[80] The prize for the most imaginative campaigning event staged so far must surely go to the group Milieudefensie in The Netherlands, which used a crane to drop a tropical log on a second hand Mitsubishi car, right in front of the European Head Office of the Corporation, which has interests in tropical timber concessions in South East Asia. Colchester, in his provocatively titled piece 'The International Tropical Timber Organization: Kill or Cure for the Rainforests', summarizes the NGO movement's arguments for restricting the tropical timber trade. He discusses the doubly negative impact of timber harvesting on tropical moist forests, destructive by itself, and, more importantly, opening up forests to further destruction. His conclusion that sustainable logging is a myth, for both ecological and economic reasons, is controversial and in direct contradiction to Poore's article in this volume. It is mainly based on the assumption that the political commitment of developing country governments to protect the interests of the weak is very limited. Colchester also critically examines a few solutions commonly advocated by other NGOs (industrial timber plantations in the tropics, higher tropical timber prices) and shows them up for what they are: two-edged swords, which may create more problems than they solve.[81]

Developed country parliaments, local and, to a lesser extent, national governments have responded to many of these campaigns. At the national level, extensive parliamentary hearings were held in both the UK and the Federal Republic of Germany, and the US House of Representatives called for a ban on teak imports from Thailand, allegedly derived from unsustainable harvesting operations organized by the military in the remote Thai-Myanmar (formerly Burma) border area. The European Parliament has called for a ban on timber imports from Sarawak, and for all tropical timber imports into the EC to derive from sustainable sources by the year 1995. There has been a proliferation of initiatives at local government level in many countries, especially in The Netherlands and Germany where many municipalities and city councils have stopped using tropical timber in their construction projects. The private sector has also responded, often belatedly, by either giving in to NGO demands (or seeming to do so) or staging counter-campaigns. The latter are based on the premise that logging is only a minor direct cause of deforestation in the tropics, and that buying timber products contributes to saving tropical forests as it gives countries an economic incentive to manage forests, rather than convert them to agriculture.[82] Recently, some NGOs have started to build bridges to the trade. The World Wide Fund for Nature is trying to convince traders and the International Tropical Timber Organization to restrict trade to tropical timber derived from sustainable sources by the year 1995, five years earlier than the ITTO target date, the year 2000.

WWF's and ITTO's targets can only be achieved if sustainably produced timber imports can be distinguished from unsustainable ones. This would imply a need for a reliable labelling mechanism, which is controversial for a number of reasons. Tropical timber producing countries are opposed to it in principle, because the cost of verification, and the fact that temperate timber has not yet been slated for similar scrutiny, would cause unfair competition.[83] They also argue that sustainability is not a simple yes/no issue.[84] In addition, there are technical problems involved in setting standards for sustainable tropical forest management, as there are many opinions on what the latter entails: ranging from sustainability of timber yields to more stringent ecological criteria.[85] There is also considerable doubt about the practical feasibility of certification, and the extent to which irregularities can be prevented.[86] Unsurprisingly, labelling by the trade itself has turned out not to be reliable: 'Virtually all green claims made by British timber traders and DIY superstores are misleading and unsubstantiated.'[87] Various tropical timber certification programmes are now operational, but they are still new, and limited in scope.[88] A complementary approach to large-scale certification is to establish direct contacts between small scale producers and Western importers.[89] The fact that production is small scale, however, does not in itself guarantee sustainability.[90] Another recent effort to assure that tropical timber in the international trade derives from sustainable sources is the restriction of trade in endangered timber species in the framework of the Convention on International Trade in Endangered Species of Wild Fauna and Flora (CITES). In March 1992, Brazilian rosewood was placed on Appendix I, allowing no further trade, whereas three others were listed under Appendix II, allowing trade under a permit system.[91] The practical results of this CITES listing are uncertain, as one observer has noted: 'CITES does not appear to have improved the status of any endangered species, prevented the endangerment of additional species, or nullified the international trade as a threat to species.'[92]

Two preparatory meetings for the renegotiation of the International Tropical Timber Agreement have been held so far. Many of the parties involved in the original agreement want considerable changes. The producer countries and the environmental NGOs are arguing for the inclusion of timber from boreal and temperate forests in the agreement, in order to prevent unfair competition with tropical timber. Consumer country governments disagree with this approach, and favour instead a Global Forest Convention based on the Forest Principles adopted at Rio in 1992 to deal with non-tropical timber. Producer countries have been critical of the role of the environmental NGOs which are observers at ITTO Council meetings, and of ITTA's focus on forest management rather than on processing and trade. They argue that the agreement should concentrate more on stimulating the latter by improving access to consumer country timber markets among others. Consumer countries have been very critical of the operational mechanisms of the ITTA, and have especially criticized ITTO for taking on too many projects and pre-project studies, a task for which the organization has little comparative advantage.[93] While the outcome of the negotiations in terms of changes in the mandate and the operational mechanisms of the ITTA is still uncertain, there seems to be a consensus on two issues. (Three words here) ITTA member countries have committed themselves to ensure all tropical timber in the international trade will be sustainably produced by the year 2000 is likely to be formalized in the preamble of the new agreement, as is the commitment to ensuring forest management will conform to ITTO's guidelines for natural forest and plantation management.

The sheer effort devoted to attempts to regulate the international tropical timber trade seems increasingly at odds with the trade's contribution to deforestation, and with the effectiveness of trade regulations in achieving environmental goals.[94] It is clear from Arnold's contribution (this volume), that tropical forests are no longer being 'pillaged by the wood-hungry North', as Jack Westoby (in his lifetime considered by many to represent the conscience of the forestry profession) said in his keynote address to the World Forestry Congress in Jakarta in 1978. It can be argued that the demand for tropical forest land to grow food and cash crops is a more powerful driving force of deforestation than the demand for forest products. A study carried out by the Dutch Forest Service demonstrated that, as far as imports into The Netherlands are concerned, tropical timber derived from selective harvesting is a less important direct agent of tropical deforestation than other imported commodities such as rubber, palm oil, coffee, cocoa, tapioca and soy bean, which require partial or outright forest clearance.[95] Imports of these commodities, rather than timber, would seem to be a more important target for people concerned with the future of the tropical moist forest.[96]

Advocates of tropical timber bans have pointed to non-timber forest products (NTFP), such as canes, resins, mushrooms and fruit as an alternative to timber for supporting forest-dependent peoples' livelihoods as well as providing their governments with much needed foreign exchange. Many NTFP are internationally traded already, and others have potential. There are, however, important limitations to this alternative. The harvesting of many NTFP is destructive (ie it causes the death of the plants or animals involved) and has resulted in economic extinction in the wild. As a consequence, NTFP such as cinnamon, the best quality rattans and some resins are now mainly or only produced in plantations, while others have disappeared from world markets altogether.[97] Still other NTFP (including dyes, glues and gums) have been outcompeted by cheap, synthetic substitutes, and are no longer in demand.[98] Furthermore, increased commercialization of NTFP may displace the poor people who depend on them.[99]

Forestry development aid

> While conservation is concerned with attaining stable balances, development is dedicated to change. (Duncan Poore)[100]

Just as non-forestry policies have contributed much more to deforestation than forestry policies (Repetto and Gillis, this volume), non-forestry development aid has had much more impact on tropical forests than forestry development aid. Projects promoting roads, large dams and other infrastructure, agricultural colonization and plantation development, have all taken their toll on tropical forests.[101] This section deals mainly with forestry development aid. The record for forestry aid is not encouraging. The largest donor initiative in recent years, the Tropical Forestry Action Plan, has not lived up to expectations in addressing major issues such as deforestation. Other large forestry donors, such as the World Bank, have experienced similar problems in dealing with social and environmental aid agendas.[102] Paradoxically, while disenfranchised groups – such as the poor, and women – depend on forest resources for their livelihoods to a greater extent than do better-off segments of rural populations, forestry aid projects have had limited success in benefiting these groups.[103]

Development aid has gone through three main phases and so has forestry aid. In the 1950s and 1960s assistance was directed to modernization, through industrialization and green revolution farming, which was expected to help developing countries 'take off' into sustained growth.[104] In the late 1960s and 1970s, when it became clear that the poorest were often not benefiting, and in some cases were suffering from this approach, donors tried to redirect aid towards fulfilling poor peoples' basic needs: food, shelter and other basic necessities. Subsequently, the linkages between development and the environment received increasing attention as a consequence of the 'limits to growth' discussions fueled by the Club of Rome report, and the first United Nations Conference on Environment and Development was called in 1972 in Stockholm to discuss these linkages. But only in the 1980s did sustainable development get full attention at the highest political levels, through the efforts of the Brundtland Commission.[105]

Douglas, in an excerpt from his book *A Re-appraisal of Forestry Development in Developing Countries*, reviews earlier forestry development approaches. Although there is now a consensus among donors that the forestry development approaches of the 1950s, 1960s (and 1970s), based on the establishment of modern, often state-owned industries (supplied by natural forests or large-scale plantations), are no longer appropriate, many forest authorities in developing countries still support the older views. His assessment of what is wrong with forestry aid in a socio-economic sense remains valid ten years hence: many forestry projects are still designed in an economic policy vacuum, with little or no consideration of income distribution effects, or of opportunity cost in relation to investment in other sectors. Moreover, aid projects involved in so-called institution building activities (strengthening the administrative framework, enhancing training and research capacity) may have been strengthening overly centralized forest authorities against pressures for necessary change.

In 1985, the Food and Agricultural Organization of the United Nations (FAO) published a document that attributed tropical deforestation to the poverty of the people living in or around them, and proposed more food production, higher incomes and greater employment through increased timber harvesting and processing as remedies.[106] Later in 1985, a task force convened by the United Nations Development Programme (UNDP), the World Bank and the World Resources Institute (WRI) published 'Tropical Forests: a Call for Action', which advocated policy changes inside and outside the forestry sector to address the multiple causes of deforestation, and increasing donor investment in tropical forest projects.[107] The subsequent merger of these two initiatives (and their divergent problem definitions) led to the constitution of the Tropical Forestry Action Plan (TFAP), sponsored by FAO, UNDP, the World Bank and WRI.[108] A permanent TFAP secretariat was recently established at FAO headquarters in Rome. The TFAP is a combination of previous development agendas (modernization, basic needs) and issues that had gained prominence more recently such as the conservation of biological diversity, the need for policy and institutional reform, and the realization that the causes of deforestation originate mainly outside the forestry sector itself. The TFAP was intended to work at two levels: internationally it was to promote consultation and coordination between the donor agencies; and at the national level it was to mobilize support for a country-level development planning process and stimulate investment to implement the resulting plans.

Priority areas of intervention identified by TFAP

- Forestry in land use. Activities aimed at the interface of forestry and agriculture and at more rational land use through community forestry, integrated watershed management and desertification control, and land assessments and forest resource inventories. Such activities would include planting of multi-purpose trees on farms to help combat declining soil fertility and shortages of poles, fuelwood and other forest products.
- Forest-based industrial development. Activities aimed at promoting appropriate forest-based industries: small-scale 'cottage' enterprises and other forest-based income generating activities in rural areas, as well as industrial plantations and the expansion of forest products exports.
- Fuelwood and energy. Activities aimed at restoring a balance between fuelwood supply and demand, by increasing production of and reducing

demand for woodfuels; as well as programmes to develop wood-based systems to generate energy.

- Conservation of tropical forest ecosystems. Activities aimed at conserving, managing and using forests' genetic resources, such as management of protected areas and of forests for sustainable production.
- Institution building. Activities aimed at removing the institutional constraints to conserving and managing tropical forests, including support for training, research, extension; greater institutional support to NGOs and the business community; the strengthening of public forestry agencies; and the revisions of laws and policies to integrate forestry more fully into national planning.[109]

As it became clear that tropical deforestation was accelerating rather than slowing,[110] the Tropical Forestry Action Plan came in for increasing criticism. A review of nine national TFAPs (of which eight were from tropical rainforest countries) sponsored by the World Rainforest Movement and *The Ecologist* even went as far as proposing a moratorium on all TFAP funding until the plan's basic flaws were amended.[111] The report concluded that the TFAP was unduly stressing commercial logging and export-oriented forest industries, while marginalizing the needs of forest dwellers and ignoring their rights, and failing to address the root causes of deforestation. The World Resources Institute, one of the original co-sponsors of the TFAP, carried out a more comprehensive review which, although noting that the TFAP process had succeeded in focusing international and national attention on tropical deforestation and in improving coordination of forestry aid among donors, was equally critical of the TFAP's achievements.[112] WRI's report notes that the original intention of the TFAP – to curb uncontrolled deforestation through more effective forest conservation and management programmes, policy reform both within and outside the forestry sector, and improved land use planning and cross-sectoral coordination – was subverted by the fact that the major players (FAO, donors and recipient governments) became preoccupied with accelerating investment in the forestry sector, at the expense of the quality control needed to guide the planning process and make the plan succeed.[113] An independent review commanded by FAO concluded TFAP implementation had suffered from major institutional and managerial shortcomings, but declined to comment on the more substantial issues the reviews cited above focussed on.[114]

It is difficult to do justice to the TFAP, which is now being implemented in more than 80 developing countries, with very varied inputs and results. Most of the criticism was levelled at TFAP exercises in tropical moist forest countries, whereas many of the participating countries are in the sub-humid and semi-arid tropics. Nevertheless, most national TFAPs seem to have a number of conceptual and procedural flaws in common. One of the basic conceptual flaws in TFAP was that it was framed as a technical planning exercise, underplaying the existence of important (political) trade-offs between national and local level interests, and between conservation and development. To assume that the contribution of the forestry sector to a country's economy and export earnings could be increased without any risk of harming the livelihoods of forest-dependent local communities and without any risk of damage to biological diversity and soil and water conservation benefits was of course wishful thinking. The TFAP process, finally, was flawed in its failure to involve sectors other than forestry (eg agriculture), and local communities and grassroots organizations.[115]

Indigenous peoples

> We do not live in the Third World; we live in our first and only world. (member of CONAIE, during a 1991 NGO meeting)[116]

> Without immediate action to resolve the tropical forest crisis, by the year 2000, indigenous peoples who have inhabited the forests for thousands of years will be displaced and, in some cases, their cultures will disappear. (The Tropical Forestry Action Plan, p 5)

The word 'crisis' has been used so often in relation to tropical forests that it has lost some of its sense of urgency. However, crisis is the right word to use when discussing the survival of indigenous peoples in tropical forests. Indigenous societies may be becoming extinct at a faster rate than the ecosystems of their ancestral lands. In Brazil alone, it was estimated that 85 Indian groups became extinct in the first half of this century.[117] With the disappearance of each indigenous group, an accumulated wealth of millennia of human experience and adaptation, and intricate knowledge of hundreds of forest plants and animals, is lost.[118]

There is no single definition of indigenous peoples that can be applied globally and in all contexts. Nevertheless, five characteristics are commonly shared by forest-dwelling indigenous peoples: long-term occupancy of forest lands; direct dependence on forests for physical and cultural well-being; political isolation from state decision-making processes; cultural differences from the country's dominant population; and vulnerability due to lack of legally recognized land rights and/or nomadic lifestyle.[119] Although worldwide celebrity status has only been achieved by Amazonian rainforest dwellers, the majority of indigenous peoples in the tropics live outside the moist forest zone.[120] The largest populations of forest-dwelling indigenous peoples are found in South and South East Asia, where they number several tens of millions.[121] Indigenous peoples' cultures have been deemed primitive and stagnant by many casual observers, because of the fact their livelihoods employ little in the way of modern technology. In reality, their land use practices are often highly sophisticated, based on an intricate knowledge of tropical forest plants (including hundreds of domesticated varieties) and animals and their ecology. Moreover, their ways of life are not static, but evolve in response to changing circumstances.[122]

Many indigenous peoples in the tropics practice swidden (shifting) cultivation; very few are hunter-gatherers and even these rely on cultivated foods for a considerable part of their diet. Although shifting cultivation has often been singled out as a major cause of tropical deforestation, this is more commonly due to pioneer slash-and-burn farming by recent immigrants than to the practices of aboriginal shifting cultivators. The latter use conservative methods involving selective felling, light burning and no tillage, and they retain a variety of high forest and fallow trees. Intercropping and relay cropping of a large variety of crops and carefully protected fallow vegetation minimize exposure of the soil to water and wind erosion. In many forest areas, there are no economically feasible alternatives to these highly adapted indigenous farming systems. In Africa, the practices of nomadic and transhumant pastoralists have been similarly blamed for causing resource degradation, although their livelihoods are based on an intricate knowledge of dryland ecosystems, and generally cause far less environmental damage than those of the farmers that displace them (Shepherd *et al*, this volume). Other observers have idealised the ways of life of indigenous peoples, and some have even used them as ammunition for political and cultural critiques of the evils of modern society.[123] There is no need to paint overly rosy pictures of the lives of indigenous forest dwellers. First, their lives are at times desperate and brutish in the extreme.[124] Second, not all indigenous peoples have cultural restraints on natural resource use, and where they do, these restraints are in some cases vanishing fast in the face of cultural erosion and hybridization (Nels Johnson, this volume).[125]

John Bodley, in an excerpt from his book *Tribal peoples and development issues*, describes the plight of small, kinship-based tribal nations occupying the last great resource frontiers in a world dominated by large, expansionist nation states. Today, several hundred million people live in recently conquered, or still autonomous tribal nations, where blatant exploitation and direct extermination by outsiders have now made way for integration efforts. As Bodley asserts, the basic assumptions underlying such integration efforts – that indigenous peoples' ways of life are materially inadequate; that integration will improve the quality of their lives; that their interest in new technology reflects a desire for integration; and that 'progress' is inevitable – are less than universal truths.

In recent years, indigenous peoples have increasingly organized themselves politically, and some governments have responded with less integrationist proposals to their address.[126] In Colombia, the government has recognized amerindian rights to large territories in the Amazon, and in the Brazilian Amazon, so-called extractive reserves for the use of Mestizo rubber tappers and indigenous peoples have been legally gazetted. Whether or not these initiatives will enable indigenous peoples to maintain their cultural identities remains an open question. De'ath and Michalenko, in their paper 'High Technology and Original Peoples: The Case of Deforestation in Papua New Guinea and Canada', explain how indigenous forest-dwelling peoples' interests are likely to be marginalized when governments try to 'develop' remote regions by doing business with socially and environmentally irresponsible transnational corporations. They also show that net economic benefits of the logging and pulping activities involved were very limited, even without taking into account the considerable environmental and social costs incurred; and that the difference in the way developed and developing country governments treat indigenous peoples is one of degree and not of principle.

Biological diversity conservation

> . . . the future of African economies will depend largely on their capacity to harness some of the advances in biotechnology and apply them in decentralized economic development strategies. The conservation of genetic resources will be crucial for such future development strategies. (Calestous Juma, 1989)[127]

The conservation of biological diversity poses ethical, technical and social problems. The question of what biological diversity we should try and conserve can be approached from either a utilitarian or an ethical angle. From a utilitarian point of view, efforts should concentrate on potentially useful species and genes, rather than forest ecosystems. Most basic ecosystem functions fulfilled by natural forests (water, carbon and nutrient recycling processes) can be performed equally well by appropriate agroforestry systems and forest plantations, except for example in the case of fragile soils or steep and dissected topography, where retaining natural forest cover is an absolute necessity.[128] Such a utilitarian approach faces the difficulty of predicting which species and varieties will be useful. But some choices are clear: eg the plants present in the fields of shifting cultivators and other farmers would take precedence over those from natural forests, except where the wild ancestors of major crop species are concerned;[129] and *ex-situ* conservation, ie maintaining selected species and genotypes in seed collections off-site, would be an important component of the conservation strategy. Obviously, *ex-situ* conservation cannot save ecosystems, but where small forest remnants are rapidly disappearing, *ex-situ* conservation of a number of potentially useful species and varieties may be the maximum achievable result.[130]

Such a utilitarian approach is also the one that would benefit poor farmers in developing countries most, if they could have guaranteed access to improved crop varieties derived from these wild genetic resources. From an environmental ethics point of view, however, it could be postulated that all species have equal rights to life, so that we must try to conserve the maximum possible number of species. The implications of this would be that conservation of intact natural tropical moist forests (*in-situ*), which are the most diverse ecosystems on earth, would take precedence over all other causes. In some cases, there may be a difficult choice between preserving natural forest biodiversity or the many varieties of useful plants contained in the fields of shifting cultivators in the same area.[131] If it is accepted that neither the expertise nor the funding is available to save all naturally occurring diversity, then priority forest areas have to be identified and choices made. This is by no means an easy task.[132]

Technically, biological diversity conservation efforts are plagued by numerous uncertainties: about the numbers of species present in different ecosystems; the numbers of species lost due to deforestation and forest degradation; the numbers of species able to survive in modified ecosystems in the long term; and the best way to conserve tropical forest biodiversity (large reserves *vs* an equal amount of land dispersed over a number of small reserves). Knowledge derived from research on disturbed ecosystems has proved most useful for the management of forest conservation areas, but such research has been restricted mainly to large game in savannah areas.[133] The resulting lack of knowledge poses serious conservation problems. Many wild relatives of agricultural crops, for instance, thrive in fairly disturbed environments and can only be conserved by maintaining the right level of disturbance.[134]

Areas intended for biological diversity conservation in developing countries face increasing problems of a social nature, mainly through pressure from competing land users outside their limits. New approaches aiming to involve the surrounding populations in the management of protected areas have been developed to address these social issues. This seems to be the only way forward: as Johnson (this volume) notes, conservation is most likely to be successful when it coincides with the self-interest of the relevant constituency. The most straightforward option is jointly to manage a natural resource with its local users, eg managing fish populations with local communities of fishermen, or forests with the villagers using them.[135] Another is the so-called buffer zone approach, where conservation is the most important objective in the core zone, and development is pursued in buffer zones which allow varying degrees of human use.[136] Although it is too early to judge the results of these new approaches, there are good reasons not to be over-optimistic about their effectiveness in the case of tropical moist forest conservation. Protected areas are often established in tropical moist forest areas of known high biological diversity, areas which are per definition sparsely populated. Resettling people from the core zone of protected areas and restricting their exploitation of the resources there, while providing alternative income generating sources in the buffer zone may have undesirable consequences. More people may be attracted into the buffer zone from outside the protected area than from the core zone, contrary to what was intended. And since there are few activities that generate higher revenues than the combination of shifting cultivation and hunting in remote tropical moist forest areas, such interventions may create permanent dependency on outside funding.

Biological diversity has three components: species diversity; ecosystem diversity (inter-species level) and genetic diversity (intra-species level). Tropical moist forest ecosystems harbour high numbers of species in small areas; other ecosystems, such as the Mediterranean heathlands of the Cape of South Africa have low species counts at that level, but are very diverse at the landscape level because they consist of many distinct plant communities. Therefore, meaningful discussions of species richness must specify scale.[137]

Some tropical plant and animal species are endemics that are restricted to very small areas, but many others occur over large areas and show considerable variation over their range due to genetic differences within the species. Genetic variation within species is essential for their ability to adapt and survive in the wild, and constitutes an indispensable raw material for domestication and further breeding of plants and animals.[138]

How many species are there in tropical forests? An oft-quoted estimate is that of Erwin, of 30 million species of plants and animals worldwide, the large majority of which consists of insects in tropical moist and seasonal forests.[139] The latest estimate by E O Wilson, an authority on the subject of biological diversity, is of 100 million species worldwide, more than 100 times Darwin's 1865 estimate of 800,000 species.[140] But, as one observer has noted, a century after the pioneering work of Wallace and other naturalist explorers the number of species contained in the world's rainforests is still 'only a matter of rough conjecture'.[141] Johnson, in his paper *'Everyone's introduction to the technical aspects of biodiversity and its conservation'*, explains how the fact that biodiversity is highly site-specific makes the results of such extrapolations extremely tentative. The reasons why some areas contain so much more biological diversity than others are still poorly understood. Attempts to generate theories to explain the diversity of life and its role in ecosystem functioning have produced mixed results and do not seem to hold great promise. As a consequence, it is difficult to predict whether or not a certain forest site will be species-rich from easily identifiable characteristics.

Estimates of species loss due to deforestation are similarly uncertain. There is no simple arithmetic relationship between tropical forest loss and species loss: a 50 per cent decrease in forest cover does not lead to extinction of a fixed percentage of all species in the area concerned. Patterns of species extinction depend on the amount of tropical forests altered or destroyed, and on the spatial arrangement of forest remnants. Many alarmist estimates of species extinction are based on extrapolation of tropical forest loss towards zero, whereas most countries now have at least some secure protected areas and/or inaccessible forests.[142] Recent estimates taking such factors into account still arrive at considerable extinction rates: 6–14 per cent of the world's species in the next 25 years.[143]

In Africa and Indo-Malaya existing protected areas include populations of almost all bird and mammal species, although they account for only 2–8 per cent of the land areas of the countries involved.[144] However, the fact that there are as yet very few documented extinctions of vertebrate animals in tropical forests is no reason for complacency. First, it is certain that fragmentation of ecosystems endangers species and genes: small, isolated populations are vulnerable to extinction from inbreeding and a number of random causes, and genetic erosion will occur long before species become extinct.[145] Second, tropical forest loss is greatest where least remains eg in West Africa and Madagascar. Third, the status of other groups of plants and animals is much less known, and may be more critical. Fourth, climate change will threaten the viability of many existing protected areas. Restoration of some of the surrounding areas will be necessary to allow for the dispersal and migration of plant and animal species as circumstances change.[146] Forests which are managed for the production of timber and non-timber products adjacent to protected areas may play an important role in this respect.[147]

Over the last 20 years, the biodiversity agenda has shifted away from social justice issues to environmental concerns. Whereas in the 1970s the debate centred on the contribution of crop genetic resources to improve the livelihoods of poor people in developing countries, the discussions since the mid 1980s have been mainly about the conservation of natural biological diversity for the benefit of humanity as a whole.[148] As a consequence, the biodiversity convention that was adopted at the UNCED conference in Rio may not be as instrumental as the developing countries that signed it had hoped.[149] Privatization of genetic resources poses various threats, such as restrictions on the use of genes of patented cultivars; counterproductive demands for increased uniformity; and restrictions on exchange of breeding material and scientific information. This danger is especially important for commodity crops, which are important generators of foreign exchange in many developing countries, but do not benefit from large publicly funded research and development programmes.[150]

Climate change

The world has become warmer by about half a degree Celsius over the past century or so.[151] The emerging scientific consensus is that this warming trend is the result of manmade greenhouse gas emissions, part of which derive from tropical deforestation. One of the difficulties in determining whether or not the greenhouse effect is real is the fact that the earth's climate fluctuates naturally, presumably due in part to cyclical variation in solar activity (sunspots) and changing tidal effects of the moon. To complicate matters further, a downward natural trend in temperatures may have masked the real warming effect of greenhouse gas emissions.[152] A model developed to estimate the impact of natural phenomena on global temperature has produced trends of up to 0.3 degrees Celsius in a century, which cannot account for the temperature rise we have witnessed. Global warming may have potentially disastrous consequences. Any sea level rise is a matter of serious concern as the majority of the world's population lives within 50 miles of the sea. Developed countries such as The Netherlands and developing countries such as India and Bangladesh would lose much of their coastline. Tropical forests are also likely to be negatively affected.

Where do greenhouse gases come from, and how do they contribute to global warming? The answer to this question is subject to major uncertainty for two reasons. First, estimates of present-day emissions of major greenhouse gases vary widely. The reliability of figures for past emissions is even worse, an issue which is important because many greenhouse gases remain active for long periods of time. (see Table I.1). Global warming, therefore, is the result of cumulative emissions up to the present day – not just of the greenhouse gases added last year. The only accurate historic records of carbon emissions are those from fossil fuel combustion, cement manufacture, and chlorofluorocarbon (CFC) production and use. But emissions of carbon due to deforestation, and of methane and nitrous oxide in general are subject to large uncertainties. Second, the translation of greenhouse gas emissions into net enhancement of the greenhouse effect is problematic because of numerous uncertainties: about the lifetimes of the greenhouse gases; the impact of chemical reactions between them; future greenhouse gas emissions;[153] and the positive or negative feedback mechanisms induced by global warming itself.[154]

Table I.1: Global warming potentials (GWPs) following the instantaneous injection of 1 kg of each trace gas, relative to carbon dioxide.

Trace gas	Lifetime (years)	Time horizon (years)		
		20	100	500
Carbon dioxide	120*	1	1	1
Methane	10	63	21	9
Nitrous oxide	150	270	290	190
CFC-11	60	4500	3500	1500
CFC-12	130	7100	7300	4500

Source: Shine *et al* 1990 (quoted in Kelly 1991)

* Approximate. NB: the GWP assigned to a particular gas is a measure of its net enhancement of the greenhouse effect (radiative forcing) over a specified period of time (the time horizon)

Deforestation leads to the release of greenhouse gases such as carbon dioxide, methane and nitrous oxide into the atmosphere, immediately through biomass burning and more slowly through carbon emissions from deforested soils and rotting biomass. The amount of carbon dioxide released due to deforestation and changing land use has been estimated at anywhere between 0.6 and 2.6 Gt carbon equivalent per year.[155] In addition, methane emissions from tropical land use changes including deforestation have been estimated at 0.05–0.1 Gt per year. All in all, clearing of tropical forests may be causing one-fifth to one-quarter of the net annual release of carbon. Difficulties in generating reliable estimates of the net contribution of deforestation are caused by uncertainties regarding deforestation rates, biomass per hectare in various tropical forest ecosystems, emission factors, land use after deforestation and soil emissions.[156] Originally it was thought that burning of biomass, especially of tropical moist forests, was a major source of nitrous oxide, which has an atmospheric lifetime of 150 years. Recent studies, however, concluded that biomass burning emits less than 7 per cent of global nitrous oxide production.[157] While the contribution of tropical deforestation and land use change to global warming is subject to large uncertainties, there is no doubt that climate change would have important negative consequences for tropical forests. Preliminary results of research assessing the impact of four climate change scenarios on global forests show that the areal extent of climates associated with moist forests would decrease, and that all forest types would undergo significant geographical shifts.[158] Moreover, many forest types may not be able to make the rapid geographical shifts dictated by changing climate, for biological reasons and because they are surrounded by farm lands.[159] Indirectly, tropical forests are likely to be affected by the impact of global warming on agriculture in developing countries.[160]

The amount of excess carbon dioxide in the atmosphere is about half the total emitted over recent centuries. The other half has been absorbed by oceans and by terrestrial sinks such as forests.[161] The creation of additional carbon sinks by means of large-scale tree planting in developing countries has received much attention from Northern governments and electricity generating companies.[162] Experts agree on the existence of several hundred million hectares of degraded lands in the tropics that could be used for large-scale reforestation.[163] Although this seems an attractive option to sequester carbon in theory, in practice it is inferior both in terms of cost-effectiveness and social desirability to investment in forest conservation, smallholder farm forestry and natural regeneration of degraded forest areas. Moreover, forestry interventions are merely stop-gap measures;[164] aggressive efforts to reduce fossil fuel emissions in both developed and developing countries are the only way permanently to reduce global warming.[165]

The Intergovernmental Panel on Climate Change (IPCC) was established by the United Nations Environment Programme (UNEP) and the World Meteorological Organization (WMO).[166] In the run-up to the 1992 United Nations Conference on Environment and Development (UNCED) in Rio de Janeiro, a UN General Assembly resolution transferred control over climate negotiations from IPCC, UNEP and WMO to an International Negotiating Committee (INC) under the auspices of the UN General Assembly itself. This resolution received overwhelming support from the developing countries, who felt the specialized international agencies had been dominated by experts from industrial nations, and did not address the broader North–South political issues (debt, terms of trade, and technology transfer) they wanted to include in the climate change debate.[167] In the meantime, UNCED has been and gone, and produced little in terms of concrete measures. The developed countries have refused to accept responsibility for their role in causing global warming, and for taking the lead in reducing it.[168] The long-awaited United Nations Framework Convention on Climate Change (FCCC) was signed by 155 states, in a much weakened form mainly to accommodate US intransigence. Whereas scientists of the IPCC had calculated that an immediate reduction of carbon dioxide emissions by 60 per cent was necessary, the convention does not commit the developed world – the main producer of greenhouse gases – to any reductions at all.[169]

The unfortunate, if understandable, reluctance on behalf of politicians in industrialized countries to convince their electorates to forego some of the joys of consumerism and, to a lesser extent, the justified desire of less developed countries for a transfer of additional capital and technology from the industrialized countries, seem to have caused considerable delays in developing appropriate responses to the global warming problem. A rational response to global warming would seem to incorporate three basic elements: improving knowledge of the sensitivity of resource use systems to variations in climate and to environmental change; gradual adjustment to climate change; and expansion of the options to respond to change. The latter implies that mitigation measures should be based preferably on actions that can be easily and cheaply implemented and reversed, and that therefore do not limit future options. And even if the threat of climate change is somewhat uncertain, many measures that would help mitigate global warming also make sense for other reasons, eg upland tropical forest conservation, which contributes to the preservation of biological diversity and prevention of soil and water erosion as well as to carbon sequestration.[170]

NOTES AND REFERENCES

1. In reality, the tropics of Cancer and Capricorn are artificial boundaries, and eg some tropical rainforests are found outside the – geographically speaking – true tropics, such as in Australia (New South Wales), the Himalaya foothills (Upper Myanmar, Assam, southern China), and in Brazil (vicinity of Rio de Janeiro) and southern Madagascar. Boundaries with subtropical and temperate rainforests are mostly ill-defined. T C Whitmore, *An Introduction to Tropical Rainforests*, Clarendon Press, Oxford 1990.
2. Even in the Asian and Latin American tropics, where closed forests constitute the majority of forest cover, one-third of all forests are open woodlands.
3. These boundaries were much more vague in Medieval times; and pigs would be taken to the woods in autumn to fatten on oak and beechmast before being slaughtered in winter. Oakmast has long ceased to be of economic significance in Western Europe, but is still practised to some extent in the cork and holm oak parklands of upland Spain and Portugal. For a history of woodland management in Britain, see: Oliver Rackham, *Trees and Woodlands in the British Landscape*, J M Dent & Sons, revised edition, London 1990.
4. Duncan Poore and others, *No Timber Without Trees: sustainability in the tropical rainforest*, Earthscan Publications Ltd, London 1989. In many ecosystems, such as African savannas and several temperate forest types, it is impossible to define the primary (or equilibrium) vegetation, because natural disturbances are too large or infrequent, ephemeral events have long-lasting disruptive effects, or climate changes interrupt any move towards equilibrium that does occur. Douglas G Sprugel, 'Disturbance, Equilibrium, and Environmental Variability: what is 'natural' vegetation in a changing environment?' *Biological Conservation* **58**: 1–18, 1991.
5. The term tropical moist forests was coined in 1976 by Sommer to refer both to tropical rainforests and tropical seasonal (or monsoon) forests. T C Whitmore, 1990 op cit, p 9.
6. It is impossible to do justice to the variety and beauty of plant and animal life in tropical forests in a non-illustrated volume such as the present one. The reader is referred to T C Whitmore's *An Introduction to Tropical Rainforests*, Clarendon Press, Oxford 1990, for a useful overview. Vernon Heywood (ed), *Flowering plants of the World*, Oxford University Press, Oxford, 1978, provides information on many tropical forest plants and on economic uses of both timber and

non-timber species. However, it does not include conifers, palms, bamboos, ferns and tree ferns, which are important constituents of many tropical forest vegetation types.

7. This is possibly the consequence of the understandable tendency of many casual observers of tropical forests to view these physically impressive ecosystems as remnants from a millenarian past – which often they are not.
8. A well-documented example is that of moist forests in western central Africa (Cameroon, Congo and Gabon) recolonizing so-called paleoclimatic savannas. D Schwartz et Raymond Lanfranchi 'Les paysages de l'Afrique Centrale pendant le quaternaire', pp 41–45 in: Raymond Lanfranchi and Bernard Clist, eds. *Aux origines de l'Afrique Centrale*, Centres Culturels Francais d'Afrique Centrale and Centre International des Civilisations Bantu, Libreville 1991.
9. The terms 'tropics' and 'developing countries' are used interchangeably in this text. Strictly speaking this is not entirely accurate: there are tropical forests in some developed countries (eg Puerto Rico in the USA and Queensland in Australia) and there are developing countries outside the tropics.
10. See eg FAO, *Household food security and forestry: an analysis of socio-economic issues*, FAO Forests, Trees and People Programme Paper, Rome 1989; and FAO, *The major significance of 'minor' forest products: the local use and value of forests in the West African humid forest zone*, FAO Community Forestry Note No 6, Rome 1990.
11. There has been relatively little attention for the conservation status of non-tropical forests until recently, except in Australia, where the environmental movement has been actively involved with the subject for almost a decade. Criticism by environmental groups from the South of one-sided condemnations of tropical timber as being unsustainably produced by Northern activists has inspired the latter to look at the beam in their own eye. See eg Nigel Dudley, *Forests in Trouble: a review of the status of temperate forests worldwide*, World Wildlife Fund, Gland 1992.
12. The third component of global environmental change is stratospheric ozone depletion. Frederick H Buttel, 'The "Environmentalization" of Plant Genetic Resources: Possible Benefits, Possible Risks' *Diversity* 8: 1, pp 36–39, 1992. For a useful introduction to global change, see: Gareth Porter and Janet Welsh Brown, *Global Environmental Politics*, Westview Press 1991.
13. Poore and others, op cit 1989.
14. In this introduction, strict protection is referred to as forest preservation, whereas the term conservation refers to any management regime that retains natural forest cover, whether or not in modified form. Forest conservation, therefore, includes sustainable utilization for timber and non-timber forest products (see also Poore, this volume).
15. See Shepherd *et al*, this volume. For an exhaustive overview of local management systems for forests and other commons, see: BOSTID, *Proceedings of the Conference on Common Property Resource Management*, National Academy Press, Washington DC 1986.
16. In this context externalities are forest benefits which, although important to some people and/or to society as a whole, are not expressed in monetary terms. Such benefits are more at risk from the unregulated pursuit of private interests, since they derive no protection from their market value.
17. Quoted in Caroline Sargent, *Defining the Issues; some thoughts and recommendations on recent critical comments of TFAP*, International Institute for Environment and Development, London 1990.
18. Anil Markandya and Julie Richardson, *The Earthscan Reader in Environmental Economics*, Earthscan Publications, London 1992, p 18.
19. Jack Westoby, *The Purpose of Forests*, Basil Blackwell Publishers, Oxford 1987, p 118.
20. Peter Glueck, 'Das Elend der Kielwassertheorie', Internazionaler Holzmarkt 5/1982, pp 15–18.
21. eg the laws of Colbert in 17th century France.
22. Poore and others, op cit 1989.
23. Poore and others, op cit 1989, p 154.

24. There is also a lack of scientific knowledge of traditional silvopastoral management systems in savannah woodlands. Bonkoungou and Catinot 1986, quoted in Gill Shepherd, *Managing Africa's Tropical Dry Forests: a review of indigenous methods*. Overseas Development Institute, London 1992.
25. Lawrence S Hamilton, *Tropical forests: Identifying and clarifying issues*. An Overview Paper on Issues for the Tropical Forests Task Force of the Pacific Economic Cooperation Council, Kuala Lumpur, September 25–29, 1990.
26. This definition is similar to that chosen by FAO: a complete clearing of (closed or open) tree formations and their replacement by non-forest land uses. Singh, K D and others, 'A Model Approach to Studies of Deforestation', DEFR 3, FAO, Rome 1990.
27. 'Rotational bush fallowing' is a more accurate description of the practices of (relatively) sedentary indigenous forest farmers than shifting cultivation, which is often used to refer to the disorganized slash-and-burn farming practised by recent colonists. Paul Richards, *Indigenous agricultural revolution: ecology and food production in West Africa*, Unwin Hyman, London 1985.
28. 80,000 square kilometres in *World Resources 1990–91*, World Resources Institute, Oxford University Press, 1990; *vs* 35,000 square kilometres in T C Whitmore and J A Sayer, *Tropical Deforestation and Species Extinction*, IUCN Forest Conservation Programme Paper, Chapman and Hall and the World Conservation Union, 1992. For a detailed technical critique of the methods used by Setzer *et al* (1988) to arrive at the original 80,000 square kilometres estimate, see: Philip M Fearnside, 'The Rate and Extent of Deforestation in Brazilian Amazonia' *Environmental Conservation,* **17** (3), Autumn 1990.
29. The establishment of causality is even more contentious where the alleged consequences of deforestation are concerned. But the fact that two events occur more or less simultaneously does not mean there is a causal link between them. Major floods such as the ones occurring in Bangladesh are caused by unusually heavy or prolonged rains that exceed the ability of soils to store and streams and rivers to drain rainwater without bank overflow, not by deforestation. Hamilton 1990, op cit.
30. Theodore Panayotou, 'The economics of environmental degradation' in: Markandya and Richardson (eds), 1992 op cit.
31. David W Pearce and Jeremy J Warford, *World without End: economics, environment and sustainable development*, published for the World Bank by Oxford University Press, 1993.
32. Matti Palo, 'Deforestation and development in the Third World: Roles of system causality and population' pp 155–72 in: Palo, Matti and Jyrki Salmi (eds) *Deforestation or Development in the Third World?*, Volume III, Finnish Forest Research Institute, Helsinki 1990.
33. It is interesting to note that our ecologist, by the same standard, could have called a cabbage field in the eastern part of The Netherlands a degraded oak-birch forest – however absurd this may seem.
34. For a discussion of how institutional interests 'generate' uncertainty in Nepal, see Michael Thompson and Michael Warburton, 'Uncertainty on a Himalayan scale, Mountain Research and Development' Vol 5: 2, pp 115–35, 1985.
35. Charles M Peters, Alwyn H Gentry and Robert O Mendelssohn, 'Valuation of an Amazonian rainforest' *Nature* **339**: 655–56, 29 June 1989. There has been much criticism of the valuation techniques used in the article by Peters *et al*, see eg Ricardo Godoy and Ruben Lubowski, Guidelines for the Economic Valuation of Nontimber Tropical-Forest Products, *Current Anthropology*, Vol 33 No 4, pp 423–33, 1992.
36. For an overview of the history of economic theory with regards to natural resources and the environment, see: Edward B Barbier, *Economics, Natural Resource Scarcity and Development: conventional and alternative views*, Earthscan, London 1989.
37. Robert Repetto, William Magrath, Michael Wells, Christine Beer and Fabrizio Rossini 'Wasting assets: natural resources in the national income accounts' pp 364–87 in: Anil Markandya and Julie Richardson (eds), *The Earthscan Reader in Environmental Economics*, Earthscan, London 1992. The International Tropical Timber Organization has recently started some pilot efforts in tropical forest resource accounting in Cameroon and Ecuador. NB. If

resource accounting is done in monetary terms, physical resource depletion may still result, if the rise in unit value of the remaining resource compensates for volume loss.

38. Alf Leslie 'A second look at the economics of natural management systems in tropical mixed forests' *Unasylva* **155**, Vol 39: 1, pp 46–58, 1987. Eg the discount rate used by Peters *et al*, 1989 op cit was a low 5 per cent. High discount rates, often blamed for making forestry investments (which generally require long periods to mature) unattractive, have a positive side as well. If ruthlessly applied, they will prevent the execution of environmentally damaging and wasteful development projects which are not capable of producing sufficiently high returns. Many of the most environmentally damaging development projects in tropical forest areas were also economically disastrous. David Burns 'Runway and Treadmill Deforestation' IUCN/IIED Tropical Forest paper No 2, International Institute for Environment and Development, London 1986.
39. Since it cannot be measured directly, existence value has to be derived from contingent (indirect) valuation methods, many of which are based on questionnaire surveys. There are important methodological problems in the field of contingent valuation. See eg D L Coursey, J L Hovis and W D Schulze 'The disparity between willingness to accept and willingness to pay measures of value' pp 92–100, in: Markandya and Richardson, 1992 op cit.
40. If adults in the developed world would be willing to pay $8 per year for the Amazon to continue to exist as is, then existence value would amount to $3 billion, or one-quarter of the entire contribution of the region to Brazil's gross domestic product. Pearce and Warford, 1993 op cit, pp 102–33.
41. Indeed, terms such as 'bequest value' and 'social discount rate' suggest moral inspiration. For an unashamedly moral plea to redefine economics, see Herman E Daly and John B Cobb Jr, *For The Common Good: Redirecting the economy toward community, the environment and a sustainable future*, Beacon Press, Boston 1989.
42. With thanks to Stephen J Bass, 1991 personal communication, for introducing me to the concept of island forestry.
43. For a comprehensive overview of the history of human views of nature, see Max Oelschlaeger, *The Idea of Wilderness: From prehistory to the age of ecology*, Yale University Press, New Haven and London 1991.
44. Jonathan Adams and Thomas McShane, *The Myth of Wild Africa*, Norton 1992. Whereas in Kenya populations of elephants have decreased dramatically over the last 30 years, in Zimbabwe they have been stable, or even increasing.
45. This question does not suggest that poor people are the most important agents of tropical deforestation; tropical forests offer economic opportunities to rich and poor people alike.
46. At the Earth Summit in Rio in 1992, a 'non-legally binding authoritative statement of principles for a global consensus on the management, conservation and sustainable development of all types of forests' was adopted, rather than the forest convention many developed countries were lobbying for. See eg GLOBE 'Rationale for a Convention for the Conservation and Wise Use of Forests' Aid-Environment, Amsterdam 1991.
47. No article on climate change and tropical forests has been included, but the pertinent issues have been summarized in this introduction instead.
48. The concept of marginality is discussed in the excerpt from Blaikie and Brookfield's *Land Degradation and Society*.
49. Michael Dove 'The Agroecological Mythology of the Javanese, and the Political Economy of Indonesia' *Indonesia* **39**: 1–36, 1985. See also David Anderson (this volume), for forests constituting a refuge for Kenyan pastoralists in the face of increasing pressure in colonial times.

    A present-day example of the powers of evasion vested in forest dwellers is that of the relations between 'pygmy' hunter-gatherers and Bantu and Nilotic forest farmers in Africa. Pygmies can subsist on hunting and gathering, but prefer to barter game and labour against farm produce and metal implements to improve their quality of life. If they perceive the terms of trade as being disadvantageous, however, they will revert to stealing farm produce and/or moving to another part of the forest. NB. The term 'pygmy' is commonly used to refer to the forest-dwelling groups of small stature whose presence in Central Africa predates the arrival of Bantu and other negro races in the region. These groups prefer to refer to themselves as Baka, Bambuti, Aka etc: the names of their individual ethnic groups.

50. Jack Westoby, at one point Director of Forestry at FAO, argues that foresters have assisted governments in achieving such aims: 'For centuries much of the work of foresters went into creating and protecting royal and princely estates, extinguishing every kind of common right in the forest and enforcing exclusive property.' (*The Purpose of Forests*, Basil Blackwell Publishers, Oxford 1987, p 254).
51. The commonly quoted rationale for the Indonesian transmigration programme was to reduce overpopulation on Java, but in view of the fact that resettlement at the height of the programme in the late 1970s and early 1980s was unable to move more than a fraction of the island's annual population growth, this is unlikely to have constituted a major reason.
52. The tree crop (mainly oil palm and rubber) settlement schemes in Peninsular Malaysia – even if they have negatively impacted on forest dwellers' livelihoods in some cases, and are likely to have caused some loss of biodiversity – were exceptional in three ways: they were economically successful; they did not provoke major land degradation; and commercial timber was systematically salvaged, not burnt.
53. Jack Westoby, 1987 op cit, page 229.
54. The use and management of fire was acquired early in human evolution, at least one to two million years ago, by *Homo erectus*. This raises the possibility that hominids used fire to clear large tracts of savanna quite early in prehistory. Karl Leopold Hutterer 'The Natural and Cultural History of Southeast Asian Agriculture: Ecological and Evolutionary Considerations' *Antropos* **78**: 169–212, 1978.
55. Eg Spencer (1966) came to the startling conclusion that all seasonal forests in continental South East Asia have been cleared by swidden cultivators at one time or another. *The Conservation Atlas of Tropical Forests: Asia*, Simon and Schuster and World Conservation Union, 1991, p 31.
56. Terborgh (1986b) ascribes extinctions of various large mammals in Pleistocene Amazonia to human hunters, quoted in J P Skorupa, 'The effects of selective timber harvesting on rainforest primates in Kibale Forest, Uganda'. Unpublished PhD thesis, University of California, Davis 1988.
57. A swidden is a temporary farm field established in a forest, through slashing and often burning of part of the forest vegetation.
58. In other areas which lost their forest cover early on forests have regenerated after the civilizations that cleared them perished, as witnessed by the jungle-covered temple complexes in eg Cambodia (Angkor Wat) and Central America (Mayan ruins). Sedentary agriculture is thought to have emerged first in the Indus valley and in South East Asia around 8–10,000 BP.
59. Many of these smallholder cash crop gardens are mixed-species stands which retain considerable biological diversity. See eg Michon and de Foresta (1990) and Alcorn (1984).
60. In many instances it is impossible to make clear distinctions between agriculture and gathering; rather there is a continuum from the definitely gathered to the definitely cultivated, through varying degrees of vegetation management and manipulation. Christine Padoch and Andrew Vayda 'Patterns of Resource Use and Human Settlement in Tropical Forests' in: F B Golley and H Lieth (eds), *Structure and Function of Tropical Forests*, Elsevier 1980. For an overview of the literature on the importance of wild plants and animals in tropical agricultural systems, see: Ian Scoones, Mary Melnik and Jules Pretty, *The Hidden Harvest: Wild Foods and Agricultural Systems*, a literature review and annotated bibliography, International Institute for Environment and Development, London 1992.
61. Some people in swiddening communities for instance are specialized forest product gatherers, such as the rattan collectors among the Dayak on Kalimantan, or hunters.
62. This is one of the main causes of the controversy about overgrazing in the African savannahs and steppes. The well known TV pictures of complete denudation near waterpoints are misleading because this is caused by hoof action and occurs only over very limited areas. Over the whole of the pasture, grass productivity falls initially in response to livestock build-up, and then stabilizes at a low level; it does not decrease to zero. Katherine Homewood and

W A Rogers 'Pastoralism, conservation and the overgrazing controversy' pp 111–28 in: David Anderson and Richard Grove (eds), *Conservation in Africa: people, policies and practice*, Cambridge University Press 1987. See also: Ridley Nelson, *Dryland Management: The 'Desertification' Problem*, Environment Department Working Paper No 8, The World Bank, Washington, DC 1988.

63. Generations of observers have been alarmed by the steady increases in human and cattle populations in Africa. But the picture changes completely if we take the history of the continent into account. At 30 million, the population of present-day Zaire has still not recovered from the onslaught of the colonial era, which reduced the population of the Belgian Congo, for instance, from 40 million in 1880 to 9.25 million in 1930. Similarly, over 90 per cent of the African cattle population was eliminated at the start of the colonial era by the rinderpest pandemic introduced to Eritrea in 1887 and reaching the Cape in 1900. R H V Bell, 'Conservation with a human face' in: Anderson and Grove, op cit 1987, p 82–83.
64. For instance, many studies quote woodfuel deficits for areas where people have long switched to agricultural residues for their energy needs. See also Leach and Mearns, this volume.
65. Michael Thompson, Michael Warburton and T Hatley, *Uncertainty on a Himalayan scale*, Milton Ash, London 1986.
66. Robert Chambers, *Rural Development: Putting the last first*, Longman, Harlow 1983, pp 10–26.
67. Many recent settlers in tropical moist forests have little interest in sustainability. Whether they are small-time slash-and-burn farmers or big-time ranchers and loggers, they have come into the forest in order to get out again, only richer. And even among indigenous peoples who have dwelled in forest areas for generations, there are some notorious examples of forest destroyers: eg part of the Iban Dayak of Borneo, and the Aouan in Cote d'Ivoire.
68. Gerald Leach and Robin Mearns, *Beyond the Woodfuel Crisis: people, land and trees in Africa*, Earthscan, London 1988, p 1.
69. For instance, some authors attribute the sudden drop in deforestation rates in the Brazilian Amazon starting in 1988 in part to the cancellation of tax credits for agricultural development (which amounted to subsidized forest clearing). Other observers maintain that subsidies, even if they were illegal, continued to be paid and that weather conditions and worsening economic conditions in Brazil were the main causes of slowing deforestation rates. Nels Johnson and Bruce Cabarle, *Surviving the Cut: Natural forest management in the humid tropics*, World Resources Institute, Washington, DC 1993, p 10.
70. See also Blaikie and Brookfield, this volume.
71. For a series of interesting forest conservation case studies, see Judith Gradwohl and Russell Greenberg, *Saving the Tropical Forest*, Earthscan, London 1988.
72. Small quantities of specialty timbers are imported from tropical dry forests and open woodlands, eg *Dalbergia melanoxylon*, blackwood, an African heavy hardwood that is used for making musical instruments.
73. The earliest documented examples of these date back to the start of the European colonial expansion in the tropics, such as the over-exploitation of island rosewoods in the Caribbean in the eighteenth century. Sara Oldfield, *Rare Tropical Timbers*, IUCN Tropical Forest Programme Paper No 6, 1989. Economic extinction refers to the situation where the market for a certain species has ceased to exist because it is very rare or restricted to inaccessible areas in the wild, and has not been brought into cultivation. Physical extinction of timber species is more commonly caused by habitat loss (eg forest conversion to agriculture) than by overharvesting, except where harvesting practices impede regeneration (eg when it is profitable to harvest small sizes).
74. It was estimated that half of all logged-over forest in the main African tropical timber producer countries, would be subsequently deforested in the 1981–85 period. FAO/UNEP, *Forest Resources of Tropical Africa*, part 1: regional synthesis, FAO Rome 1981.
75. *ITTO Guidelines for the Sustainable Management of Natural Tropical Forests*, ITTO Technical Series No 5, International Tropical Timber Organization, Yokohama 1990. No producer countries have yet completed national forest management guidelines based on this document,

as was intended. Most primary forests outside protected areas in South East Asia, which provides 90 per cent of the international tropical timber trade, will have been logged over by the year 2000. As various observers have noted, for some of the world's forests this target will be too late, even if it is achieved. See eg Pearce and Warford, op cit 1993, p 293.

76. Trade creation was the most important reason for producer countries to sign the ITTA. The removal of tariff and non-tariff barriers to tropical timber imports into developed country markets would have generated an additional trade of $150 million and $690 million respectively in 1976 and 1980. I J Bourke 'Trade in forest products: a study of the barriers faced by the developing countries' FAO Forestry paper No 83, Rome 1988.
77. For an overview of the strategic importance of wood in supporting the expansion of major civilizations from 4000 years BP (Mediterranean basin) until last century (USA) see: John Perlin, *A Forest Journey, the role of wood in the development of civilization*, W W Norton and Co, New York and London, 1989, p 25–31.
78. Duncan Poore and others, *No Timber Without Trees: sustainability in the tropical rainforest*, Earthscan, London 1989. The 0.125 per cent statistic has often been quoted out of context. The report qualified the figure by adding that many of the remaining forests, especially in remote areas, are logged very selectively and that a substantial proportion of these, although not under formal management, may not be deteriorating. The reason for not granting such 'log-and-leave' forests the predicate 'managed', however, is that logging standards have not been established, or are not controlled, and that they are vulnerable to encroachment by pioneer farmers (Poore and others, op cit pp 191–193).
79. A recent study of the forest history of Burma (Myanmar) found that although these conditions had not been fulfilled, the country had managed its forest remarkably well for more than 100 years. Alex Mc Gillivray, unpublished MSc thesis.
80. See eg 'A Hard Wood Story' 'Timber!' 'The Good Wood Guide' (published by Friends of the Earth); 'ITTO Position Papers 1–3' 'Timber from the South Seas' (published by the World Wildlife Fund); various articles written by the World Rainforest Movement and the Rainforest Information Centre and published in the magazine *The Ecologist*. For an overview of tropical timber campaigns and the government and private sector responses to them, see Robert Goodland 'Race to Save the Tropics' Island Press 1990.
81. According to Colchester, the argument that forests can be saved only by making them more lucrative is entirely spurious; higher tropical timber prices are likely to lead to faster mining of forests. In the present context, this may well be the case. On the other hand, where a reliable management system is in place, strong forest product markets can enhance management, eg in Uganda where forest managers were able to increase the share of marketable timber species by selling trees unsuitable for timber to charcoal burners. FAO 'Management of Tropical Moist Forest in Africa' FAO Forestry Paper No 88, FAO, Rome 1989.

    The tropical timber plantation record has been succinctly summarized as follows: 'At their best, industrial plantations can become a major asset to local development by providing raw materials, infrastructure, employment, income and environmental and recreational services. At their worst, plantations, usually imposed from a "top-down" perspective and ignoring local needs, values and rights, have monopolized land in times of food shortage, degraded wild animal and plant populations, and destroyed habitats and landscapes.' Caroline Sargent and Stephen M J Bass (eds), *Plantation Politics: Forest plantations in development*, Earthscan, London 1992.

    For a standard work on tropical timber plantations, see: Julian Evans, *Plantation Forestry in the Tropics*, Oxford University Press, 1991 (revised edition).
82. Eg the 'Real Wood Guide' published by The UK Timber Trade Federation as part of its Forest Forever campaign; the leaflet 'Save the rainforest – buy tropical timber products' published by the Danish Timber Trade federation and the Association of Danish Furniture Industries. For a more balanced view from within the trade the reader is referred to the monthly bulletin 'Tropical Timbers: market information for producer nations'.
83. Only a small part of the tropical timber trade consists of unique products such as decorative veneers and cabinet timbers. Most tropical timber products are utility items which are

vulnerable to substitution by temperate timber and non-timber (aluminium, PVC) products.

84. In response to this, the Rainforest Alliance's 'Smartwood' programme offers sustainability certificates at different levels. Richard Donovan, 1992 personal communication. Similarly, the 1990 ITTO mission that visited Sarawak to assess the state's forest management practices distinguished four elements of sustainability (sustainable timber yields, sustainable catchment management, biological diversity, economic sustainability), each of which were separately assessed and scored. *The promotion of sustainable forest management: a case study in Sarawak, Malaysia*, International Tropical Timber Organization, Yokohama 1990.
85. Nels Johnson and Bruce Cabarle, *Surviving the Cut: Natural forest management in the humid tropics*, World Resources Institute, Washington, DC 1993, p 43.
86. One of the problems is caused by the fact that tropical forest resources are very heterogeneous. Therefore, even large-scale processing enterprises holding vast forest concessions regularly have to buy in timber from small logging enterprises to feed their mills. This makes verification extremely difficult and costly.
87. Fred Pearce 'Timber trade stripped of its green veneer' *New Scientist* 134: p 4, 1992.
88. Eg 'Smart Wood' from the Rainforest Alliance, and 'Green Cross' from Scientific Certification Systems. Johnson and Cabarle, 1993 op cit, pp 42–43.
89. Examples of this approach are: Ecological Trading Company (timber); Traidcraft (finished wood products); and the Body Shop (non-timber forest products).
90. See eg Johnson and Cabarle, 1993 op cit, pp 28–29, on the flaws in the silvicultural system applied by small-scale logging cooperatives (ejidos) in Quintana Roo, Mexico.
91. Johnson and Cabarle, 1993 op cit, p 18 Listed on Appendix I: *Dalbergia nigra* (Brazilian rosewood); on Appendix II: *Pericopsis elata* (afrormosia), *Swietenia mahogani* (Central American populations of true mahogany) and *Guaiacum officinale* (lignum vitae). Intense lobbying by producer countries caused four other species proposed for Appendix II to be withdrawn: *Swietenia macrophylla* (American mahogany), *Schinopsis* spp (quebracho), *Gonystylus bancanus* (ramin, a widely traded species) and *Intsia* spp (merbau).
92. The quote continues: 'There are inherent problems in using trade controls to achieve species conservation aims. The variability of the trade, the informational requirements involved in effectively implementing trade controls, and the difficulties facing implementors attempting to detect and penalize trade in violation of CITES' goals have combined to divert participants' attention away from CITES' original conservation mandate . . . Moreover significant regulatory costs are imposed on largely legitimate portions of the wildlife industry.' Mark C Trexler 'The Convention on International Trade in Endangered Species of Wild Fauna and Flora: Political or Conservation Success'. Unpublished paper, 1989.
93. Bass, Stephen, Duncan Poore and Bart Romijn 'ITTO and the Future in Relation to Sustainable Development' a report prepared by AIDEnvironment and the International Institute for Environment and Development, Amsterdam and London, 1992.
94. 'Trade measures may be an extremely blunt, and in many cases an inappropriate, tool for eliciting the proper incentives for sustainable timber management . . . If combatting global deforestation is the goal, then multilateral measures should focus less on trade interventions and more on designing appropriate international environmental agreements to deal with this problem directly.' Edward B Barbier, Bruce A Aylward, Joanne C Burgess and Joshua T Bishop 'Environmental Effects of Trade in the Forestry Sector' Paper prepared for the Joint Session of Trade and Environment Experts, (OECD Paris), International Institute for Environment and Development, London 1991, p 44.
95. K F Wiersum (ed) 'Nederlandse aktiviteiten en Regeringsbeleid ten aanzien van tropische bossen' ('Dutch activities and government policies concerning tropical forests'), Staatsbosbeheer, Utrecht 1985. This is despite the fact that The Netherlands had the highest per capita consumption of tropical timber of all the countries in the European Community at the time of the study.
96. But see Pearce and Warford, 1993 op cit, p 293–94, for a critique of the attribution of deforestation to the 'cassava connection'.
97. Geneviève Michon 'De l'homme de la forêt au paysan de l'arbre: agroforesteries

indonésiennes' Unpublished PhD thesis, USTL Montpellier, 1985.
98. F L Dunn *Rainforest collectors and traders: a study of the resource utilization in modern and ancient Malaya*, Monographs of the Malaysian Branch of the Royal Asiatic Society 5, 1975.
99. Julia Falconer 'Non-timber forest products in Ghana's high forest zone' ODA 1992.
100. Poore *et al*, 1989 op cit, p 15.
101. For an overview of the impact of non-forestry development aid on tropical forests see the IUCN Conservation Atlases of Tropical Forests on Africa and Asia, and Susanna Hecht and Alexander Cockburn *The Fate of the Forest: Developers, destroyers and defenders of the Amazon,* Penguin Books 1990. For an excellent introduction to agroforestry development projects and the reasons for their success or failure, see Paul Kerkhof *Agroforestry in Africa: A survey of project experience,* PANOS, London 1991.
102. '. . . rural development or social forestry projects and environmental forestry projects have proven difficult to formulate and implement. Projects of this kind tend to accentuate intrinsic sociopolitical and institutional weaknesses and if so, they compromise sustainability in the long term.' *World Bank, Forestry: The World Bank's experience* A World Bank Operations Evaluations Study, Washington, DC 1991.
103. Raymond Noronha and John S Spears 'Social Variables in Forestry Project Design' pp 227–266; and Michael M Cernea 'Alternative Units of Social Organization Sustaining Afforestation Strategies' pp 267–293; in Michael M Cernea (ed) *Putting People First: Sociological Variables in Rural Development* World Bank and Oxford University Press, 1985. On the importance of forest resources for the livelihoods of poor people and women, see: Robert Chambers, Czech Conroy and Melissa Leach *Trees as Savings and Security for the Rural Poor* International Institute for Environment and Development, London 1993; and Irene Dankelman and Joan Davidson *Women and Environment in the Third World: Alliance for the future* Earthscan/IUCN, London 1988.
104. 'The idea of a "take-off into sustained growth" was introduced in 1960 by professor William Rostow of the Massachusetts Institute of Technology in a book subtitled *A non-communist manifesto*.' David Burns, Runway and Treadmill Deforestation, IUCN/IIED Tropical Forest Policy Paper No 2, London 1986.
105. World Commission on Environment and Development *Our Common Future* Oxford University Press, Oxford 1987, is commonly referred to as the Brundtland report.
106. FAO Committee on Forest Development in the Tropics, Draft Proposals for Action Programmes in Tropical Forestry, FAO: FDT/85/3, April 1985.
107. *Tropical Forests: A Call for Action* Report of an International Task Force convened by the World Resources Institute, the World Bank, and the United Nations Development Programme, Washington, DC 1985.
108. *Tropical Forestry Action Plan* Food and Agricultural Organization of the United Nations, Rome 1985 (second edition 1987).
109. *Tropical Forestry Action Plan* Food and Agriculture Organization of the United Nations in cooperation with United Nations Development Programme, World Bank and World Resources Institute, Rome 1987.
110. For example, from the FAO's 1988 update of its 1980 tropical forests assessment.
111. Marcus Colchester and Larry Lohmann *The Tropical Forestry Action Plan: What Progress?* World Rainforest Movement and *The Ecologist*, Penang and Sturminster Newton 1990.
112. Multilateral and bilateral funding for tropical forestry was estimated to have risen from $0.6 billion in 1984 to $1.0 billion in 1988. Caroline Sargent *Defining the Issues: some thoughts and recommendations on recent critical comments on TFAP* International Institute for Environment and Development, London 1990.
113. Robert Winterbottom *Taking Stock: The Tropical Forestry Action Plan after five years* World Resources Institute, Washington 1990.
114. Ullsten, O, Salleh Mohamed Nor and M Yudelman 'Tropical Forestry Action Plan: Report of the Independent Review' Kuala Lumpur 1990.
115. Winterbottom, op cit 1990.
116. 'Vivimos en nuestro primer y unico mundo'. CONAIE stands for Confederacion de

Nacionalidades Indigenas de Ecuador, an Ecuadorian indigenous peoples' organization. Some authors refer to tribal peoples as first peoples, and to tribal areas as the Fourth World, see eg Julian Burger *The Gaia Atlas of First Peoples: A future for the indigenous world* Anchor Anchor Books, Doubleday 1990.

117. Ribeiro (1970), quoted in Darrell Posey 'Intellectual Property Rights and just compensation for indigenous knowledge' *Anthropology Today* **6**: 4, August 1990.
118. Some of the knowledge of extinct indigenous groups does survive, however, eg among the 'caboclos', descendants of detribalized indigenous peoples in the Amazon. Clay (1988) cited in Posey 1990, op cit.
119. Definitions of the term 'indigenous peoples' have been proposed in the United Nations Working Group on Indigenous Populations Document E/CN.4/SUB.2/1986/7, and the International Labour Organization Convention 169. 'Indigenous Peoples and the Tropical Forestry Action Plan' unpublished paper, World Resources Institute, Washington, DC 1990.
120. Eg highland farmers in Latin America and nomadic pastoralists in Africa. Strictly speaking, the term 'indigenous peoples' can in Africa only be applied to the very small remaining groups of hunter-gatherers (pygmies and San bushmen), since all other human populations are relatively recent immigrants. Nevertheless nomadic and transhumant pastoralists are often called indigenous peoples because of the many characteristics they share with them.
121. African indigenous peoples number 25 million, of which only 150,000 are forest dwellers: the Aka, Baka, Mbuti and Twa (Efe?) hunter-gatherers. South American indigenous peoples number 15 million, of which 1 million are in the Amazon, and 0.5 million dwell in other (sub-) tropical forest areas. Central American indigenous peoples number 13 million, the majority of which are highland farmers. South and South East Asia and Oceania contain the majority of indigenous tropical forest dwellers, several tens of millions. Burger, 1990 op cit.
122. Julie Sloan Denslow and Christine Padoch '*People of the Tropical Rainforest*', University of California Press and Smithsonian Institution, 1988.
123. Turnbull (1961) ascribed the fact that certain pygmy groups in the Ituri forest practised bow-and-arrow in small groups rather than net hunting in large groups to the influence of capitalism, whereas Ichikawa (1983) demonstrated that this was due to differences in the forest environment of the localities concerned. C M Turnbull *The Forest People* Triad/ Paladin, Grafton Books, London 1961. Ichikawa 'An examination of the hunting-dependent life of the Mbuti pygmies, Eastern Zaire' *African Study Monographs* **4**: 55–76.
124. To quote one researcher: 'Mortality and morbidity among children are very high. Many children do not survive to puberty, and few adults reach the age of 50 or 60. Malaria, infant diarrhea, bilharzia, filariasis, intestinal worms, amoebic dysentery, tropical ulcers, ringworm, pneumonia, leprosy, tuberculosis, sleeping sickness, and sickle-cell anemia are just some of the diseases that sap the strenth of all forest dwellers.' David S Wilkie 'Hunters and farmers of the African forest' chapter 7 in: Denslow and Padoch, 1988 op cit.
125. For a well-documented example of the loss of the conservationist ethic through the influence of modern scholarization, see Mac Chapin's article on the Kuna indians of Panama 'Losing the way of the Great Father' *New Scientist*, 10 August 1991.
126. Major aspirations and rights claimed by indigenous peoples worldwide, such as rights of communal ownership to ancestral lands, legal recognition of indigenous organizations, and rights to self-determination, are summarized in Document W/CN.4/SUB.s/1986/7 of the United Nations Working Group on Indigenous Populations and Conventions 107 and 169 of the International Labour Organization. Lynch, 1990 op cit.
127. Calestous Juma '*The Gene Hunters: Biotechnology and the scramble for seeds*' Zed Books, London 1989, p 238.
128. Duncan Poore and Jeffrey Sayer 'The Management of Tropical Moist Forest Lands: Ecological Guidelines' IUCN Forest Conservation Programme Paper No 1, Gland 1991 reprint.
129. Many of these are perennial crops. For an overview, see: Nigel J H Smith, J T Williams, Donald L Plucknett, P Greening and Jennifer Talbot *Tropical Forests and Crop Genetic Resources* Princeton University Press 1991.

130. Ex-*situ* conservation of plant genetic material has proved useful in tropical forestry; eg 40 per cent of the source areas of *Swietenia* mahogany collected in Puerto Rico have since been converted to agriculture, and many Nicaraguan pine provenances exist now only in seed collections and plantations (John Palmer, personal communication 1991).
131. In Europe, where there are no true natural forests left, ancient agro-ecosystems and silvicultural systems are the main reservoir of biological diversity. Although the reverse is true in most of the tropics, there is much useful biodiversity in tropical agro-ecosystems. For an account of the biodiversity in Indonesian agroforestry systems, see Genevive Michon and Jean-Marie Bompard 'Agroforesteries indonésiennes: contributions paysannes à la conservation des forêts naturelles et de leurs ressources' *Rev. Ecol. (Terre Vie)* **42**: 3–37, 1991. Crop populations growing and breeding in adverse circumstances (ie rainfed fields) are a key source of plant genetic material that can be used for breeding to mitigate the impact of climate change on future agricultural harvests. John Palmer, personal communication 1991. See also Martin Parry *Climate change and world agriculture* Earthscan, London 1990.
132. A first round of analysis would involve recognition of global priority areas through the measurement of both species richness and their taxonomic distinctness; within these areas a second round of analysis would identify a network of reserves to contain all local taxa and ecosystems. R I Vane-Wright, C J Humphries and P H Williams 'What to Protect – Systematics and the Agony of Choice' *Biological Conservation* **55**: 235–54, 1991.
133. For a general review of the relevance (or lack of it) of ecological research for the management of protected areas, see: Craig L Shafer *Nature Reserves: Island theory and conservation practice* Smithsonian Institution Press, Washington and London, 1990.
134. This is part of the more general problem that genetic resources present in protected areas are often poorly known and documented, and therefore of limited utility. Robert and Christine Prescott-Allen '*Genes from the Wild*' Earthscan, London 1988, p 94. Many wild species similarly thrive on some degree of disturbance, eg in the African tropical moist forest biome, the late secondary stages of forest succession are thought to be more diverse than the climax forest. FAO *Management of Tropical Moist Forests in Africa* FAO Forestry Paper No 88, Rome 1989.
135. Michael Brown and nn, 1993; Shepherd *et al* (this volume).
136. Jeffrey Sayer *Rainforest Buffer Zones: Guidelines for Protected Area Managers* IUCN Forest Conservation Programme Paper No 2, Gland 1991.
137. See Whitmore, 1990 op cit, pp 28–32, for a detailed discussion of plant diversity in tropical forest ecosystems.
138. R and C Prescott-Allen, 1988 op cit, p 91. An example of the range of genetic variation in the wild is the bark of certain Bolivian varieties of *Cinchona* which contains up to 13 per cent quinine, whereas the bark of other varieties contains no more than 2 per cent of this anti-malaria drug. Catherine Caulfield *In the Rainforest* Heinemann 1985, pp 209–17.
139. Erwin, T L, 1982 'Tropical forests: their richness in Coleoptera and other arthropod species' *The Coleopterists' Bulletin* 36, pp 74–75. His estimate is based on the extrapolation of the species numbers he obtained from sampling the insects living in 19 trees over three seasons in a scrubby tropical seasonal forest in Panama. His figures were later re-interpreted by Stork (1988, quoted in T C Whitmore and J A Sayer *Tropical deforestation and species extinction* IUCN Forest Conservation Programme Paper, Chapman & Hall and the World Conservation Union, 1992), who arrived at an estimate of 20–80 million species worldwide.
140. E O Wilson, in Whitmore and Sayer 1992, op cit.
141. Whitmore, 1990 op cit, p 58.
142. Whitmore and Sayer, 1992 op cit.
143. W V Reid 'How many species will there be?' in: Whitmore and Sayer, 1992 op cit.
144. Whitmore and Sayer, 1992 op cit.
145. Whitmore and Sayer, 1992 op cit. See also the *Databook on endangered tree and shrub species and provenances* FAO Forestry Paper No 77, Rome 1986.
146. The only published example of tropical forest ecosystem restoration is the work of Dan Janzen in the tropical dry forests of Costa Rica. See references in Johnson, this volume. On

the importance of dispersal pathways, see: *Nature Conservation: The Role of Corridors* (Proceedings of a Conference at Busselton, 11–15 1989), Surrey-Beatty, Sidney 1990.

147. Many species of vertebrates are able to persist in logged-over forests, especially if ecologically sustainable modes of harvesting and silviculture are applied. Andy D Johns 'Species conservation in managed tropcial forests' chapter 2 in: Whitmore and Sayer, 1992 op cit. See also: Jill M Blockhus, Mark Dillenbeck, Jeffrey A Sayer and Per Wegge (eds) *Conserving biological diversity in managed tropcial forests* IUCN Forest Conservation Programme Paper No 9, IUCN/ITTO, Gland 1992.
148. Buttel, 1992 op cit, attributes this 'environmentalization of plant genetic resources' to the fact that calls for social justice evoked little response from the neoconservative regimes that ascended to power in nearly all member states of the OECD in the 1980s.
149. More than 150 countries signed the biodiversity convention; the only important country not among the signatories was the USA. Their absence may not make much difference, as 'US companies are already drawing up deals to fulfill the convention's purpose, providing Western funds for conservation in return for access to Third World genetic resources.' Fred Pearce 'How Green Was Our Summit' *New Scientist*, 27 July 1992.
150. R and C Prescott-Allen, 1988 op cit, pp 68–70.
151. Annual averages of global surface air temperatures in the 11 years from 1980–1990 were 0.1–0.4 degrees Celsius higher than the 1950–79 average. The year 1990 was the warmest on record worldwide, and it was the fourth time in ten years that the global temperature record had been broken. *Tiempo, Newsletter on Global Warming and the Third World* Vol 1: 1, International Institute for Environment and Development, London and University of East Anglia, Norwich, April 1991.
152. 'An assault on the climate consensus' by John Gribbin, pp 26–29 *New Scientist* 15 December 1990.
153. pp 2–3 in 'The Allocation of Responsibility' unpublished paper by P M Kelly, School of Environmental Sciences, University of East Anglia, Norwich 1991.
154. Feedback occurs when an indirect consequence of a certain process reinforces (positive feedback) or decreases (negative feedback) its impact. Increased water-vapour content of the atmosphere as a result of global warming is expected to further increase temperature, whereas the impact of increased cloud cover is still poorly understood 'Positive about water feedback' by Robert D Cess *Nature* Vol 349, 1991.
155. Watson *et al* 1990, cited in Kelly 1991. NB. 1 gigatonne is one thousand million tonnes. The Intergovernmental Panel on Climate Change (IPCC) has estimated carbon dioxide emissions due to tropical deforestation at 1.0 to 2.0 Gt of carbon equivalent per year, based on an annual tropical moist forest deforestation rate of 10 plus or minus 5 million hectares. Intergovernmental Panel on Climate Change *Tropical Forestry Response Options to Global Climate Change*, Conference Proceedings Sao Paulo, January 1990.
156. IPCC, op cit 1990.
157. Collins, 1991 op cit.
158. T H Smith, H H Shugart and P N Halpin 'Progress reports on international studies of climate change impacts section four: global forests' p 214 in: David Howlett and Caroline Sargent *ONEB/ITTO Technical Workshop to Explore Options for Global Forestry Management* (Bangkok 24–30 April 1991), International Institute for Environment and Development, London 1991.
159. George M Woodwell 'The role of forests in climatic change' pp 75–91, in: Narendra Sharma (ed) *Managing the World's Forests: Looking for balance between conservation and development*, Kendall-Hunt, Dubuque 1992.
160. Martin Parry *Climate change and world agriculture* Earthscan, London 1990.
161. This does not mean that only half of last year's emissions have remained in the atmosphere whereas the other half has been instantly absorbed by carbon sinks; in fact most of last year's carbon dioxide exhaust is still with us. The absorption of carbon dioxide by the oceans depends on the concentration of this gas in the atmosphere, and the latter depends on global emission history rather than on last year's additions. P M Kelly, 1991 op cit.

162. The Noordwijk Declaration on Atmospheric Pollution and Climatic Change, which was adopted at the Ministerial Conference on Pollution and Climatic Change held in Noordwijk, The Netherlands, 6–7 November 1989, stated that a world net forest growth of 12 million hectares a year early next century should be considered as a provisional aim. SEP, a Dutch consortium of electricity generating companies, has committed US$ 300 million over the next 25 years for reforestation projects in Bolivia, Ecuador and Indonesia aiming to offset carbon emissions from two new electricity plants in The Netherlands.
163. 'For the tropics as a whole, there may be as many as 579 million hectares of degraded land, formerly covered with forests or woodlands, available to be planted, and managed as, plantations'. R A Houghton, J Unruh and P A Lefebvre 'Current land use in the tropics and its potential for sequestering carbon, pp 297–310 in: David Howlett and Caroline Sargent (eds), 1991 op cit.
164. When planted, conserved or more intensively managed forests reach their natural biomass productivity equilibria over a period of 30–100 years, they are in balance in terms of gas fluxes, and theoretically serve neither as sinks nor as sources of greenhouse gases. FAO 'Climate Change and Global Forests: Current knowledge of potential effects, adaptation and mitigation' pp 213 in Howlett and Sargent (eds), 1991 op cit.
165. Mark Trexler 'Estimating tropical biomass futures: a tentative scenario' pp 311–17 in: Howlett and Sargent (eds), op cit.
166. The IPCC was recognized by the UN General Assembly Resolution 43/53 on Protection of global climate for present and future generations of mankind.
167. Collins, 1991 op cit.
168. For a strong plea for the developed countries to accept responsiblity for the greenhouse effect, see Anil Agarwal and Sunita Narain *Global warming in an unequal world: a case of environmental colonialism* Centre for Science and Environment, New Delhi, 1991. The paper was written in response to the section on climate of the 1991 World Resources Report, which ascribed an important part of greenhouse gas emissions to the developing countries. *World Resources 1990–91*, The World Resources Institute and Oxford University Press, 1991.
169. European countries, however, have agreed to reduce greenhouse emissions on a voluntary basis. Fred Pearce 'How green was our summit' *New Scientist*, 27 July 1992.
170. 'Climate hazards, climatic change and development planning', by William E Riebsame, pp 288–296 in: *Land Use Policy*, October 1991.

PART I
POLICY AND MANAGEMENT

Chapter 1

The Sustainable Management of Tropical Forest: The issues

Duncan Poore

There can be few people now who are not aware that the tropical forest is in danger. This has become one of the clarion calls of environmentalists, a call which is now echoed in the statements of political leaders in both the developing and the developed world. It is one of the main issues identified in the report of the World Commission on Environment, the Brundtland Commission,[1] and there is no dearth of international action. But there is still much misunderstanding about the nature and scale of the problem and, above all, about what might be done about it.

This widespread concern about tropical forests is based on a number of issues: that these forests are disappearing at an alarming rate; that the loss of so much forest has potentially disastrous environmental effects – on soil, water, climate, the genetic richness of the globe and the supply of possible future economic products; that the uses to which the land is being converted are often not sustainable – that the forest in fact is being destroyed for no ultimate benefit, and that forest-dwelling peoples are being arbitrarily displaced. All of these are partly true, and to a greater or lesser degree in different parts of the world. And, as usual, it is drama and disaster that seize the headlines.

But there is another side of the coin. Successes can be chalked up here and there, examples of good planning and management, of conservation. They too make the news – with the better correspondents. But they are still pitifully puny and infrequent in relation to the scale of the problem; and often, good though they may be, they are of only local relevance. The challenge is to extract from these local successes some general principles and to apply these widely and rapidly at a scale which will have some real impact.

First, though, it is necessary to find out what the real issues are, to separate what is important from the verbal froth. How vital really is it to preserve forest cover in the tropics? How much forest should be preserved intact and where? Is it possible to use some parts of the forest for economic purposes while maintaining environmental values? How might this be done? Where have planning and management failed? What are the reasons for failure and success? What can genuinely be done to solve these problems in the prevailing political and economic context of the tropical countries in which the forest is found?

The book *No Timber Without Trees: sustainability in the tropical forest* from which this first chapter is taken (Poore, 1989), does not set out to be a comprehensive account of tropical forest destruction. Instead, it concentrates on the management of tropical forest for the sustainable production of timber. It cannot do this, however, without trying to set this very important use – the production of timber – in a more general setting and examining some of the truths and myths of the tropical forest story.

It is justly pointed out that in the past much forest has been cleared elsewhere in the world, indeed that development has apparently often been based on forest clearance. Why should tropical countries be denied the same opportunity? But this clearance has been mostly in those climatic zones of the world in which agriculture has proved readily possible – the temperate regions, Mediterranean climates, the semi-arid zone, the subtropics; only in the boreal zone has most of the land surface remained under forest. Even in tropical and equatorial climates fertile alluvial soils were deforested centuries ago for irrigated agriculture. What then is so different about the remaining tropical forests?

Three points can be made in relation to this. First, it is now recognized that the deforestation in many of the long-established seats of agriculture was a very destructive process which led to much unnecessary loss of soil; if it were possible to develop those areas now with all the advantages of present knowledge, any wise government would hope to do it very differently. Second, there are good reasons why most of the remaining areas of tropical forest remain as forest: many, though not all, are on lands which have been inherently difficult to settle or cultivate, either because of disease or the infertility of the soil. Third, the forests of the wet tropics are exceptionally easy to destroy utterly, with all the plants and animals they contain. In most other parts of the world, forests can be destroyed but most of the tree species survive; even many of the other woodland organisms can exist outside the forest or in small fragments of it. This does not seem to be the case in the wet tropics, so exceptional care is needed if it is wished to develop these lands in ways which do not waste their resources.

There is another consideration. Everyone is the product of the age to which he or she belongs. Developments which occurred in previous centuries took place in a climate of opinion and balance between population and potential resources which were very different from those of today. Liberties were taken with human rights and with the environment which would no longer be acceptable to world opinion. It is mistaken to suggest that these concerns are confined to the developed world, and that it is solely those from developed nations who are attempting to force their values on other poorer countries. It is evident that similar concerns are being voiced spontaneously by people in developing nations too and that these changes are signs of an inevitable progression in human awareness.

It is as a result of this new consciousness that the Tropical Forest Action Plan (TFAP) has come into being and that the International Tropical Timber Agreement (ITTA) contains the historic clause about 'sustainable utilization and conservation of tropical forests'.

What then is necessary for the sustainable utilization and conservation of tropical forest lands? The following section of this chapter will examine some of the concepts involved and try to set out the questions that need to be asked, and the standards that should be set, if sustainable utilization and conservation are to become realities in practice.

SOME DEFINITIONS

In order to make the study as exact and as useful as possible, some definitions are necessary. Much confusion and misunderstanding can be, and have been, caused by the imprecise use of terms and we wish to avoid this as far as possible. Even in the subject of the study there are two terms that may be misinterpreted: 'natural forest' and 'management for sustainable timber production'. Some explanation is given below of the sense in which certain key terms are used in this book.

Natural forest

This term is used in contrast to forest plantations (forest crops raised artificially either by sowing or planting), which are in general areas in which the naturally occurring tree species have been totally replaced by planted trees. Natural forest includes a range of types which have been subjected to varying degrees of modification by man and which grade almost imperceptibly into one another. The following list illustrates this range, starting with those that are least modified:

- virgin forest (compare with b, c and d below), essentially unmodified by human activity, will contain gaps caused by the natural death and regeneration of trees and may include areas or phases which have been affected by natural events such as landslides, typhoons or volcanic activity;
- forest, similar to the above, the composition and structure of which may have been modified by the hunting and gathering activities of indigenous peoples;
- forests which have been subjected to a light cycle of shifting cultivation or in which cultivation has been abandoned, so that a full tree cover of indigenous species has been able to develop;
- forests which have been subjected to various intensities and frequencies of logging, but which still remain covered with a tree or shrub cover of indigenous species. These may be of two kinds: those in which new tree growth is entirely derived from natural regeneration and others where this has been supplemented by 'enrichment planting'.

The first two of the above categories (ie a and b) will be referred to as *primary forest*, the second two (c and d) as *secondary forest*.

The following, both of which can make an important contribution to timber supplies, will be excluded from the definition of natural forest: areas which have been so intensively modified by cultivation, fire or other disturbance that they remain covered with grass or non-forest weeds – *degraded forest lands*; and forest plantations (as defined previously), whether of native or introduced species.

Sustainable timber production

This phrase requires considerable explanation, both in itself and in relation to the more general term, 'sustainable management'. The relation between these two will be explained later.

When primary forest is first logged it normally contains a high standing volume of timber, a variable proportion of which is marketable, depending upon composition and market demand. Because this standing volume has accumulated over a long period, the commercial timber is likely to be of a quality and volume that will probably not be matched in future cuts (because it contains slow-growing specimens and species, large diameters etc) unless the logged forest is closed to further exploitation for a century or more. In this sense the first crop is, in practical terms, not repeatable.

If production of timber is to be genuinely sustainable, the single most important condition to be met is that nothing should be done that will *irreversibly reduce the potential of the forest to produce marketable timber* – that is, there should be no irreversible loss of soil, soil fertility or genetic potential in the marketable species. It does not necessarily mean that no more timber should be removed in a period of years than is produced by new growth; overcutting in one cycle can, at least in theory, be compensated for by undercutting in the next or by prolonging the cutting cycle.

But even this description of sustainable production begs several questions. For example, markets will certainly change between one phase of logging and the next – new species will become marketable and current fashions may decline – so the timber production to be sustained will not be the same from one cutting cycle to the next but will contain a different mix of species. One consequence of this is that valuable species may be over-exploited initially for economic reasons.

Moreover, the forest will certainly alter somewhat in composition as a result of selective harvesting and there may even be some loss of soil fertility.[2] But potentially damaging changes of this kind can be compensated by new investment – in these instances by the application of fertilizers or by enrichment planting. Whether to invest or not is essentially an economic decision.

The best silvicultural practice requires the calculation of the volume of timber which may be cut in one year in a given area (the 'annual allowable cut' – AAC), a volume which should be set at a level that provides the maximum harvest while ensuring that no deterioration occurs in the prospects for future sustainable harvests.

When an area of virgin forest is cut the AAC depends upon the volume of marketable timber in the area which may be cut while leaving enough stems on the ground for the next crop. This is calculated from an inventory of the standing stock and an estimation of the length of the cutting cycle – the period between successive loggings. If the forest is well stocked, this figure may be high. But the situation alters for subsequent cuts, because the rates of growth of the remaining trees are changed when the first crop is removed. At this later stage it should be possible, by measurement of permanent sample plots, to determine a stable AAC which corresponds to the annual growth in volume of the forest in question. This is likely to be different from, and often but not always lower than, the AAC from the primary forest.

Sustainable production per unit area and sustainable supply

There is sometimes also confusion between these two terms because of the use in trade statistics of 'production' in quite a different sense from that defined above – to mean, in effect, the supply of timber from whatever source. Representatives of some countries use the term 'sustainable timber production' to mean continuity of supply from the natural forest, implying that when one source is exhausted, another will be found. It need hardly be remarked that this usage is dangerous, for it need not include any provision for continuity of production on sites exploited and can lead to the total destruction of the resource. In fact, in this sense, supply is sustainable until it runs out; then it is someone else's problem.

Management and its intensity

Management in the broadest context can be defined as taking a firm decision about the future of any area of forest, applying it, and monitoring the application. Dr Synnott, in his account in chapter 4 of *No Timber Without Trees* writes:

> The term 'management' is sometimes used loosely. Some activities are called 'management projects' which actually consist of demonstrations and trials. Equally it is often said that management would be uneconomic when it is meant that silvicultural treatments would be. Here, management is understood to include many possible components: silviculture is one component of management, but it is possible to have effective management with few or no silvicultural interventions, and with only a few of the more important management activities.
>
> Here we distinguish between the characteristic tools of *silviculture* . . . regulation of shade and canopy opening, treatments to promote valued individuals and species and to reduce unwanted trees, climber cutting, 'refining', poisoning, enrichment, selection . . . and *management* objectives, yield control, protection, working plans, felling cycles, sample plots, logging concessions, roads, boundaries, prediction, costings, annual records, and the organization of silvicultural work.

The tropical forest can be managed for the sustainable production of timber at a number of different levels of intensity. This is often misunderstood and it is assumed that, if the forest is not being managed intensively, it is not being managed at all. We may take five different levels as examples, starting at the lowest:

Wait and see
Where forest is remote and there is, as yet, no market for logs, the most effective management may be to demarcate the forest and protect it from encroachment until it becomes worthwhile to extract timber.

Log and leave
In this 'extensive' form of management, after logging the forest is closed and protected from encroachment or further logging. The speed of recovery and the volume of the future crop will depend upon the kind of forest, the nature and standard of the first logging and the length of time that the forest remains closed.

Minimum intervention
Marked trees (up to the limit defined by the AAC) are removed with minimum damage to the remaining stand and according to a well-research silvicultural system, leaving behind an adequate stocking for the next crop to be taken after a defined number of years. Removals are confined to stems intended for sale and defective stems of marketable species; no other species is interfered with. The area is then closed and protected without any further tending until the next logging is due.

Stand treatment
Logging is carried out in the same way as in the previous system, but the growth of the remaining stems or regeneration is enhanced by various treatments, which may include poisoning of unwanted stems or species, poisoning of lianes, weeding etc.

Enrichment planting
The treatment is the same as 'stand treatment' above, but saplings of desirable species are planted where the stocking of residual stems is low; special treatment is given to encourage these saplings.

All of these, if properly applied, would constitute management for sustainable timber production. Both the benefits, in terms of timber sales, and the costs, in terms of protection, tending, planting stock and chemicals, increase from the first option to the last. The decision about which to use at any time is essentially a matter of policy which is likely to be strongly influenced by prevailing costs and likely benefits.

In assessing the reports from the countries surveyed in this study, it is very important to recognize that these different levels of management may exist, at least potentially, and that all may be justifiably termed sustainable. On the other hand, any system, however good, may produce results that are unsustainable, if it is not applied conscientiously and consistently. Many good systems have failed because they are so complicated that there is neither the will nor the ability to operate them properly for even a few years, far less for as long as a scientifically based cutting cycle or even for the length of a rotation.

THE CONTEXT

The management of natural forest for the sustainable production of timber cannot be isolated from other aspects of national life, and these inevitably have an influence on the priority that is afforded forest management in national policy and on the way it is regarded by various sectors of the government and of the people. This study has established that the success or failure of natural forest management depends upon a great variety of different factors, but that technical constraints, although they certainly exist, are much less important than those that are political, economic and social.

There are a number of issues surrounding natural forest management which need to be aired. These concern the trade and the future sources of its timber, the national economic context of forest management, sustainability and environmental quality and the social setting. These are far from being academic questions: they are all of great importance in considering whether natural forest management and trade in tropical hardwood timber have a real future.

The trade and its future sources of timber

Tropical timber can come from a number of sources:

(1) from previously unlogged forest which it is intended should remain as forest;
(2) from logged forest which is being logged again, either with or without a plan of management;
(3) from unlogged or logged forest which is being converted to another use in either a planned or an unplanned manner;
(4) from secondary regrowth;
(5) from trees planted in association with agriculture, roadside and canal-bank trees etc; and
(6) from forest plantations.

Accurate statistics to determine the relative amounts which come from these different sources are lacking. There is little doubt, however, that most of the present supply to the market comes from the first cut of previously unlogged forest (1); substantial amounts also come from both (2) and (3), though, as we shall see, very little of that from logged forest is under an operational plan of management; and small amounts so far come from regrowth (4), agroforestry (5) and plantations (6).

There are many uncertainties about the future of these categories and the boundaries between them are by no means distinct. For example, in category (1) much forest that is planned to remain as forest in reality does not remain so, and in (2) little logged forest is under a plan of management which is effective and strictly applied. Forest in (6) is likely therefore to be deteriorating in quality and potential productivity; relogging is generally determined by the availability of a market rather than by considerations of sustainable silviculture; if light conditions prevail, it may be sustainable but, if not, it is likely to lead to progressive degradation of the resource. In (3) clear felling, carried out when the forest is being converted to another use, results from a perfectly legitimate land-use decision, but permanently removes the forest from timber production, and in (4) there are no accurate trend statistics about what is happening to degraded forest or to sparsely populated or abandoned land; according to circumstances, it may degrade further or recover naturally and become capable of providing a supply of fast-growing, light-demanding timber species. Both (3) and (6) have potential, but at present make a very small contribution to tropical timber supply.

In any country, as the supply declines from the logging of land which is designed for conversion to agriculture and from the first cut of previously unlogged forest, the future must lie in four possible sources: managed natural forest; managed secondary regrowth; agroforestry; and plantations. Meanwhile the immediate market reaction to shortages or very steep price increases in timber from traditional sources is likely to be a movement away from the countries where supply has declined to those which have a largely untapped forest resource. Grainger, for example, has predicted a strong shift from South East Asia and Africa to South America.[3] The temptation for these newly favoured producing countries will be to follow the downhill path of resource mining which has been pursued by many of their predecessors, largely because they have a resource which they consider to be infinite.

As previously unlogged forest is logged and forest designed for agriculture is progressively cleared of timber, it is highly unlikely that the supply of tropical timber can be sustained at expected rates by plantations alone, however efficient. *Either* the management of natural forest for the sustainable production of timber must be practised widely over an extensive forest estate, *or* unprecedented steps must be taken to promote production in secondary growth, in plantations and in farm forestry. The decision about what should be the right mix is one for each individual producer country but the nature of these decisions is of considerable moment for the timber trade, for the International Tropical Timber Organization (ITTO) and for each individual producer country.

The national economic context of forest management

There seems to be general agreement that most governments seriously underestimate the *economic* worth of forests (both as productive resources and for the services that they provide) and that, conversely, they do not appreciate the cost of transforming the capital of natural forest into other forms of capital. There seems to be little doubt, at least in the view of one school of economists, that if the full non-market benefits of forests were to be taken into account, the 'sustainable utilization and conservation' of natural forests would prove convincingly economic.[4]

Apart from these economic considerations, there are very great differences in the way in which forests are treated in *financial* terms, differences which make comparison difficult.

At one extreme – and this situation is relatively common in Asia – profits from the sale of timber are expected to generate large amounts of foreign exchange for investment in national infrastructure or other forms of profitable enterprise. This is in addition to covering the cost of the government forestry service and providing funds devoted, in theory at least, to reforestation and forest management. In Indonesia in the 1970s, for example, timber was the largest earner of foreign exchange and is now the third; and in Honduras the forestry service is a self-financing corporation which, in addition, makes a contribution to general revenue.[5]

At the other extreme, in Queensland, Australia, the exploitation of tropical forest provides profit for all involved in logging and the wood industries, but the cost of the forest service is borne from general revenue, justified, presumably, on the grounds of supply of a valuable product, the employment provided by the industry and the environmental benefits of retaining forest on the ground. In between there are a number of variants: separation of functions between forest service and forest enterprise; different methods of partition of royalties, with huge differences in the amounts charged; different ways of providing (or not providing) for the costs of forest management etc.

Under these varying circumstances it is nearly impossible from accessible information to judge whether natural forest management is financially viable (or indeed to assess what is meant by this term!).

Sustainability and the environmental context

The International Tropical Timber Agreement is concerned with the managing and harvesting of natural forest to meet the conditions of sustainability, and indeed this is one of its most important purposes. Something has already been said about the meaning of 'sustainable timber production', but this is in fact only a small part of the more general concept of environmental sustainability – an idea which is now being treated with the utmost seriousness in all informed international debate.[6] It cannot any longer be ignored in considering the use of tropical forests for timber production.

One expression of the conditions for an environmentally acceptable policy for tropical forest lands is that set out by the IUCN (International Union for Conservation of Nature and Natural Resources).[7] This emphasizes that it is essential, if the conditions of long-term sustainability are to be met, that forestry for timber production should be part of a land-use policy that makes full and proper provision for the conservation of biological diversity and for the protection of those forests which are critical for environmental stability (on erodible and infertile soils or where the removal of forest would have harmful effects on local climate or even global climate). Some further explanation of this ideal pattern of land use and management is included later in this chapter.

The sustainability of management of any area of forest can be assessed from two different points of view related respectively to the desired product and to the state of the forest: (a) to maintain the potential of the forest to provide a sustained yield of a product or products; (b) to maintain the forest ecosystem in a certain desired condition. Consideration of these two will sometimes lead to the same result, but not always. Both, however, depend upon further definition and clarification: in the first case it is necessary to answer the question 'what product(s)?'; in the second, 'what conditions?'

This point can perhaps best be illustrated by examples.

What product?

The most natural interpretation of this from the point of view of the timber trade is the sustained production of specified timber products. It should be possible to produce a silvicultural system to meet a certain mix of products. But if the aim is to maintain the potential to meet a market that may have changed during the period of the rotation, the system may have to be modified to cope with this uncertainty. There is also an important distinction between harvesting the mean annual increment of a certain chosen species and maintaining the potential of the forest to provide a sustained yield. The situation becomes even more complicated if the yield of other forest products is to be taken into account, rattan for example or game animals.

What condition?

Environmental arguments are likely to concentrate more on the condition of the forest than on its products. These arguments will hinge upon what condition it is desired to maintain, and therefore on the proper position of any particular forest in national policies for the allocation of forests to different uses. Let us take as examples two extreme cases.

At one extreme it may be considered sufficient to preserve the 'forest condition' – the function of the forest in relation to soil conservation and water circulation. In this respect a well-planned rubber plantation with cover crops may be an acceptable substitute for natural forest, and any system of extraction and silvicultural management that provides adequate soil protection and continuous vegetation cover will equally prove acceptable.

At the other extreme, if the purpose of management is to maintain the forest as near as possible to a virgin state for genetic conservation and scientific study, this condition is only sustainable by leaving the forest untouched.

Between these two extremes are various intermediate degrees. For example, if the successful establishment of a certain timber tree depends upon birds that are themselves dependent on a non-timber tree, the forest can only be maintained by natural regeneration if the second species of tree is permitted to remain; but it would also be maintained by the artificial planting of the first species.

Thus, if one is to be strictly accurate, sustainability can only be defined in relation to a specified set of products and a specified condition. It may, however, be possible to design a system that is an acceptable compromise between a number of objectives. For example a 'minimum intervention' system of silviculture may provide a reasonable and sustainable harvest of timber and retain a very considerable degree of the biological richness of the forest; but the more intensively a forest is managed for timber production the more its composition is likely to be simplified and impoverished in variety of species and of genotypes. (Even a style of harvesting which is of very low intensity but highly selective – high-grading – is considered by many to cause genetic deterioration of the stock.) An ideal land-use policy would compensate for the intensification of timber production in some areas by more generous provision for species conservation in others.

This ambiguity in the definition of sustainability and the need for a nation to have forests which meet a number of different objectives emphasize the cardinal importance of careful land allocation in land-use policy, and the need for an exact specification of management objectives in relation to each tract of forest.

Indeed, it is questionable whether one universal definition of 'sustainable management' is useful, because it will lend itself to different interpretations by different interests. To avoid misunderstanding it is essential to be very clear which kind of sustainability we are talking about in each instance.

There are also wider questions relating to sustainability which should not be ignored. It seems certain that the next century will see some degree of global warming. How this will affect the tropical moist forest is not clear. But it is evident from the forest fires in Kalimantan, Sabah and Brazil that even small climatic fluctuations can make forests which have rarely, if ever, burnt before, much more vulnerable to fire. This should be taken into account when determining standards of management. Extensive deforestation also certainly has an effect on the carbon dioxide balance and, perhaps, on regional climates. But in this respect any continuous tree cover is likely to be at least nearly as beneficial as natural forest.

Another issue of great significance will be the selection and management of protected areas against a background of climatic change. The near certainty that this will take place should clearly have a considerable effect on the criteria used for selection, the choice of boundaries and the relation between the management of the protected area itself and that of the areas which surround it.

The social context

The regulation of forest management has, in many countries, been considered to be exclusively the province of government, with exploitation being carried out by government itself or by its licensed contractors. In carrying out these functions the customary rights of local peoples have or have not been respected to varying degrees. In Papua-New Guinea, for example, the great majority of the forest is considered to belong to the local people, whereas in some other countries customary rights are largely ignored.

It is generally true, however, that the benefits from government-promoted forest exploitation have seldom accrued to those who live in or near the forest. For hunter-gatherers timber extraction may seriously damage their environment, though the light extraction in some African countries apparently may have the opposite effect. But it is often the case that farming peoples seldom see the advantages of retaining forest as rather than converting it to another – and to them more lucrative – use. The sustainable management of government-run forests depends, therefore, upon the resolve of governments, not only to manage the forest consistently for long periods, but to protect it from encroachment by a largely unsympathetic populace.

But there are other possible models for sustainable management. This study has shown that there are small-scale local successes in forests which are managed by private commercial concerns and by local communities; there may be possibilities in extending the use of these models much more widely. What is crystal clear is that no system works unless there is long-term security for the managed forest and unless those concerned can see that the enterprise is profitable to them. It must also, of course, be profitable to the country.

THE SUSTAINABLE DEVELOPMENT AND CONSERVATION OF TROPICAL FOREST LANDS

What does it mean?

This is a question that should be asked – and answered – for the answer is by no means self-evident, and different people provide different interpretations. But it is important to reach some level of agreement, for governments and organizations will otherwise work at cross-purposes, and the objective will be that much more difficult to accomplish.[8]

The easy way of answering the question is to quote the various axioms and goals of the World Conservation Strategy.[9] This is certainly a good starting-point, for these are both valid and wise. The main axiom of the World Conservation Strategy – that conservation depends upon development, and that lasting development is impossible without conservation – is an attempt to distil in a few words the essence of a very complex set of ideas. Like all such axioms it can easily be twisted by the unscrupulous to fit their own ends. Conservation *can* be an obstacle to development; and development *can* be very destructive. The goal must be to bring about development, where development is needed, *with* conservation.

Take an extreme example. Two of the three objectives of the strategy are to maintain essential ecological processes and to preserve genetic resources. The best way of accomplishing these in the tropical forest would be to maintain the *whole* forest untouched and inviolate. Soils would remain intact, the rivers would be moderate in behaviour, the water clear and the wildlife abundant and in natural balance. Technological man would have no place. For these two objectives, preservation is the best form of conservation; there can be little doubt of this.

But such a course of action is no longer possible anywhere. Hence the third objective of the strategy: to ensure the sustainable use of species and ecosystems. This immediately throws up the questions: how much should be used, and how much preserved?

There are other crucial issues associated with the word sustainable. Land and water can be used in a variety of ways that are ecologically sustainable. It is

possible to exploit the natural forest itself more, or less, int
can be replaced by an artificial ecosystem (agro-ecosystem) a
be exploited sustainably at different levels of intensity, accord
amounts of energy and skill that are invested in caring for it. The more
intensive forms of management usually also carry their own costs in higher
of energy, increased risks of pollution and so on. We must use *all* these different
ways of managing natural resources if even the basic needs of growing populations in tropical countries are to be satisfied; further intensification cannot be avoided if the quality of life of these peoples is to be raised. Tropical forest conservation must therefore be concerned, first and foremost, with finding an acceptable balance between these different intensities of use. How much land should be used in each of these ways? Are there limits beyond which it is unacceptable to go? These are questions that must be faced in any tropical forest policy. They are not easy questions to answer.

It is perhaps easier to understand the problem – though no easier to solve it – if one appreciates that the three objectives of the World Conservation Strategy are different in kind. The maintenance of essential ecological processes and life support systems is concerned with keeping the whole machine running smoothly and preserving the natural capital of productive soil; the sustainable use of species and ecosystems, while maintaining the capital of natural systems that are cropped; and the preservation of genetic resources, with an insurance for the future. Choices must be made all the time about how much should be invested in each of these.

The only way of providing for conservation effectively and securely is by leaving natural ecosystems intact, and then only if the intact areas are very large. Once they are managed they are changed. The slightest manipulation alters the natural balance of populations and species, and as intensity increases, so does the probability of irreversible change. Populations of plants and animals are affected first, next whole ecosystems go, then soil coverage is lost. So conservation is concerned not only with the balance between different kinds of use but also with the way in which one use is changed into another, and with the standards of management at all levels of intensity.

In the last analysis land, resources and people *will* reach a balance; sustainable use is the only ultimate possibility. They may reach this equilibrium in a planned, humane manner and in such a way that relatively rich resources remain at the disposal of humanity, or by a series of cataclysmic plunges that cause untold hardship and leave an impoverished world. The first is the only sensible course to follow.

The balance of use

Is it possible to say anything useful about balance between different uses – about how much of the natural ecosystems should be preserved, how much exploited in a sustainable way, and how much converted to other sustainable uses? Guidance is needed at both the national and the global levels.

The management of its natural resources is the responsibility of each nation.

resources in the total context of its economic
e present relation between resources and
ndard of living and distribution of wealth,
all of these, and national social and economic
ds to work out a balance for the conservation and
nds that is acceptable in this context. There can
answer to the question of balance which is valid for
nor is there a single path to reach that balance.
ld perspective, for many national actions have world-
equired is a set of national policies for the balanced use
of trop ds that fit together, like the pieces of a jigsaw, into a
sensible wor e of tropical forest conservation.

Conservation: the moving target

If the resources of tropical forest lands are to be effectively and lastingly conserved, it will not be enough to design answers that only fit the circumstances of today. There will be great changes even in the next few decades, and more in the next century. While conservation is concerned with attaining stable balances, development is dedicated to change. Policies should be imaginative and designed to adapt to changing circumstances in the balance between populations and resources, in economic well-being, in world energy policies, in the balance of trade and in the attitude of people to environmental issues. Many land-use policies are obsolete before they are implemented. Conservation policies for tropical forest lands must, therefore, look forward, be integrated as far as possible with policies for population and for all sectors of the economy, and aim to hit a moving target.

What is the scope of tropical forest conservation?

Tropical forests are very varied in character. In the wetter, equatorial and tropical regions there are true rainforests. These are very rich in species, frequently occur on poor soils and, if cleared or mismanaged, can prove very fragile. Once the forest cover is destroyed, very few indigenous species can survive. Destruction or damage is being caused by agricultural expansion, planned or unplanned, and by bad forest exploitation; fuelwood cutting is rarely a problem until the forest has already been reduced to remnants. As the climate becomes drier the rainforests give way to semi-evergreen forests, then deciduous forests which gradually become lower in stature as the rainfall decreases. These drier forests have been much more extensively modified than the rainforests – by agriculture, grazing and burning. They have been replaced over large areas by savannas, grasslands, degraded scrub or wasteland. In such areas, shortage of fuelwood is often a problem. The mountain forests near the Equator have characteristics that are not shared by those outside the tropics (strong daily fluctuations of climate but small seasonal differences); those in the subtropics have much in common with the forests of more temperate regions. In all these forests there are great variations due to biogeographic history, and local differences in climate, geology, topography and soil. They are very heterogeneous.

It is important to realize that there is great variety in forest type included within the term 'tropical forest' and that the problems facing these forests differ widely from place to place. Most generalizations break down in face of this variety. For example, in the regions where there is exploitation of tropical timber there is rarely a fuelwood problem.

Neither can attention be confined to those areas that are covered by forest. In many places the forest has been degraded to grassland or scrub. Some of these secondary communities have their own flora and fauna and are worth preserving because of this and some could and should recover to forest or be used for the development of new tree plantations. The problems of deforestation often lie outside the forest, for example in policies concerned with the expansion of agriculture. Some countries, such as China, India and Peru, are only partially countries of tropical forest; what happens to the forest can depend on priorities determined outside the tropical region. The policies for tropical forest conservation are therefore concerned, in the first instance, with the forest, but beyond that, with the future of tropical forest lands and the various factors and pressures that influence the future of these lands.

What is the starting-point?

In a very few countries, such as Gabon, Brazil and Cameroon, where the population is low and much of the land is still covered by intact forest, it may still be possible to develop an ideal policy, but in most there is strong pressure to bring forest land into economic use or to clear it for agriculture. Where there is not yet pressure of population or of economic development on resources, the forest and its soils can and should be looked upon as an asset upon which to build future national economic development; where there is severe pressure, there may be little choice but to clear to produce more food and fuel. In extreme cases, and such are not uncommon, the resource of the once-forested land is so degraded and the pressures so great that a gigantic task of restoration is required.

While there is undoubtedly value in defining the ideal conservation policy and the limits of the unacceptable, in many countries things have deteriorated so far that the only politically possible course of action may be to reduce the amount of the unacceptable and edge gradually in the right direction; the ideal may always remain elusive.

The elements of the ideal

Suppose, then, that one were given a free hand to design the ideal policy for a country which had almost all its forest intact but where there was a need for economic development to meet the reasonable requirements of a gradually expanding population: what should be the main elements?

Conservative land husbandry should be looked upon as a spectrum ranging from land that is totally protected to land that is used very intensively. In between there may be many shades, different degrees of exploitation or restraint from use, according to the values that it is thought important to preserve.

Protection

The natural forest is a bank of land and of resources, secure and maintained at little or no cost. Until it is needed for other purposes, all land should be kept under forest. The burden of proof should be on those who wish to alter or remove the forest. If it is necessary to intensify land use, areas which are not under forest should be chosen first if such are available.

Certain areas should be permanently dedicated to protective uses. If these were to be lost the loss would be irreversible. Absolute protection is not always required, however, to maintain the value of these.

Living room for indigenous peoples

In many tropical forest areas there are indigenous populations who are in harmony with the ecosystems in which they live. Provided that this harmony is maintained and that these people wish to go on living in this way, the areas which they inhabit should be protected for their continued use.

Fragile soils

Measures for the conservation of all fragile soils, and the resulting protection of water resources, should take priority over any form of use or development which would cause erosion or loss of fertility.

This means the mapping of all areas with fragile soils in order that no use or development which would cause erosion or loss of fertility should be allowed in them. In very fragile areas there should be a permanent ban on exploitation. In others, no use or development should be permitted until it has been established that this could be carried out, and later managed, in such a way that no soil deterioration would take place.

Genetic resources

Special measures need to be taken to preserve the intra-specific variation of species of economic importance. These would include protecting some areas specially for this purpose and applying management constraints in chosen areas elsewhere.

Samples of ecosystems

These include measures for the preservation of representative samples of widespread ecosystems (climax, sub-climax and extra-zonal) and examples of exceptional ecosystems.

The full range of species

Many species will be protected by the full range of measures above. These should be supplemented to include: areas of special interest because of their exceptional richness or unusual character; the feeding, roosting and nesting sites of migratory birds; sufficient territory to encompass full populations of wide-ranging mammals and birds.

Forests for sustainable production

A sufficiently large area should be allocated for the production of timber, wood and other forest produce to meet foreseeable domestic requirements for the next 50 years at least; further areas should be set aside to satisfy export markets if there is no shortage of forest in the country. If this area of forest is likely to prove insufficient to meet domestic needs when managed in an environmentally satisfactory manner, the shortfall should be made up by intensively managed, highly productive plantations (wood farming).

Land for food

Where not required for conservation purposes or for productive forests, the best soils should be held available for agriculture; but they should not be opened up until there is a real need. The first call on this new land for agriculture should be for those in greatest need, especially those who have no alternative to overusing and degrading the land on which they live.

If pressure for essential food or wood makes it necessary to turn forest on fragile or vulnerable land into farmland, this should be developed with combinations of food and tree crops that retain as far as possible the protective characteristics of the forest structure. Where land is allocated to agriculture or wood farming it should be made as productive as possible to help to reduce pressures of overuse on other areas.

Other environmental safeguards

In addition, there should be appropriate environmental safeguards covering any major developments, such as large reservoirs, irrigation systems, urban settlements, and communications to ensure that they are in harmony with this pattern of land use and with the essential requirements of conservation.

Economic and social considerations

The above are the main ecological elements of an *ideal* policy for the sustainable development of tropical forest lands. But ecological sustainability is only one facet of the problem. If solutions are to be lasting, they must be economically viable and socially acceptable. It is necessary to develop towards the ideal in steps that are practicable from the situation which exists today in many tropical forest countries. The development of a policy and the adoption of the measures necessary for tropical forest conservation depend upon the evolution of appropriate social and economic mechanisms. At the national level, they need to be politically acceptable and to fit readily into the framework of sensible economic policies. At the local level, patterns of land use need to be acceptable to the people they affect and not to intrude harshly into the harmonious development of local communities. They should be based on a broad understanding of the ecological circumstances and of the social setting.

There are many obstacles to this harmonious development. Central, perhaps, is a lack of understanding that the intact forest and its soils *are* a capital resource. This is reflected in the plundering of this capital and a failure to invest in the protection and maintenance of the resource. But there are many other obstacles: political pressures, social unrest, considerations of national security; failure to treat land-use questions in a manner that is integrated and socially sensitive; a lack of trained staff and the knowledge that can be derived from well-conceived research.

As indicated above there are many different starting points. Some fortunate countries have the resources to plan immediately for the ideal. Others may plan to power their economic development by depending upon timber and on the cash crops that can be planted on good forest soils. They should be encouraged to examine how far these objectives are consistent with the ideal (and therefore how far they are sustainable) and to move towards an ideal pattern of land use. Where the pressures to meet everyday needs are urgent and immediate, it may be totally impracticable to think in terms of all elements of the ideal; in these cases soil and water conservation, and the growing of food, fodder and fuel, are bound to take precedence in domestic priorities over genetic conservation. Here international planning may bridge the gap: it may be possible to preserve the same resources somewhere else or, if this is not possible, to compensate for national hardship. There are cases too where the sustainable use of resources is likely to remain a pipe dream unless there is an investment of international resources.

Planning considerations

It was stressed earlier that the conservation target is constantly moving. At present we tend to think in terms of a pattern of development that will retain a substantial proportion of the rural people in the countryside. Where this is the case, it will often be necessary to plan an attractive package of measures to combine conservation with development in an acceptable way. Examples include a protected area with production forest around it, both of which provide local employment; the development of agroforestry and the intensification of agriculture; some forest plantations and local industry. As well as developing thriving local communities, these will serve to buy time.

But there may be countries where development will bring about a rapid change in the pattern of rural settlement and in social priorities. This has already happened in some places. It is necessary for conservation policies to anticipate these changes and plan accordingly, for example, depopulation of rural villages can lead to a reduction in demand for wood and fodder in the countryside but a growth in interest in recreation and rural amenities from a more affluent urban class. The change can be very rapid.

International dovetailing

There is an important international element in policies for tropical forest conservation. The effective conservation of tropical genetic resources is a matter for international planning and co-operation, as regards the choice and safeguarding of protected areas. The same is true of measures to protect species. But this in no way lessens the duty of each nation to manage its own resources in a responsible way; indeed it increases the responsibility by adding an international dimension.

However, good land-use planning and proper sustainable use can become very difficult unless there are international policies which encourage them, or at least do not positively discourage them. International markets and policies can very readily encourage unwise use of resources, for instance the movement into cash crops rather than food crops, over-exploitation of tropical forests caused by incentives to satisfy foreign markets, and unwise decisions about land use as a result of the instability of commodity markets.

A very important (and difficult) element in determining national policies in relation to tropical forest conservation is to strive for the harmonization of international policies, particularly on trade and aid, so that these provide a real incentive to the sustainable use and conservation of the resource. The Tropical Forestry Action Plan is working towards harmonization of aid while ITTO is concentrating on trade.

DEFINITION OF ISSUES IN THE ITTO STUDY

In order to define the status of the 'sustainable utilization and conservation' of natural forests and, in particular, the status of management for the sustainable production of timber, it is necessary to find answers to these general questions:

- Over what areas is natural forest managed at an operational scale for the sustainable production of timber?
- Where such management has been undertaken successfully, what are the conditions that have made this possible?
- Where such management has not proved possible or has been attempted but failed, what have been the constraints that have made it difficult or impossible to apply? (The conditions for success and the constraints tend to be mirror images of one another.)

In analyzing the situation some more detailed questions proved to be useful, all of which are relevant to successful management. These came under the following headings.

Policy

Is there a national land-use policy? Is there a national policy for the sustainable management of a permanent forest estate? If not, why?

Extent

What area of natural forest is managed for the sustainable production of timber?

Allocation

Is there a satisfactory system for choosing, demarcating and protecting those areas that will be used as production forest? If not, why?

Is there a satisfactory system for choosing, demarcating and protecting those areas that will be used as protection/conservation forest? If not, why?

Are there pressures from other sectors or interests to remove productive forest from use? What measures are being taken to counter or divert these pressures?

Sociological and economic conditions
In what ways do the various people who have an interest in or are affected by the management of the forest benefit from this management or suffer from mismanagement (people dwelling in or near the forest, loggers, middlemen, wood processors, small industries, consumers generally, the forest authority, and other parts of government)? Are the benefits adequate to provide an incentive to good management? Is there equitable distribution of these benefits? If not, why?

Management
Are the objectives of management conducive to sustainable production? Are the management prescriptions appropriate for the particular forest type? Are they rigorously applied and reviewed? If not, why?

Pre-exploitation survey
How comprehensive and adequate is the pre-exploitation survey: choice and marking of trees for felling; analysis of trees to remain unfelled; existing regeneration; environmental conditions; routing of extraction roads? If inadequate, why is this so?

Choice of exploiters
Does the choice take into account the best long-term interests of the forest? How?

Conditions of exploitation
Do these bring reasonable benefits to the various parties concerned: government revenues, any reforestation fund, the logging companies, local contractors, logging labour, those with customary rights in the land?

Are the conditions of exploitation such as to encourage long-term investment in the sustainable management of the forest? Are there reasonable incentives to encourage good management? What proportion of revenues are returned to forest management? If the conditions are unsatisfactory, what prevents their improvement?

Quality of exploitation
Are there guidelines on the siting, construction and maintenance of extraction roads; on weather conditions in which exploitation should not take place; on the equipment to be used; on directional felling and cutting of lianes etc? Are such guidelines followed? If not, why?

Are the above conditions monitored during and after exploitation? How? How well?

Post-exploitation survey and treatment
Are there guidelines? Are they sensitive to different forest types? Are they adhered to? Is later performance monitored? How? If not, why?

Control and follow-up
Is there effective control of operations at all stages? If not, why? Are there

arrangements for monitoring and reviewing prescriptions? If not why?

Research

Is research designed to support sustainable timber production from natural forest? Does it provide the necessary information to answer the questions set out above? Are there permanent sample plots to provide the data upon which sustainable yield can be calculated? Are the data processed and made available to management within a reasonable time?

Education and training

Are enough trained staff at all levels being produced with qualifications in the skills needed in natural forest management?

The answers to these questions form the basis of the following four chapters in *No Timber Without Trees: sustainability in the tropical forest.*

This chapter was taken from Poore, D (1989) *No Timber Without Trees*, Earthscan Publications, London.

NOTES AND REFERENCES

1. World Commission on Environment and Development *Our Common Future* (Oxford, 1987).
2. Palmer, J, in Chapter 6 of *No Timber Without Trees*, expresses doubt whether the depletion of nutrients is ever a significant factor.
3. Grainger, A, *Tropform: a Model of Future Tropical Timber Hardwood Supplies, Proceedings of the CINTRAFOR Symposium in Forest Sector and Trade Models* (Seattle: University of Washington, 1987).
4. Burns, D, *Runway and Treadmill Deforestation*, IUCN/IIED Tropical Forestry Paper No 2 (London, 1986); HIID *The Case for Multiple Use Management of Tropical Hardwood Forests*, Study prepared for ITTO by the Harvard Institute for International Development (Cambridge, Mass, 1988); Repetto, R and Gillis, M (eds) *Public Policy and the Misuse of Forest Resources* (Cambridge: Cambridge University Press, 1988).
5. For further figures see HIID, op cit, Figure 2.2.
6. See World Commission on Environment and Development, op cit.
7. Poore, D, and Sayer, J *The Management of Tropical Moist Forest Lands: Ecological Guidelines* (Gland, Switzerland: International Union for Conservation of Nature and Natural Resources, 1987).
8. This section is based on a text prepared for IUCN and is reproduced with acknowledgement to IUCN.
9. IUCN *World Conservation Strategy* (Gland, Switzerland: International Union for Conservation of Nature and Natural Resources, 1980).

Chapter 2

Public Policies and the Misuse of Forest Resources

Malcolm Gillis and Robert Repetto

PRINCIPAL FINDINGS

Some low income nations are confronted by a shortage of wood; the world is not. The world is, however, facing a decline of natural tropical forests. Even as policy-makers begin to understand the value of natural tropical forests, they are rapidly shrinking and deteriorating. Many scientists have clearly established the extent of forest decline and likely economic, social, and environmental consequences (Brown *et al* 1985; Eckholm 1976; Fearnside 1982; Grainger 1980; Lanly 1982; Myers 1980, 1984, 1985; Spears 1979). These and other studies (Allen and Barnes 1985; Bunker 1980; Ehrlich and Ehrlich 1981; Plumwood and Routley 1982; Tucker and Richards 1983) have also established the principal causes of deforestation. They have identified three major outgrowths of population growth and rural poverty – shifting cultivation, agricultural conversion, and fuel-wood gathering – as threats to natural forests in the Third World, along with the impacts of large development projects. Virtually all previous studies have also focussed upon commercial exploitation, including logging and land-clearing for ranches, as major sources of the problem.

The authors support these findings. In addition, they have identified government policies that have significantly added to and exacerbated other pressures leading to wasteful use of natural forest resources including those owned by governments themselves. It is important to emphasize the policy dimension, because changes in policy can substantially reduce resource wastage. The authors' emphasis on wasteful use in the economic sense does not ignore or minimize the importance of non-economic objectives underlying forest policies. Rather, research undertaken implies that policies leading to economic waste have also undermined conservation efforts, regional development strategies, and other socioeconomic goals.

Separating the effects of government policies from those of other causes is impossible, because policy-induced exploitation interacts with other pressures on natural forests: forests opened by loggers encouraged by liberal concession terms are more accessible to shifting cultivators, government-sponsored settlements also attract spontaneous migrants; development policies that worsen rural poverty lead to more rapid encroachment on forest lands. The indirect effects of government policies are complex and substantial. Despite this, important conclusions can be drawn.

One such conclusion is that wastage of publicly owned natural forests has been widespread and long-standing. To an extent heretofore unappreciated, these outcomes have been the avoidable consequences of government policies.

The policies in question include not only those formulated by government agencies nominally responsible for oversight of forest utilization, but a wide range of other policies designed to serve broader governmental goals (ie non-forestry policies) but which nevertheless have badly undermined the value of natural forest assets.

Forestry policies include those pertaining to the terms of timber harvest concessions, such as their duration, permissible annual harvests and harvest methods, levels and structures of royalties and fees, policies affecting utilization of non-wood forest products, and policies toward reforestation. The 12 case studies included in Gillis and Repetto, 1988 (from which this chapter is drawn), including four in West African countries and three in the Malaysian states), furnish ample evidence that forestry policies have provided strong incentives for wasteful use of natural forest resources. Inadequate use of royalties and other charges to collect the economic rents potentially available to harvesters of mature timber have set off timber booms and scrambles for short-term profits. In most of the countries studied in the aforementioned publication, including Indonesia, Malaysia, the Philippines, Ivory Coast, and Ghana, the result has been rapid, careless timber exploitation that outran both biological knowledge and administrative capacity for sustainable forest management.

Timber concessions have generally been too short in duration to allow loggers to conserve forest values even if they were inclined to do so. Harvesting methods, particularly the variants of selective cutting prescribed or allowed by most tropical countries, have undermined forest quality. The structure of timber royalties has promoted excessive mining (high-grading) of forest value and yielded too few revenues for governments as resource owners. Where reforestation policies aimed at regeneration and restoration of commercial mixed forest stands have been tried, they have proved largely ineffective.

In many countries non-forestry policies have caused greater forest destruction than misdirected and misapplied forestry policies. Non-forestry policies prejudicial to forest conservation may be arranged on a continuum that ranges from self-evident to subtle. Most obvious are the effects of policies leading directly to physical intrusion in natural forest areas. These include agricultural programmes that clear forest land for estate crops such as rubber, palm oil and cacao, for annual crops, and even for fish ponds. Closely related are investments in mining, dams, roads, and other large infrastructure projects that incidentally result in significant, once-and-for-all destruction of forest resources. Many such projects are politically driven and of questionable economic worth, even apart from the forest and other natural resource losses they impose.

Further along the continuum are tax, credit, and pricing policies that stimulate private commercial investment in forest exploitation, whether in logging or timber processing. Such policies have induced timber harvesting in excess of rates that otherwise would have been commercially profitable. One step removed are policies that stimulate private investments in competing land uses, such as ranching, farming, or fish culture. The principal instruments of fiscal and monetary policies contributing to forest destruction are generous tax treatment, heavily subsidized credit, and direct government subsidy.

Next on the continuum are land tenure policies that encourage deforestation. Of these, the most direct are tenurial rules that assign property rights over public forests to private parties on condition that such lands are 'developed' or 'improved'. Such rules have facilitated small-farmer expansion into forested regions, but in some countries have been used by wealthier parties to amass large holdings. A few countries have demonstrated that this policy works in reverse, by awarding private tenures to *deforested* public wastelands on condition that they be reforested.

A more indirect tenurial policy has been the centralization of proprietary rights to forest lands by national governments, superseding traditional rights of local authorities and communities. Although intended to strengthen control, such actions have more often undermined local rules governing access and use, removed local incentives for conservation, and saddled central governments with far-flung responsibilities beyond their administrative capabilities.

Finally, the furthest points on the continuum represent those policies that appear at first glance to have few implications for forest use, but which ultimately prove to be significant sources of policy-induced forest destruction. Included here are all domestic policies that further impoverish households living close to the margin of subsistence, especially in rural areas. These include pricing policies and investment priorities biased against the agricultural sector, development strategies that depress the demand for unskilled labour, and agricultural policies that favour large farms over smallholders. These policies retard the demographic transition, make rural populations more dependent on natural forests for subsistence needs, and increase the concentration of agricultural landholdings.

This emphasis on the policies of national governments responsible for their public forest lands does not imply that external agencies have been blameless in the long-term misuse of natural forests in developing countries. It has long been apparent that trade barriers protecting wood-processing industries in the United States, Japan, Europe, and Australia have usurped many of the benefits from forest-based industrialization that could have accrued to poor countries, while inducing Third World governments to take strong, sometimes excessive, countermeasures. And multinational timber enterprises from industrial countries were among the primary beneficiaries of log extraction in tropical nations well into the 1970s, although these large multinationals have since steadily withdrawn from the forests of developing countries. By 1980 they had completed divestment of virtually all logging operations in South East Asia and West Africa, their places having been taken by smaller multinationals from developing countries and by firms owned by domestic entrepreneurs. The image of the now-departed multinational firms, popularly identified as the principal engines of forest destruction for so long, no longer obscures the part played by domestic government policies.

There is little doubt that policies of governments have been inimical to the rational utilization of valuable forest resources. Why have these policies been adopted, and why do they persist? In many countries the policies have been deliberately intended to reward special interest groups allied with or otherwise favoured by those in power. The existence of large resource rents from harvesting mature timber has attracted politicians as well as businessmen and women to the opportunities for immediate gain.

That is not the whole story, however. To a considerable degree, the policy weaknesses identified in Gillis and Repetto (1988) arose despite well-intentioned development objectives. The shortcomings have been failures of understanding and execution. Distillation of the lessons of this publication provide six reasons that help to explain why government policies have erred in the direction of excessive depletion of natural forest resources. The six are first presented in summary form, then discussed at greater length.

- The continuing flow of benefits from intact natural forests has been consistently undervalued by both policy-makers and the general public.
- Similarly, the net benefits from forest exploitation and conversion have been overestimated, both because the direct and indirect economic benefits have been exaggerated and because many of the costs have been ignored.
- Development planners have proceeded too boldly to exploit tropical forests for commodity production without adequate biological knowledge of their potential or limitations, or awareness of the economic consequences of development policies.
- Policy-makers have attempted – without much success – to draw on tropical forest resources to solve fiscal, economic, social, and political conflicts elsewhere in society.
- National governments have been reluctant to invest the resources that would have been required for adequate stewardship of the public resource over which they asserted authority.
- National governments have undervalued the wisdom of traditional forest uses and the value of local traditions of forest management that they have overruled.

Natural forest endowments remain undervalued in all countries studied, not only by the general public but by governments as owners or as regulatory authorities and by international institutions. An asset that is undervalued is an asset that will inevitably be misused.

Forest exploitation has concentrated on a relatively few valuable commodities, neglecting other tangible and intangible values. Natural forests everywhere serve protective as well as productive functions. Assigning monetary values to the protective services is much more difficult than estimating the market value of timber harvests. This difficulty accounts for much of the worldwide tendency to undervalue natural forest assets, but, in addition, potential production has also been undervalued. Forests in the tropics have generally been exploited as if only two resources were of any significance: the timber and the agricultural land thought to lie beneath it. A third resource has been overlooked virtually everywhere: the capacity of the natural forest to supply a perpetual stream of valuable non-wood products that can be harvested without cutting down trees. In the tropics, these include such commodities as nuts, oils, fibres, and plant and animal products with special uses. In advanced temperate zone countries, forests' recreational value is often underestimated.

Second, the timber and agricultural products expected to flow from log harvests and clearing of natural forests have been overvalued. The common expectation that tropical logs could be harvested every 35 years in cut-over stands was based on over-optimistic expectations of the rate and extent of regeneration and has been grossly unfulfilled in Indonesia, the Philippines, and other countries studied. Assumptions about the agricultural potential from land underlying tropical forests have been even more optimistic, and results have been even more disappointing, particularly in the Indonesian transmigration programme and in the Brazilian schemes for promoting Amazonian development through large-scale cattle ranching.

Policy-makers have usually overestimated the employment and regional development benefits associated with timber industries, infrastructure investments, and agricultural settlements in tropical forests. Where such initiatives have not been economically sound to begin with, they have not induced further development or even been able to sustain themselves without continuing dependence on government subsidies. However, one result of such inflated expectations has been that governments, both as holders of property rights in forests and as sovereign taxing authorities, have often allowed even the timber that is readily valued in world markets to be removed too cheaply. This problem is reflected in persistently low timber royalties and licence fees and unduly low – sometimes zero – income and export taxes. Only in Sabah (and then only after 1978) have governments been moderately successful in appropriating a sizeable share of the rents available in logging. Elsewhere, sizeable rents available to timber concessionaires have generated destructive timber booms and pressures for widespread, rapid exploitation.

The employment benefits expected from forest utilization have also not been realized. The wood products industries in tropical countries have provided some employment to be sure, but with the single exception of Gabon, the timber sectors in tropical wood exporting nations have typically provided jobs for less than 1 per cent of the labour force, a figure only as high as in the diversified United States economy and half as high as in Canada. The drive to expand employment, domestic value-added, and foreign exchange earnings has led to strong protection of domestic processing by banning log exports and imposing high export taxes on logs but not on timber products. Stress on forest-based industrialization has also lead to tax and credit incentives for sawmills, plymills, and even pulp and paper projects that are inappropriate for short-fibred tropical hardwoods. Jobs in the forest-based sector were indeed created, but at very great cost to the nation. Large amounts of taxes and foreign exchange earnings were dissipated, and some of the domestic industries that were sheltered became inefficient claimants to more of the forests than foreign mills could ever command.

Along with overestimated benefits, there has been a pervasive tendency to underestimate not only the economic but also the social and environmental costs of forest exploitation. Although some of these costs have by now been well documented, they have nowhere been taken as an explicit offset to putative benefits from forest use. The destruction of habitat that threatens myriad little-known species endemic to tropical forests and the displacement or disturbance of indigenous communities have been especially neglected costs. In addition, the costs of 'boom-town' development that often arise from intensified logging and processing activities were overlooked in the drive to promote development of lagging or backward regions. In Indonesia, Brazil, and elsewhere, neither the infrastructure costs of providing for large inflows of immigrants to timber provinces in the early stages of timber booms, nor the costs of maintaining excessive infrastructure in post-boom periods, were viewed as offsets to the private economic benefits flowing from the opening of the natural forest. Instead, they were viewed as investments in regional development that could be financed through timber sales.

In some areas, large-scale extractive activity in natural forests has imposed heavy environmental costs. Misused, fragile tropical soils have been seriously damaged over large areas, for example. In Indonesia and the adjacent East Malaysian state of Sabah, these hitherto unforeseen costs reached calamitous heights in 1983. In that year of severe drought, fire in the moist tropical forests in both countries burned an area about one-and-a-half times that of Taiwan. Previous droughts had brought fire, but on a far smaller scale and with minimal damage. But extensive logging in both areas had, by 1983, predisposed even the wet rain forest to disastrous damage from fire, while unlogged forests suffered far milder damages. In Indonesia alone, losses in the value of the standing stock of trees exceeded $5 billion, and the costs of ecological damage are still unknown.

These miscalculations of costs and benefits demonstrate the third underlying reason referred to above: governments have proceeded without adequate biological or economic knowledge of tropical forest resource management. Little is known about the potential commercial value of all but a very few of the tropical tree species, so most trees are treated as weeds and destroyed during logging operations. Much remains to be learned about potential regeneration of currently valuable tree species and successful management of heterogeneous tropical forests for sustained yields. Without this knowledge loggers have blundered through the forests, extracting the few highly valued logs and severely damaging the rest. Even less is known of the potential value for agricultural, scientific, or medical purposes of the millions of other plant and animal species, despite clear indications from previous discoveries that unknown treasures may exist in the forests. Consequently, forest habitat is recklessly cleared for commodity production of marginal economic value.

Large-scale agricultural settlements and livestock operations have been encouraged without adequate study of land use capabilities. Painful and costly failures have driven home the lesson that the lush tropical forest does not imply the existence of rich soils beneath it. It is now recognized that most underlying soils are too nutrient-poor for sustained crop production without heavy fertilization, and that the better soils – along rivers, for example – are probably already being used by shifting cultivators. Similarly, massive conversions to monocultures and ranches have taken place without prior attention to potential problems of plant and animal diseases or to pest and weed management, with costly and sometimes disastrous consequences. Only now has serious attention been paid to the capabilities of tropical soils and the development of sustainable farming and livestock systems suited to them.

Governments have pushed ahead with forest exploitation not only in advance of ecological knowledge but even before understanding the likely consequence of the policy instruments with which they hoped to stimulate development. Several countries awarded concessions for the majority of their productive forest estates before enough time had elapsed to assess properly the adequacy of their forest management system, or the impact of forest revenue systems on licensees' behavior. Governments have stimulated large domestic processing industries before evaluating the appropriateness of the levels of protection afforded them, their technical and economic efficiency, and the costs and benefits to the national economy of the incentives provided. Similarly, governments have gone ahead with large-scale conversions of tropical forests before adequately evaluating the economic viability and worth of the alternative uses.

Such precipitate actions are related to a fourth reason: governments have tended to grasp at tropical forests as a means of resolving problems arising elsewhere in society. Migration to forested regions has been seen in many countries as a means of relieving overcrowding and landlessness in settled agricultural regions, whether those conditions sprang from rapid population growth, highly concentrated land tenures, or slow growth of employment and opportunities for income generation. Rather than modifying development strategies to deal more effectively with employment creation and rural poverty, or to tackle the politically difficult problem of land reform, many countries have in effect used forests as an escape valve for demographic and economic pressures.

Sale of tropical timber assets has been seen as a ready means of raising government revenues and foreign exchange. Governments have found that drawing down these resources has been easier in the short run than broadening the tax base and improving tax administration, or reversing trade policies that effectively penalize nascent export industries, although ultimately the assets and easy options are exhausted and the underlying problems remain.

In addition, governments (and development assistance agencies) have failed to devote the resources and attention to the forest sector that would have been necessary to ensure proper management and stewardship. Development spending on the forest sector, for example, has been a tiny fraction of that allocated to agriculture. While most sizeable countries in the Third World have built up substantial agricultural research programmes (including tree crop research in some countries), none have developed appreciable research capabilities or activities focussed on natural forest ecology and management.

Despite the enormous value of the resource, including billions and in some countries even trillions of dollars in timber alone, and the large sums of money represented in the annual log harvest, governments have not built up adequate technical and economic expertise or effective management and enforcement capabilities. As countries with large petroleum and other mineral resources have found, the cost of such expertise and managerial capability is small relative to the value of the resource, and is an investment that is returned quickly in increased earnings to the national economy and the government treasury. Gillis and Repetto (1988) provide much documentary evidence of the needless waste of forest resources from inappropriate policies and widespread evasion of justifiable but poorly enforced stipulations. Yet the record of policy analysis in the forest sector is sparse, and infusions of funds and technical assistance to train, staff, equip, and monitor forest administration agencies have been inadequate.

Finally, while national governments have overestimated their own capabilities for forest management, they have underestimated the value of traditional management practices and local governance over forest resources. Local communities dependent on forests for a wide variety of commodities and services, not just timber, have been more sensitive to their protective functions and the wide variety of goods available from them in a sustainable harvest. Moreover, when provincial and national governments have overruled traditional use-rights to the forests, local communities and individual households have been unable, and unwilling, to prevent destructive encroachment or overexploitation. Conversely, some governments have found that restoring or awarding such rights to local groups has induced them to attend carefully to the possibilities of sustainable long-term production from forest resources.

IMPROVING THE POLICY ENVIRONMENT FOR UTILIZATION OF NATURAL FORESTS

Those with important interests in the natural forests form a wide and diverse constituency. Included are not only public and private proprietors and enterprises based on forest resources, but also those who benefit from the protection that forests provide to soils, water, and wildlife. With respect to the tropical forests this constituency is worldwide.

Conflict is frequent between competing interests over forest policy and is sometimes seen as inevitable. In particular, conservationists and developers are supposedly unalterably in opposition. When conservation interests are external to the region or nation seeking to develop, conflicts of interests can be acute. However, an important implication of the book from which this chapter is taken is that such conflict is often more apparent than real. Policies that have led to wasteful exploitation of forest resources, both in developed and developing countries, have been costly not only in biological terms, impoverishing the biota and soils, but equally costly in economic terms. Uneconomic investments have been promoted, assets have been sacrificed for a fraction of their worth, and government treasuries have been deprived of revenues and foreign exchange earnings sorely needed for genuine development purposes.

Reforms of public policies toward publicly owned forests can save both natural and financial resources. Rather than a win–lose predicament, opportunities for policy reform present win–win situations for nations with natural forest endowments. For example, Brazilian tax and credit incentives for conversion of Amazonian forests to livestock pastures has resulted both in ecological and fiscal disasters: the incentives have made social waste privately profitable at great expense to the government budget. In the United States, the Forest Service has subsidized timber production at a large cost to the treasury on lands economically unfit for the purpose, sacrificing potential superior recreational uses along with hundreds of millions of dollars. Since the central concern of economic policy is the efficient management of scarce resources, policies that promote wasteful use of forest resources are rarely economically justifiable.

Opportunities for policy improvement notwithstanding, there is a vein of truth in the assertion heard in the Third World that rich countries, having pre-empted a large share of the world's resources, wish developing countries to bear the costs of conservation. Since the clientele for tropical forest conservation is worldwide, it is in the rich nations' interests to help defray some of the costs of policy reforms affecting far-away forest endowments. For example, worldwide interests are clearly furthered when Brazil, Indonesia, Malaysia, or Gabon sets aside new areas in parks and forest reserves. Survey and exploration of natural forest resources, for wood and non-wood products as well as genetic resources, are costly but essential for better management. International interests are served by forest conservation and research; international financial and scientific support for them is therefore sensible.

The following sections of this chapter identify a series of measures that should be undertaken by tropical country governments, by industrial country governments, and by international agencies to improve economic utilization and conservation of natural forest assets. This list is not intended to be an exhaustive agenda of needs and opportunities for slowing deforestation. Such a list would include recommendations for accelerating botanical, genetic, and ecological research, for improving agroforestry, plant breeding, and selection, and for strengthening forestry practices in responsible government organizations.

The focus is complementary to these latter concerns, although the concerns of the authors are the same. Needs and opportunities have been identified which would facilitate policy reform by altering incentives affecting decisions about natural forest resource use. For convenience of presentation, the reforms are grouped into two categories: those that are essentially the responsibility of governments as owners of forests and as regulators of activities in them, and those for which responsibility falls to agencies representing the worldwide constituency for forest conservation. Neither category is intended to be airtight; for example, the need for joint responsibility for financing reforms is particularly great. Further, the international community should help to provide better and more complete information on which policy reforms by owners can be based. This requires expanded policy research on several forest-related topics.

Policy reforms by national governments of tropical countries

In most cases, two objectives can serve as adequate guidelines for policy reform: (1) more efficient development of the *multiple* uses of natural forests; and (2) improved financial returns to governments. At first glance these objectives may seem to be too narrow, as they may appear to ignore non-economic considerations such as preservation of indigenous communities and ecosystems. However, steps to achieve these two objections will generally also contribute strongly to these non-economic goals as well, because of the complementarities discussed previously. Thoroughgoing reform is required both in forestry and non-forestry policies.

Forestry policies

Royalties and related charges on private contractors for harvesting and sale of government-owned timber have been deficient on two principal counts. First, in all cases studied in this volume, charges have been maintained at levels well below the stumpage value of the timber. This has resulted not only in lost timber rents for the governments, but in enormous pressures from business and political interests to obtain timber concessions and the large, short-run profits they offer. Combined with other flaws in timber policy, this rent-seeking syndrome has led to over-rapid, wasteful exploitation, including the harvest of timber in marginal stands, often in ecologically vulnerable sites such as slopes and critical water catchment areas. Second, in many of the developing countries the structure of royalties has, usually in combination with inappropriate selection systems, exacerbated loggers' proclivities for high-grading (or mining) forest stands, thereby causing needless damage to forest quality. Sensible reform calls both for increases – sharp increases in many countries – in royalty levels and for modification of defective royalty structures. Inflexible, undifferentiated, specific charges based on the volume of timber harvested should be replaced by *ad valorem* royalty systems, based on export prices properly discounted for costs of harvesting logs and transporting them to ports. Valuation of logs for royalties should be equivalent for log exports and for those delivered to processing mills, for the measure of the opportunity cost of a log is usually its f.o.b. value. Countries should use differentiated *ad valorem* rates, with lower rates for so-called secondary species than for the most valuable primary species, to the extent that their forest services are sufficiently well-trained and administered to enforce them. But, if the forestry service is undermanned and untrained, flat-rate *ad valorem* royalties constitute the best of inferior options, and can be adequate if set at moderate levels and combined with other measures to capture resource rents.

Governments have generally proven reluctant to enact royalty reform. Of the countries studied by Gillis and Repetto only Sabah (1978) and Liberia (1979) have implemented sharp increases in royalty levels in recent years. China has sharply increased log prices administratively and by permitting market transactions, and proposals that were pending in China in 1987 will raise stumpage fees further. Elsewhere, including Malaysia, the Philippines, Indonesia, the Ivory Coast, Ghana, and Gabon, royalties continue well below true stumpage values and should be raised. In the United States, although royalties are bid (usually competitively) and approximate private stumpage values, the government's absorption of substantial logging costs means that timber with negative stumpage values is routinely harvested. The simplest remedy for this is the imposition of minimum acceptable bid prices high enough to recover the government's full separable costs of growing and marketing timber.

The effect of these changes would be to restrict harvesting of un-economic forest lands, to slow the pace of timber exploitation in Third World countries to one more in line with the growth of forest management capabilities, to permit more complete concession areas, and thus to reduce wastage, infrastructure costs, and the forest disturbance that opens the way for secondary clearance and agricultural conversion.

Policy reforms in the administration of concession agreements would also further both conservation and development goals. Realignment of concessions policies requires changes not only in duration of concessions but in the level of license fees charged on concession areas. Prior to the Second World War, many tropical timber concessions were granted for periods of up to a century. Newly independent governments in the post-war period tended to view such arrangements as vestiges of colonialism. Consequently, the duration of concessions was steadily compressed. By 1987, concession periods were typically 5–10 years in length even for large tracts; few governments allowed concessions longer than 20 years.

Given the long growing cycles of commercial tropical hardwoods, short-term concessions are inconsistent with sensible resource management. Logging firms have no financial interests in maintaining forest productivity under such circumstances. Instead, they have grounds for repeated re-entry into logged-over stands before expiration of their concessions, compounding damages arising from the initial harvest.

Tropical foresters have long called for extension of concession periods to at least 70 years, so as to provide loggers with at least two rotations of 35 years each. Governments have not heeded such proposals, partly because pre-war experience with longer concessions showed little evidence of conservation practices by loggers. Therefore any movement to extend the life of concessions must necessarily be accompanied by appropriate safeguards to defend the public interest. Among the safeguards proposed has been periodic review of concessionaire performance, with renewal of logging rights conditional on adherence to prescribed practices. In practice, as the experience of the Philippines shows, such arrangements are difficult to police without a substantial improvement in administrative capacity. Further, firms, especially the multinational companies, have grown wary of all long-term contracts, including timber concessions, because so many have been abrogated by governments. A complementary approach is to structure policies from the outset to provide strong incentives to firms for rational forest use.

For example, governments could use area licence fees much more effectively to promote conservation, rational harvesting, and more complete rent capture. Fees based on the area awarded in concessions have been found to be extremely low in most countries. Ideally, logging concessions should be auctioned competitively, as off-shore oil leases are in the United States, ensuring that governments capture virtually all the available resource rent. Auction or competitive bidding systems work well only where the owner has enough information about the resources in particular forest tracts to enable him or her to ascertain their approximate value, and where the number of bidders ensures competition. Acquiring the necessary information about resources is difficult in tropical timber stands, which are typically inaccessible and far more heterogeneous in species composition than temperate forests. Nevertheless, successful examples of auction systems have been reported for both Sarawak and Venezuela. The investment by governments in more detailed exploration and inventory of forest resources is likely to have an immediate pay-off in improved revenue capture.

Auction systems can be combined with higher reservation or minimum bid prices to guard against bid-rigging, or to discourage firms from entering regions that are ecologically sensitive or are better kept back for harvesting at some future time. Where auction systems are not feasible, concession contracts should employ much higher licence fees per hectare than is now common in the tropics. Higher licence fees serve two conservation goals, as well as revenue objectives: they discourage exploitation of stands of marginal commercial value, and when combined with royalty systems differentiated with respect to stumpage values, they encourage economic utilization of timber stands by providing logging firms with incentives to recover more volume and more species per hectare.

In nearly all ten of the tropical forest cases studied in Gillis and Repetto (1988) serious problems have been identified in selection systems governing harvest methods. Selective cutting systems are used almost everywhere in mixed tropical forests. Notwithstanding some evidence that careful selective logging can be done with minimal damage to residual stands, the selective cutting methods actually practised in tropical forests yield unsatisfactory ecological results, principally because of heavy incidental damage and poor regeneration of harvested species, and inferior long-term economic results because of deterioration of the quality of the stand.

It is unfortunately true that much remains to be learned about ecologically sound selection systems in tropical forests. Virtually no one supports clear-cutting (whole-tree utilization) as a silviculture method in tropical forests except where natural forests are to be cleared for other land uses. There is therefore little basis for recommendations on harvest systems, other than careful research in specific forest regions. But better enforcement of concession terms to avoid excess damage to soils and remaining trees is possible, and the policy reforms discussed above would provide stronger incentives to concessionaires to reduce logging damage and waste.

The most plausible alternative to current methods is one or another of the so-called uniform cutting systems, which also involve selective harvests but generally take more stems per hectare. Uniform cutting systems yield greater immediate volume but also entail greater costs and ecological impacts. The most sensible long-run approach to improving harvest methods is more research on the ecological, silvicultural, and economic implications of alternative selection methods, to find attractive variants of uniform cutting systems or more appropriate forms of selective logging.

Reforestation requirements in the temperate forests of North America and China are well understood, even if costs or other constraints have sometimes limited reforestation programmes on public lands. The situation is different in tropical countries. If by reforestation is meant the restoration of logged-over stands to something closely resembling their natural states, with similar species frequency, age distribution, and density, then existing knowledge of the ecology of the tropical forest severely constrains the success of such efforts. If by reforestation is meant enrichment planting of primary species in cut-over stands, current policies have brought little success, despite encouraging results in Indonesia and a few other areas.

Despite the general lack of success in regenerating cut-over tropical forests, some reforestation policies have served other purposes. Indonesia's reforestation deposit ($4 per cubic metre harvested) did capture additional timber rents and helped discourage logging on marginal stands, although it was not nearly high enough to induce firms to undertake significant reforestation activity. The same can be said of similar charges in Malaysia and West Africa. Nor have regeneration programmes mounted by governments as owners had much success, whether financed by earmarked forest taxes or directly from national treasuries. Most of these have been either under budgeted (the Philippines, Ghana, Gabon), ill-designed (Indonesia and the Philippines), or rendered ineffective by institutional constraints (Sabah).

A viable set of regeneration policies in logged-over areas would have the following features:

1. prime focus on cutting methods favourable to natural regeneration, coupled with enrichment planting of native species;
2. carrot and stick incentives for firms to undertake regeneration (the carrot could be a one-year extension of the concession period for every prescribed number of hectares of the concession area in which regeneration is under way, and the stick could be a sizeable deposit of at least $5 per cubic metre of *primary* species harvested, refundable to firms as regeneration proceeds, on proof of their expenditures for this purpose;
3. more budgetary and scientific resources for regeneration research programmes;
4. more support for government regeneration efforts by international lending institutions and aid donors.

Where reforestation is defined as replacement of forest cover on cut-over tracts by non-indigenous trees (such as pines) other than those grown for tree crops, some success has been recorded in tropical countries, including Indonesia, Malaysia, the Philippines, and Brazil. Plantation programmes, especially for conifers, are well established in China, the United States, and other temperate countries such as Chile.

Plantations are an essential part of any programme to conserve natural tropical forests, because they create an alternative source of supply to meet growing demands for wood products. Tropical areas under tree plantations (other than those established for industrial tree crops such as palm oil) have grown substantially in the post-war period. They covered an estimated 11 million hectares in tropical countries (excluding China) by 1980, and their area is expected to double by the end of the century (Spears 1983).

Policies that promote investment in tree plantations can help in reducing demands on natural forests, especially for domestic needs. A number of countries have adopted incentive programmes, including tax benefits, concessionary credits, and guaranteed markets to stimulate private investment in plantations. The studies underlying Gillis and Repetto (1988) have not evaluated such policies, because others have already done so in some detail (Berger 1980; Gregersen and McGauhey 1985; Laarman 1983; Matomoros 1982). However, impediments to plantation investments still exist in tropical countries. For example, potential foreign investors in Indonesia cannot own forest land in fee simple nor can they obtain leases for a sufficiently long period of time. Such limitations discourage investment commitments. Even with increasing investments in plantations, deforestation exceeds reforestation by large margins in most tropical countries, and depletion of timber supplies greatly exceeds additions from all sources. Consequently, positive incentives for plantation investments, while important, are insufficient. Adverse policies that exacerbate losses of natural forest resources must also be changed.

Neglect of non-timber forest products is one such adverse policy, based largely on ignorance. Most tropical country governments do not even collect information on the annual value of production or export of dozens of valuable non-wood products that can be harvested without damaging the complex forest ecosystem. Of all the cases studied in by Gillis and Repetto, only Indonesia supplies relatively complete and up-to-date data on production and export of these items, which now exceed 10 per cent of gross log export value. Complicated new policies to protect non-timber products are not needed, but better information would be useful, export taxes on them should be removed, and export controls relaxed except, of course, for trade in sensitive or endangered species. If governments realize how valuable such products can be, they will view their loss with more concern.

Traditionally, local communities have been more sensitive to these benefits, while central governments have tended to regard tropical forests as immense timber warehouses. Even then, central governments have not been effective resource managers, nor have provincial or state governments done much better. Transfer to private hands of permanent rights to natural forests is not a realistic option in most countries (both for constitutional and other reasons), and in Brazil, where it has taken place on a large scale, many large investors have been drawn to opportunities for short-term or speculative gains (in large part due to policy-induced distortions in investment incentives). The record of experience thus far, however, does suggest that local communal or collective ownership of forest property rights has been more consistent with conservation of *all* forest values than has central government ownership.

Accordingly, a larger share of rights and responsibilities over natural forests should be returned to local jurisdictions, where long-standing traditions of forest use exist, or to county governments (such as Indonesia's *kabupatens*), or to village cooperatives, as in China. Central governments would retain sovereign taxing power, and therefore would not give up forest revenues, although revenue-sharing with local communities is not only equitable but also effective in ensuring their interest in resource management. Reversion would mean principally that more forest management decisions, about the location, size, and length of logging concessions, and about land-clearing for agriculture, would be in the hands of local groups with a continuing stake in the multiple benefits that natural forests provide.

Non-forestry policies

The experiences examined in Gillis and Repetto (1988) demonstrate that government forestry policies have resulted in more resource depletion than would have occurred had governments tried to minimize their effects on private decisions over forest utilization. Not that *laissez-faire* would have been the desirable policy stance: the point is that, with market outcomes as a standard of comparison, government policies, on balance, have leaned toward more rapid exploitation rather than more conservation. Economic policies affecting the forest sector have had non-neutral effects that have exacerbated deforestation in almost every country investigated. In several, these effects have been extremely large.

Three forms of government subsidies have strongly affected forest use in many countries:

1. revenues lost by failure to collect resource rents from timber harvesting through appropriate royalties and taxes;
2. revenues foregone through income tax incentives for investment in logging, timber processing, and land-clearing activities for estate crops and cattle ranching;
3. indirect subsidies in the form of artificially cheap credit for investments in those activities.

The timber booms in the Philippines (1950–1970), West Malaysia (1960–1975), East Malaysia, and Indonesia (1967–1980) were partially fueled by generous income tax exemptions for logging and processing firms, as well as by enormous uncollected rents. Some tax holidays were stretched illegally to a dozen years. Fortunately, tax incentives for logging firms are now relics of the past in South East Asia, although they are still found in a few African nations, including Gabon and the Ivory Coast. Tax holidays largely explain the poor performance in capturing timber rents in most log-exporting countries, particularly before the late 1970s. They are unnecessary, and should be rescinded where they still survive.

Moreover, in Brazil and other countries, tax credits, provisions for loss write-offs, and generous depreciation provisions joined with tax holidays to convert socially wasteful land-clearing projects into privately profitable ones. Policies offering tax subsidies for extractive activities where severe environmental costs are involved, as in logging, are particularly misguided. Taxes should reflect the social costs not borne by the private investor. Therefore, tax subsidies create perverse incentives. All tax incentives for logging and agro-conversion in tropical countries should be removed.

Credit subsidies have added to perverse incentives for forest investments in Ghana, the Philippines, and especially Brazil. The artificially low interest rates and long grace periods available on loans for alternative land uses involving forest clearing have by themselves produced very powerful incentives for forest destruction. When coupled with generous tax subsidies, the incentives for very large-scale forest clearing have proven irresistible, even where the alternative land uses were intrinsically uneconomic. Virtually all kinds of subsidized credit programmes have been proven wasteful. Credit subsidies for irreversible destruction of forest assets cannot be justified on any economic or environmental grounds, and they should be abolished.

As much as half of the mass of a log becomes sawdust, woodchips, or other residue when the log is processed. Residue can be used for fuel or to make such products as particle board, but is much less valuable than lumber and plywood. Domestic processing of logs therefore offers substantial potential saving in shipping and manufacturing costs. Countries with forest endowments and low labour costs enjoy comparative advantages in forest-based industry; over time most timber processing should gravitate to the raw material source. Governments in industrial countries that have built processing industries on imported logs have tried to delay this process by trade barriers discriminating markedly against imported wood products, while governments in log-exporting countries have adopted policies to force exports of processed timber products rather than logs, and to encourage investments in sawmills, plymills, and other wood products industries.

While the net effect of these conflicting international trade policies has been to increase fiscal levies on processed and unprocessed tropical timber, and so restrict world consumption somewhat, severe distortions in investment patterns and losses in economic efficiency have also resulted. In industrial countries, capital and labour have been retained in declining industries. In exporting countries, log export bans have led to evasion, corruption, and the construction of high-cost processing facilities as a means of ensuring access to logs. And the tax and tariff protection provided to wood-processing industries has been so high as to weaken competitive pressure and to undermine incentives to minimize costs. Consequently, very sizeable amounts of timber rents have been destroyed; in some countries these losses have run into billions of dollars. Moreover, the low recovery rates in timber processing that have resulted from heavy protection have intensified demands on natural forest endowments. In time, forest-based industries established under the umbrella of heavy protection may achieve higher recovery rates, especially if the degree of protection is gradually reduced. But even then, the processing capacity induced by government incentives will retain a strong claim to raw materials supplies, because governments will not be inclined to allow closure of forest-based firms during the periodic slumps characteristic of world markets for wood products and will provide the logs to keep domestic mills going even if economic or ecological considerations argue against it.

Large industries have been established in many of the countries studied, including Indonesia, West Malaysia, the Philippines, Brazil, and parts of West Africa. Replacing log export quotas and log export bans in these countries with increased taxes on log exports would result in higher government revenues and send signals to processing industries of the need to modernize and raise efficiency. This trade liberalization should be negotiated in an international forum in exchange for reduced rates of protection for the timber-processing industries of log-importing countries. The result should be a gradual transfer of most tropical wood-processing industries to countries with large forest endowments, and substantial modernization and improved efficiencies of existing processing industries.

The costly lessons derived from the experiences of these countries can still be put to good use in those countries where forest-based industries have yet to emerge on a large scale, as in East Malaysia, Gabon, and others (Zaire, Cameroon, Papua New Guinea, and Burma, for example). For such nations, gradualism rather than haste is indicated in policies for promoting forest-based industrialization. Higher export taxes on logs than on lumber and plywood are superior in all respects to bans or quotas on log exports. Export taxes furnish whatever degree of protection is desired, raise government revenues, and also make redundant any income tax or credit incentives for sawmills and plymills. Similarly, log exports bans and quotas are inappropriate means of preventing logging in sensitive or protected areas. Such areas should given protected status, and no logging concessions (for domestic use or exports) should be awarded.

Many ambitious forest-based industrialization programmes have been founded on wishful thinking. The same can be said about large-scale resettlement programmes that have moved people into forested regions. In general, it has been wishful thinking that sponsored out-migration could relieve pressures of rural poverty, land scarcity, or even environmental degradation in the areas of origin. The costs of resettlement have been too large, and the numbers resettled too small to make such a strategy viable without dealing with the root problems of rapid population growth, rural inequality, and underemployment in the place of origin. Similarly, those in need of resettlement because they have been displaced by large infrastructure projects, such as dams, have often been victims of misapplied analysis that produced inflated estimates of projected returns on such investments. Most such investments have experienced serious cost and schedule overruns and have produced benefits only a fraction as large as anticipated.

The best way to ensure against such over-optimistic undertakings and the human, environmental, and economic losses they entail is to make the expected beneficiaries financially responsible for most of the costs of the projects. In all developing countries, spontaneous migrants outnumber government-sponsored migrants several-fold and typically bear all the costs of their movement as an investment toward an anticipated better future for themselves and their offspring. If resettlement programmes were based on the principle of cost recovery, the need for sound planning would be reinforced, and there would be checks on the scale and pace of implementation. Similarly, if agencies responsible for large-scale development projects, such as dams, were financially responsible for recovering operating and most capital costs from project-generated revenues, the tendency toward inflated projections of investment returns would be brought down to earth.

Where resettlement programmes are carried out, needless damage to natural forests can be reduced in several ways. The most obvious, of course, is *not* to locate new communities in natural forests at all. In many countries, ineffective or inappropriate land and agricultural tax policies allow large landholders, including governments themselves, to retain huge estates that are used to only a small fraction of their agricultural potential. In such countries, because smallholders typically achieve much higher per hectare yields and incomes through intensive cultivation, land redistribution with compensation can be brought about, provided that the government is able to reform rural taxes to induce large landowners to sell part of their holdings and is willing to finance land purchases by small farmers on long-term mortgage credit. In addition, there are degraded public forest lands and wastelands in many Third World countries that could be transferred to landless peasants, to be used for mixed farming systems involving agroforestry. The main problem here is that government agencies are typically as reluctant to disgorge any of their estates as any other large landholder.

If any agricultural resettlement projects are sited in forest areas, secondary and logged-over tracts should be selected in preference to intact forests. Individual holdings must be large enough to allow sustained support so that settlers are not forced to encroach on adjacent forest areas. Colonies should be based largely on such tree crops as rubber, which are more suitable to available soils, and on agroforestry systems. Further, collection, storage, and marketing facilities for non-wood forest products near resettlement sites would encourage relocated families to use forests sustainably. Finally, award of sites must irrevocably convey all property rights to the awardee, including clear rights of transfer. This not only enables resettled peoples to offer land as collateral in borrowing for improvements, but reinforces the incentive to protect and enhance value of the property.

This discussion of domestic policy measures to reduce wastage of forest resources can be suitably concluded by drawing attention to the connection between deforestation and failed development policies. Pressures on natural forests are reduced when towns and villages achieve adequate growth in rural output and employment, begin to convert from wood as an energy source to electricity and petroleum products, and experience declining population growth rates as birth and death rates decline and young workers are drawn off into expanding urban industries. Conversely, economic stagnation and impoverishment of rural and urban populations inevitably accelerate deforestation, as the experience of Ghana during the 1960s and 1970s illustrates so clearly.

Not much is known about government policies that can accelerate economic development, except those that promote a stable, broadly-based political and economic system that rewards enterprise and productive investment by households and enterprises alike. Much painful experience has accumulated, however, about economic policies that retard development. Salient among these are market distortions, such as deeply over-valued exchange rates, negative interest rates, and flagrantly distorted commodity prices. These not only result in serious market disequilibria, scarcities, and rationing throughout the economy, but also create perverse investment incentives and generate speculation and corruption. Such policies also often imply heavy penalties on the rural sector, as the relative prices of agricultural outputs are lowered relative to industrial prices and investments are redirected toward the urban sector. If, at the same time, governments expand their direct role in the economy beyond their managerial capabilities, creating expensive state-operated industrial and agricultural white elephants, economic stagnation is nearly assured.

Reversal of policies clearly inimical to growth will not immediately stem forest destruction arising from the spread of shifting cultivation and from the search for fuelwood. But the recent experiences of relatively prosperous countries such as Gabon, Venezuela, and the western part of Malaysia reveal little deforestation arising from either shifting cultivation or fuelwood demands.

Policy changes by industrial countries and international agencies

Of course, economic development in tropical countries is influenced by policies of the industrial countries. The 1980s have seen stagnation and decline throughout much of the Third World as a consequence of worldwide recession in the early 1980s, extremely high real interest rates, and reductions in net capital flows to developing countries on international capital markets, as well as growing protectionist pressures in the industrial countries. While beyond the scope of this study, the connection must be pointed out between economic stagnation leading to deforestation in the Third World and policies in industrial countries that restrict outflows of capital to developing countries and inflows of commodities exported from developing countries.

More specifically, industrial country trade barriers in the forest products sector have been partially responsible for inappropriate investments and patterns of exploitation in Third World forest industries. Within the context of either the General Agreement on Tariffs and Trade (GATT), the International Tropical Timber Agreement (ITTA), or some other international forum, negotiations between exporting and importing countries should result in (a) reduced tariff escalation and non-tariff barriers to processed wood imports from the tropical countries, and (b) rationalization of incentives to forest industries in the Third World.

While international assistance agencies have become increasingly concerned with forest sector problems over the past 15 years, their involvement has primarily taken the form of support for discrete development projects. These have ranged widely from reforestation and watershed rehabilitation projects to support for fuelwood and industrial timber plantations and to funding for wood-processing industries. Associated with them, of course, has been a significant amount of technical assistance in a wide variety of forestry-related subjects.

It must also be said that development assistance agencies have, in the aggregate, provided huge amounts of funding for projects that lead directly and indirectly to deforestation, including roads, dams, tree crop plantations, and agricultural settlements. Greater sensitivity is required in ensuring that such projects are, wherever possible, sited away from intact forests and especially away from critical ecosystems; in ensuring that such investments are, in fact, economically and ecologically sound once the non-market costs of forest losses are weighed in the balance; and in ensuring that such projects are executed in accordance with appropriate safeguards to minimize unnecessary damages.

Only to a much lesser extent have development assistance agencies been involved with the kinds of sectoral policy issues discussed by Gillis and Repetto. The Food and Agricultural Organization of the United Nations (FAO) has indeed studied forest revenue systems and published manuals and related material for policy-makers. Multilateral and bilateral assistance agencies such as the Inter-American Development Bank and USAID have examined policies that would encourage private investment in forestry. In recent years, the World Bank has carried out forest sector reviews in several countries that address some of the broader policy issues identified in Gillis and Repetto (1988), and new directions in World Bank policies indicate growing attention to tropical forest values. Yet, it is clear that still more emphasis on policy reform in the forest sector is required. Especially as international capital flows become increasingly divorced from specific projects and linked to broad macroeconomic and sectoral policy agreements, the international development agencies must identify and analyze the effects of tax, tariff, credit, and pricing policies, as well as the terms and administration of concession agreements, on the use of forest resources. Working in cooperation with host country agencies, they can help to a greater extent in identifying needs for policy changes and options for policy reform.

More generally, governments of industrialized countries and international agencies should act upon the recognized community of interests among all nations in conservation of tropical forests. During 1987 the FAO, the World Bank, the United Nations Development Programme and the World Resources Institute collaborated with non-government organizations, professional associations, and governments around the world on a tropical forestry action plan (FAO 1987). Intended not as a blueprint but as a framework for coordinated action, this plan identifies priority action needed on five fronts: forest-related land uses; forest-based industrial development; fuelwood and energy; conservation of tropical forest ecosystems; and institution building. Its recommendations are complementary to and compatible with the policy implications of this study.

The call for action sounded by participants in that process serves equally well as the conclusion to Gillis and Repetto (1988):

> Above all, action is needed now. Hundreds of millions of people in developing countries already face starvation because of fuel and food shortages. The forest resource potential of these countries can and must be harnessed to meet their development needs. Properly used and managed, the tropical forests constitute a massive potential source of energy, a powerful tool in the fight to end hunger, a strong basis for generating economic wealth and social development, and a storehouse of genetic resources to meet future needs. This is the promise and the challenge. (FAO 1987: 3)

This chapter was taken from Gillis, M and Repetto, R (1988) *Public Policies and the Misuse of Forest Resources,* Cambridge University Press, Cambridge.

REFERENCES

Allen, Julia C, and Barnes, Douglas F. 1985. 'The Causes of Deforestation in Developing Countries'. *Annals of the Association of American Geographers*, Vol 75, No 2: 163–184.

Berger, Richard. 1980. 'The Brazilian Fiscal Incentive Act's Influence on Reforestation Activity in São Paulo State'. PhD diss, Michigan Stage University, East Lansing.

Brown, Lester, *et al*, 1985. *State of the World*. Worldwatch Institute. New York: W W Norton.

Bunker, Stephen G. 1980. 'Forces of Destruction in Amazônia'. *Environment*, Vol 20, No 7 (September): 14–43.

Eckholm, Eric. 1976. *Losing Ground*. Worldwatch Institute. New York: W W Norton.

Ehrlich, Paul R, and Ehrlich, Anne H. 1981. *Extinction: The Causes and Consequences of the Disappearance of Species*. New York: Random House.

Fearnside, Philip M. 1982. 'Deforestation in the Amazon Basin: How Fast Is It Occurring?' *Interciencia*, Vol 72, No 2: 82–88.

FAO, World Resources Institute, World Bank, and UN Development Programme. 1987. *The Tropical Forestry Action Plan*. Rome, June.

Gillis, Malcolm and Repetto, Robert. 1988. *Public Policies and the Misuse of Forest Resources*. Cambridge University Press.

Grainger, Alan. 1980. 'The State of the World's Tropical Forests'. *The Ecologist*, Vol 10, No 1: 6–54.

Gregersen, Hans M, and McGauhey, Stephen E. 1985. *Improving Policies and Financing Mechanisms for Forestry Development*. Washington, DC: Inter-American Development Bank.

Laarman, Jan G. 1983. *Government Incentives to Encourage Reforestation in the Private Sector of Panama*. Panama: USAID, September.

Lanly, Jean-Paul. 1982. *Tropical Forest Resources*. FAO Forestry Paper 30. Rome: FAO.

Matomoros, Alonso. 1982. *Papel de Incentivos Fiscales para Reforestación en Costa Rica*. Centro Agronomico Tropical de Investigación y Enseñanza. Costa Rica, May.

Myers, Norman. 1980. *Conversion of Tropical Moist Forests*. Washington, DC: National Academy of Sciences.

Myers, Norman. 1984. *The Primary Source: Tropical Forests and Our Future*. New York: W W Norton.

Myers, Norman. 1985. Tropical Deforestation and Species Extinction: The Latest News. *Futures*, Vol 17, No 5 (October): 451–463.

Plumwood, Val, and Routley, Richard. 1982. 'World Rainforest Destruction – The Social Factors.' *The Ecologist*, Vol 12, No 1: 4–22.

Spears, John. 1979. 'Can the Wet Tropical Forest Survive?' *Commonwealth Forestry Review*, Vol 57, No 3: 1–16.

Spears, John. 1983. 'Sustainable Land Use and Strategy Options for Management and Conservation of the Moist Tropical Forest Eco-Systems'. International Symposium on Tropical Afforestation, University of Waginengen, Netherlands, September 19.

Tucker, Richard, and Richards, J F, eds. 1983. *Global Deforestation and the 19th Century World Economy*. Durham, NC: Duke University Press.

Chapter 3

Managing the Forest: The conservation history of Lembus, Kenya, 1904–63

David Anderson

In colonial Africa, as elsewhere, conservation has invariably been linked to the dynamics of political life. The colonial state in Africa set down the parameters within which conservation policies were defined. Attitudes to the African environment evolved as the colonial period progressed, and conservation accordingly took on new forms and new roles. Although historians have been able to mark out 'conservation eras', to monitor the rise of public awareness of particular issues and to chart the emergence of technical expertise in the general field of conservation management, they have also stressed the conflicts of interest present at every phase of the evolution of conservation policies (Powell, 1976; McCracken, 1982; Anderson, 1984; Beinart, 1984; Ofcansky, 1984; Helms and Flader, 1985; Anderson and Millington, 1987). This chapter takes up these themes in an examination of the colonial history of conservation in the Lembus Forest of Kenya.

Lembus was awarded to a commercial company for the development of a timber industry while the British conquest of the region was still incomplete. It was one of the largest and most favourable land concessions made to Europeans in Kenya. The subsequent administration of this concession, and the political battles that ensued for control over the management of Lembus, are the central concern of this chapter. The political struggle that emerged was three-sided, between Africans who wished to continue to utilize the resources of the forest, the commercial company who wished to exploit the forest without interference, and the colonial Forest Department, who sought to control and conserve the forest in order to maintain a financial return while also sustaining the forest resource. Events in Lembus serve to illustrate the manner in which political considerations weighed heavily upon government attitudes towards land use and conservation. Contradictions between private and public interests were clearly exposed in the problems of managing the Lembus Forest, and exacerbated by the need to reconcile the often conflicting aims of commerce and conservation.

THE LEMBUS FOREST

The Lembus Forest covers an area of approximately 130 square miles (see Figure 3.1). It is made up variously of valuable economic forest land, some scrub and bush areas, and a large number of glades. The glades, varying from

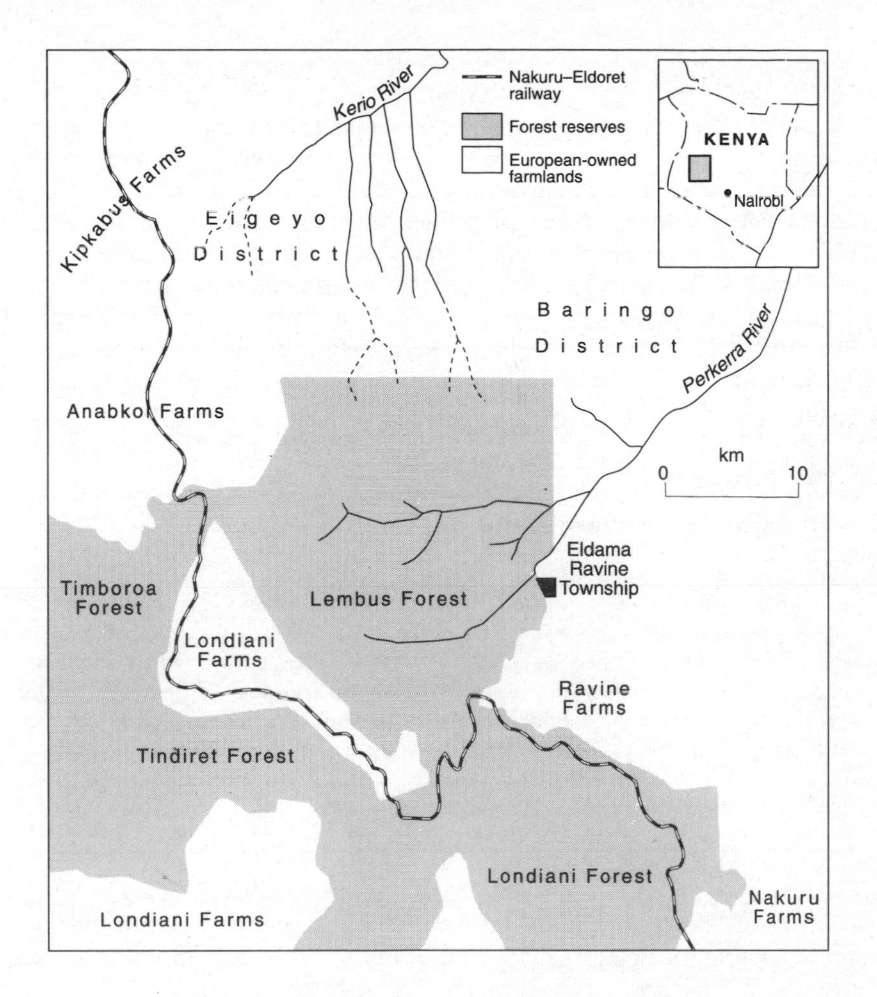

Figure 3.1 Lembus Forest

30 to 300 acres in size, offer valuable upland grazing to herders inhabiting the surrounding lowlands. Large areas of the forest-proper are dense, consisting mainly of Podo and Cedar woods, with considerable quantities of bamboo on the higher parts. Although earlier estimates had been consistently more optimistic (Johnston, 1902; Eliot, 1905), surveys completed in the 1950s revealed that around 55 per cent of Lembus then comprised economic forest, which might be reserved and exploited as such.[1] Aside from commercial aspects, the management of the ecosystem of Lembus has far-reaching importance for adjoining areas, the forest lying across the watershed dividing the river systems of the Rift Valley and western Kenya. The Perkerra River, which feeds lake Baringo, rises in Lembus, as do several important tributaries

of the Kerio River (Hutchins, 1909; Hughes, 1949; Kenya, 1950; ALDEV, 1962).

From the human aspect, a number of different groups have made use of the forest in the recent past. Uasin Gishu Maasai, Nandi, Elgeyo and Tugen peoples have each made seasonal or irregular use of the forest resources, gathering forest produce for themselves, using the forest glades to graze their animals, and clearing and cultivating areas of scrubland on the forest fringes. Various Africans have taken up residence in the forest including a small number of Dorobo (Blackburn, 1974). To them all the forest has had considerable strategic importance, and all were in some sense dependent upon access to it (Matson, 1972; Anderson, 1982; Waller, 1985). The forest was not, then, a 'tribal' land during the colonial period (or previously). Rather, it was a zone of ethnic indeterminance, over which no single group exercised exclusive control. As we shall see, the demands of colonial administration helped to bring this situation to an end, with the Tugen of Baringo District asserting their dominance over the forest.

Commerce and conservation

> 'The forests of the Mau, Nandi, Plateau, and the slopes of Mt Elgon contain hundreds of thousands of magnificent conifers – juniper and yew. The timber of the juniper is to all intents and purposes like cedar-wood. The mere thinning of these woods which is necessary for their improvement, and which might be carried out concurrently with the establishment of European settlements, would provide millions of cubic feet of timber, which would find a ready market on the east coast of Africa.' (Johnston, 1902: 291–2)

Sir Harry Johnston's description of the potentials of forestry was typical of the optimism that brought European settlers to East Africa in the early years of this century. Seldom substantiated by anything more than the passing glance of the traveller's eye, large tracts of East Africa were heralded as being rich in the resources that, with some initial investment, could support a new white settler colony. Among the greatest supporters of this view of the future was the High Commissioner for East Africa from 1900 to 1904, Sir Charles Eliot. Aware that investment on a large scale would be necessary to establish the new colony on a sound footing, Eliot used his position to encourage 'men of capital', by offering land concessions on highly favourable terms (Sorrenson, 1968). Among the early proposals this attracted was one for the establishment of a timber industry in the Lembus Forest. Responding enthusiastically to this proposition, Eliot pressed quickly ahead with the negotiations of a concessionary lease, without seeking advice from within his own Forest Department. By the time of Eliot's resignation in 1904, the government found itself committed to the granting of a substantial forestry concession to a small group of businessmen, with Canadian and South African connections, headed by Ewart Grogan.[2]

In his haste to secure forestry investment in the colony, Eliot had placed substantial powers in the hands of the concessionaires, largely ignoring the requirements of the Forest Department and giving government very little control over the working of the concession. The Forest Department now foresaw huge difficulties in administering any programme of reafforestation in Lembus under the terms of the lease, and bemoaned the loss of revenue that would attend the working of the concession at the extremely low royalties stipulated. From 1904 to 1916 the Forest Department conducted protracted negotiations with Major Grogan, the concessionaire, in a belated attempt to recoup the situation. Grogan naturally wished to retain the powers granted in the original agreement, and shrewdly maintained his position. The government found itself in an unpleasant predicament. While keen to encourage investment in the development of a timber industry, the government found that the forest concession lay at the centre of a tangled web of land dealings involving Grogan, none of which seemed to accrue any great benefit to the government or the colony. The most worrying of these was Grogan's claim to 100 acres of valuable wharfage at Kilindini Harbour, Mombasa, which had been secured by Grogan as part of the forest concession. While Grogan traded one proposal against another, and the Forest Department pressed for a foreclosure of the lease, the government, aware of consequences broader than the considerations of the Forest Department, sought to find the best political accommodation.[3]

When the renegotiated lease was finally signed in March 1916, the considerations of good forestry practice were indeed sacrificed to bring a settlement that would encourage investment in the timber industry. Grogan won exceedingly favourable terms for his concession, not substantially altered from the original lease, giving him considerable advantages over other sawmillers in East Africa and leaving his operations in Lembus largely beyond the control of the Forest Department.[4]

The Conservator of Forests was appalled by the freedom given to the concessionaire under the terms finally agreed.[5] The reasons for his outrage were partly commercial and partly conservationist. Colonial forestry in Kenya, as throughout British Africa, was initially modelled on the example of India (Unwin, 1920; Stebbing, 1937). During the second half of the nineteenth century Indian forestry had been established on a sound commercial base, raising revenue for government while also being able to finance extensive programmes of reafforestation (Stebbing, 1922). Forestry in India paid for itself, and from this position of strength gained support in government circles for a conservationist strategy that would ensure continued revenue surpluses. In short, commercial viability facilitated conservation. Indian-trained foresters were recruited to the African colonies in the hope of achieving the same result, but it became apparent that the lower yield of merchantable timber in the majority of African forests made it impossible to generate sustained revenue surpluses. Unable to realize their revenue raising potential, and in need of subsidy in order to mount programmes of reafforestation and preservation, forest departments came to be seen as a drain on the limited financial resources of the colonial state in Africa (Stebbing, 1941; Brasnett, 1942; Ofcansky, 1984). In these circumstances, the Lembus concession was particularly galling to the Kenya Forest Department. Not only was the Department being denied income because of the ludicrously low rates of royalties fixed under the lease, but the preservation work of the Department was being hampered by the manner in which Grogan's agents and contractors were exploiting the forest (Nicholson, 1931). To understand the attitude of the foresters we need to examine more closely the financial arrangements under the Lembus forest least, and the working of the lease by the concessionaire.

An important feature of the early negotiations over the Lembus concession was the agreement that the normal royalties payable to government on all merchantable timber extracted from the forest would be set at a reduced rate. Therefore, despite the protests of the Conservator of Forests, the final agreement of 1916 set royalty payments at only two rupees per 100 cubic feet of timber, with an additional sum payable on each acre clear-felled. These rates were less than half those payable by sawmillers working other Kenyan forests. After 1919, when sawmilling concessions were granted only by competitive tender, the disparity between the terms operating in the Lembus concession and elsewhere in the colony became even more marked. In 1928, Grogan's sawmillers were paying only six cents per cubic foot of timber, while sawmillers elsewhere in Kenya paid 50 cents per cubic foot. Furthermore, Grogan's agents Equator Sawmills Ltd (ESM) paid a fixed annual licence fee of KShs 12,000/- (initially 6000 rupees), against which royalty payments were offset. Timber to the royalty value of KShs 12,000/- had therefore to be felled before the Department saw any revenue on the timber extracted from the concession.[6] By the mid-1920s ESM were extracting more timber than all other Kenyan sawmillers put together, yet were paying considerably less for the privilege (Kenya Forest Department, 1919–28).

Forest Department antagonism towards ESM was deepened further by the suspicion that the company was infringing the already generous terms of its licence. The commercial working of the Lembus concession had begun prior to the First World War, but due to shortages of finance and staff, it was not until 1921 that the Forest Department began to monitor the activities of ESM (Kenya Forest Department, 1913–21). Even at this stage the only detailed map of the forest was held by Major Grogan, having been drafted by a surveyor in his employment in 1905. This gave the concessionaire a much clearer knowledge of the potential of the forest than had the Forest Department.[7] The unwillingness of ESM to furnish the Department with working plans for Lembus raised the suspicions of the Conservator of Forests, and by 1923 he had uncovered evidence of irregularities in ESM's calculation of royalties due and of areas clear-felled. Certainly, ESM were stretching the terms of the lease to their absolute limit in order to delay royalty payment and seemingly, to avoid the supervision of the Forest Department. On their own admission, the policy of the company was 'to pick the eyes out of the forest', rather than to clear-fell.[8] Consequently, when the Department sought to reafforest areas that they believed ESM to have clear-felled, they invariably discovered that trees had been left standing. By claiming that these areas were not in fact clear-felled

ESM avoided making any payment, while also preventing the Forest Department from getting on with its programme of reafforestation (under a clause that gave the company rights to re-enter any areas *not* clear-felled for a further 20 years after their original working).[9]

The Forest Department relentlessly pursued ESM over these practices throughout the 1920s, while pressing for the Attorney-General to declare foreclosure on the concession, in view of the company's infringements of the lease. As a result of this pressure, ESM were compelled to submit monthly returns on their activities in the forest, to settle outstanding royalties, and to hand over a number of clear-felled areas to the Department for replanting. But, while it was widely acknowledged by the late 1920s that the forest concession had been a serious error, the government was not prepared to challenge Major Grogan by attempting to terminate the agreement.[10] Further opportunities to terminate the lease occurred during the 1930s, as economic recession descended upon the timber trade and upon Kenya (Kenya Forest Department, 1930–34). Although the advantages of low royalties allowed the Lembus concessionaires to weather the depression better than their competitors, from 1933 financial difficulties prevented them from making full payment of the annual licence fee.[11] Non-payment of the licence fee was a specific clause under which government had power to terminate the lease, but wider political concerns again prevailed over the demands of the Forest Department: in the circumstances of the depression, the Kenya government was more intent on persuading businesses to remain in the colony than on seeking ways of expelling them. The Forest Department had succeeded in asserting a degree of control over the management of Lembus, yet one of Kenya's potentially richest forests continued to add only negligibly to the revenues of the Department (see Table 3.1).

While the commercial arrangements in the Lembus remained of grave concern to the Forest Department, the broader prospects for forest conservation improved considerably during the 1930s. Inspired by a variety of motives, public opinion in the colony became more conscious of the need for government to take an active role in implementing conservation measures (Anderson, 1984). With greater public attention focused upon questions of land use and environmental degradation, the Forest Department immediately benefited from the emphasis given to the importance of forest cover in relation to fears about soil erosion. Lively public debate on matters of conservation, much of it conducted through the newly formed Arbor Society, improved the image and status of the Forest Department, so that by the end of the 1930s opinion was very much in favour of stronger policies for forest protection (Ward, 1937; Anderson, 1982, 1984; Murray, 1982). These concerns were reflected in the prominent role played by the Forest Department in the rural development plans implemented after 1945, many of which were organized on the basis of catchment areas. One such scheme was centred on the Lembus Forest, at the head of the Perkerra River catchment. Protection of Lembus, and of the other forests around it, was accordingly given the highest priority in

Table 3.1 Equator Sawmills Ltd: timber milled and royalty payments, 1912–36

Year	Timber milled (cubic feet)	Royalty payment (incl licence fee)[a]
1912	56,450	6,000 rupee
1913	52,758	6,000 rupee
1914	143,166	6,000 rupee
1915	282,703	6,000 rupee
1916	151,175	6,000 rupee
1917	99,020	6,000 rupee
1918	90,701	6,000 rupee
1919	188,262	6,000 rupee
1920	176,926	6,000 florin
1921	222,186	13,313 sh
1922	462,249	14,103 sh
1923	47,590[b]	12,000 sh
1924	121,111	12,000 sh
1925	223,260	14,467 sh
1926	397,148	25,150 sh
1927	582,102	36,163 sh
1928	475,279	30,363 sh
1929	548,723	33,594 sh
1930	498,239	30,500 sh
1931	167,809	12,000 sh
1932	83,324	12,000 sh
1933	111,600	12,000 sh
1934	109,357	12,000 sh
1935	?	12,000 sh
1936	?	12,000 sh

[a] From 1912 to 1920 the licence fee was 6000 rupees pa In 1920 it was 6000 florins, and from 1921 onwards it was 12,000 shillings pa
[b] The mills were shut down for part of 1923 because of a collapse in the timber market
Source: Kenya National Archives, various Forest Department and Attorney-General files for the period 1902–39

planning the future pattern of African land use and husbandry throughout the entire catchment. This gave the Forest Department the political authority after 1945 to exercise much greater control over commercial activities in Lembus (Kenya, 1950; ALDEV, 1962).

The enhanced status of the Forest Department was also bolstered in hard financial terms by the boom in timber trading during the Second World War. Timber production in the colony climbed from 19,750 Hoppus tons in 1938 to 116,500 Hoppus tons in 1945, and in the Lembus Forest alone a further six mills came into operation. Although this brief period of rampant exploitation left 'chaotic conditions' in many forests, the revenue gained by the Department was

Table 3.2 Kenya Forest Department: war-time revenue, 1938–45

Year	Revenue (£)	Expenditure (£)	Surplus (£)
1938	41,550	31,323	10,227
1939	43,702	31,051	12,651
1940	57,170	30,800	26,370
1941	75,136	29,473	45,663
1942	119,020	36,608	82,412
1943	140,492	45,646	94,846
1944	142,079	60,920	81,159
1945	156,314	74,363	81,951

Source: Kenya Forest Department Annual Reports, 1938–47

substantial (see Table 3.2) (Kenya Forest Department, 1945–7). Concern over forest conservation worked to strengthen the financial position of the Department at this time with the establishment of a Forestry Sinking Fund, an initiative unique to Kenya. Taking the surplus revenue raised for 1940 as the base line, all annual revenues raised by the Department above the 1940 figure were placed in the Sinking Fund, to be used to replenish and develop the forests exploited during the wartime boom. By the end of the war the Sinking Fund stood at over £300,000 (Gardner, 1942; Graham, 1945; Kenya Forest Department, 1945–7). The Forest Department was in a financially and politically healthy position for the first time, largely due to the emergence of broader concerns over environmental protection (Logie, 1962).

It might be said, then, that this marked the victory of the conservationist aims of the Forest Department over the commercial aims of the timber company. However, the important relationship between the successful commercial exploitation of the forest resource during the war and the subsequent implementation of improved forestry protection measures must be stressed. There was no necessary contradiction between commerce and conservation, so long as the foresters were able to utilize the former to achieve the latter. From the Forest Department's viewpoint, the Lembus Forest concession was undesirable because it made this difficult. After 1945 the operation of the lease continued, and although the controls enforced by the Department were improved they were still by no means complete. The reassertion of forest conservation in Lembus had only been made possible by a growing political will in government to protect the environment. But the conservation policies enforced as part of the post-war development effort in Kenya were not always greeted enthusiastically by the Africans on whose lives they impinged. As far as many Africans were concerned, conservation for the public good meant only the restriction of their private rights; rights to graze their animals, to cultivate, and to cut timber and fuelwood. Africans making use of the forest resources of Lembus were particularly energetic in defending those rights and it is this aspect of the conservation history of Lembus that will now be considered.

Conservation and African rights

The question of 'native rights' in Lembus had been raised during the early negotiations between the government and Major Grogan. The government was concerned to ensure that Africans living within the area of the concession would be permitted to continue to utilize the forest for their livelihood. As Major Grogan was of the opinion that the forest was 'virtually uninhabited', other than by a small number of Dorobo families, he raised no objection to a clause being included in the terms of the lease protecting the rights of these few 'traditional' forest dwellers (Kenya Land Commission, 1934).[12] However, during the final stages of the re-negotiation of the lease – June 1913 to March 1916 – dispute arose over the precise extent of such rights. A compromise was eventually agreed, that the Governor should, in the near future, 'be required to endeavour to ascertain and define the nature and extent of free grazing rights and other such customary rights as may have been exercised in the [forest] prior to the dates of the concession'.[13] This clause was not immediately acted upon, but once Grogan's agents began to work the forest more intensively it became apparent that the numbers of Africans occupying Lembus were much greater than had been supposed. While the timber contractors and the Forest Department shared the view that these were 'unauthorized persons' who should be removed from the forest, the District Administration at Eldama Ravine insisted that little could be done until a formal definition of native rights in the forest was proclaimed by the Governor.[14]

In December 1923, after a census of the forest had been conducted, Governor Coryndon issued his definition of native rights in Lembus. This document, known as the Coryndon Definition, laid down 11 specific rights, including the rights to construct dwellings, to graze animals, to cultivate, and to gather forest produce. These rights were granted to all those Africans who were able to satisfy the administration that they had enjoyed rights within the forest 'according to native law and custom', prior to the initial signing of the lease. Appended to the Coryndon Definition was a full list of all African 'Right-Holders', as they now became known. The list was viewed with horror by both Major Grogan and the Forest Department, for it identified no fewer than 485 Tugen and 11 Dorobo right-holding families, and further stipulated that such rights were to be passed down to the descendants of each right-holder named. With the children of listed right-holders already numbering 650, and around 40 per cent of the adult males listed either unmarried or married but as yet without children, it was clear that the future management of the forest was going to be problematic.[15]

The Coryndon Definition remained a bone of contention throughout the colonial period. To the Africans of Lembus it was a charter of undeniable rights to be exercised in perpetuity; an absolute guarantee of their security in the forest. To the sawmillers and to the foresters it represented a serious hindrance to the economic and ecological management of the forest. The scene for political confrontation was set, with both the timber company and the Forest Department, though with differing motives, seeking to exploit other clauses in the lease to place firmer controls upon the right-holders and ultimately to press for new legislation that would override the terms of the Coryndon Definition. There was also discontent among the Africans living on lands surrounding Lembus regarding the implications of the Definition, for by securing the rights of those families listed in the Coryndon Definition the Governor had effectively closed the forest to all others. From being a zone of ethnic indeterminance, the forest had become, *de jure*, the rightful home of a clearly defined section of the Tugen. Many other Tugen and Elgeyo who made periodic use of the forest, but were not listed as right-holders, now found themselves legally excluded from the forest.

This effective exclusion of the occasional and irregular users of the forest – mainly other Tugen and Elgeyo inhabiting lands adjacent to Lembus – came at a time when three related factors combined to make access to the resources of the forest increasingly important. Firstly, during the mid-1920s the administration completed the final demarcation of the Native Reserves, the parcels of land within which it hoped to confine each African group. African populations either remained resident in their designated Native Reserve, or took temporary labour contracts on European farms. These moves towards the establishment of boundaries and the control of the movement of Africans and their livestock across those boundaries served to accentuate the role of the forest as a place of refuge (Murray, 1982). Africans wishing to move livestock around from one area to another, and particularly those with labour contracts seeking to smuggle extra cattle onto European farmlands, did so under cover of the forests. Strategically placed at the heart of the White Highlands between several European farming areas, and coupled with the advantages of concealment and good grazing, the Lembus Forest became a particularly important entrepôt for Africans and their livestock. Secondly, between 1925 and 1936 a series of droughts affected the Baringo Plains, to the east of Lembus, and over the same period locust invasions damaged grazing in parts of the Elgeyo Reserve to the north of Lembus, and in southern Baringo (Anderson, 1984). Herders hard pressed for grazing in these areas made greater use of the forest glades, particularly during the dry seasons, and there is evidence that stockowners commonly 'loaned' animals to friends and relatives among the Lembus right-holders. At times of drought the forest was a crucial resource to pastoralists on the neighbouring plains (Anderson, 1982). Thirdly, in 1929 the colonial administration in Baringo initiated a programme for the reconditioning of pasture on the plains to the east of Lembus. This was prompted by anxieties over land degradation, believed to be caused by the overstocking of the area. The success of the reconditioning programme depended upon the restriction of the number of animals allowed back into reconditioned areas once the reseeding of pasture was complete. To accomplish this cattle counts had to be conducted, and attempts made to encourage the Tugen to accept the necessity of destocking. The Tugen herders naturally distrusted a process that seemed effectively to result in the enforced reduction of their livestock holdings. Their response was pragmatic. While their spokesmen accepted the pronouncements of the administration, each Tugen herder was busy depositing as many livestock as he could around his kinsmen in those locations as yet unaffected by reconditioning (Anderson, 1982). The Lembus Forest, with its population of right-holders, was the perfect destination for livestock displaced from southern Baringo by reconditioning schemes and the threat of destocking.

At a time when government policy was aimed at removing Africans from the forests, the legal residents of Lembus were therefore a striking anomaly. With right-holders legitimately entitled to graze cattle in the forest, it was difficult for the administration to detect the presence of cattle belonging to non-right-holders. Suspicion that the forest was heavily used in this way by non-right-holders was confirmed by occasional police raids, one such raid on Torongo Glade in 1929 uncovering more than 400 head of cattle herded illegally in the forest.[16]

All of this placed the Lembus right-holders in something of a predicament, for while the Coryndon Definition protected their status it also threatened the viability of the regional economy. The forest economy was based upon the cultivation of cereals within the forest, and the dynamics of movement between the forest glades and the surrounding lowland grazing areas. Forest grazing and forest cereals were critical dry-season reserves for the peoples over the surrounding region. Short-term (ie seasonal) and longer-term fluctuations in the utilization of the forest could therefore be considerable, with forest resources being under greatest pressure during prolonged periods of drought. These were the conditions prevailing in the years following the Coryndon Definition, the enforcement of which implicitly demanded the breaking of economic linkages between the forest and its neighbouring areas. While the right-holders were happy to accept the guarantee of their security, they could not isolate themselves from the local economy. The influx of people and livestock into the forest throughout the 1930s were stimulated both by the importance of the forest within the regional economy, and by the privileged status enjoyed by the right-holders. The predicament this held for the right-holders was exposed towards the end of the 1930s, as the Forest Department mounted wider programmes for reafforestation and forest preservation.[17]

By the early 1930s regular complaints from ESM of shortages of grazing for their working oxen, and reports from the Forest Officer that the acreage under cultivation was steadily increasing, indicated that the Africans of Lembus were exploiting the Coryndon Definition to the full.[18] The efforts of the Forest Department to remedy this situation stirred the Tugen of Lembus into political action. Initially, forest officers simply tried to close parts of the forest to African grazing and cultivation, but the District Administration pointed out that this infringed the rights set out in the Coryndon Definition.[19] By 1938, strengthened by the growing conservation lobby in the colony, the Forest Department challenged the legality of the Definition by framing rules under the Forest Ordinance that gave them the power to restrict the numbers of animals in particular forest glades, control which areas could be cleared for cultivation, and remove people and livestock from areas where reafforestation programmes were to go ahead.[20] These measures were similar to regulations then being applied in other parts of Kenya in connection with soil conservation and grazing control, but once again the District Administration stepped in to uphold the special status of the Lembus right-holders. By this time the Tugen of Lembus were well aware of the significance of the Coryndon Definition, and the administration were accordingly concerned that any attempt to curtail forest rights in Lembus would have serious political repercussions. These differences of opinion between the Forest Department and the District Administration were immediately apparent to the Tugen, who found that orders and advice from one wing of government were contradicted and countermanded by another. Skilfully playing one side against the other, the leaders of the Lembus Tugen were able to exploit the politics of conservation in the forest. Facing opposition from the District Administration, and hostility from the Tugen, the Forest Department was unable to implement its conservation programme in Lembus.[21]

This postponement was prolonged by the Second World War, during which non-right-holders enjoyed a further period of unhampered access to the forest. By 1946 the Lembus Tugen had taken advantage of the interregnum in administrative decision-making brought about by the war to organize themselves against the threat posed by the Forest Department. Two main issues had galvanized the right-holders towards a more organized defence of their position. The first concerned the failure of the Forest Department to provide shops, schools, medical dispensaries and other services within the forest, facilities for which the Lembus residents paid a supplementary tax. Because of the unwillingness of the Department to provide these services in the area worked under the concession, by the mid-1940s the right-holders still lacked *any* of the services that their taxes were supposed to pay for, and that were by then common in other African locations. This failure was deeply resented by the right-holders. The second issue concerned the arrangements to be made for Lembus on the termination of the forest lease in 1959. The right-holders were determined to avoid full control of the forest passing to the Forest Department, whose activities they now viewed as entirely hostile to their own interests. Also, the right-holders wished to ensure that their rights would be recognized beyond the expiry of the lease. These issues formed the basis of a campaign now initiated by the right-holders for the excision of the Lembus Forest from the forest reserve, to become part of Baringo District.[22]

To assert firmly their claim to Lembus, the Tugen right-holders embarked upon this campaign by exercising their rights in the forest to the full. Between 1945 and 1948 the 'alarming' increase in the level of cultivation in Lembus prompted the Forest Department to threaten Tugen cultivators with prosecution under the 1941 Forest Ordinance. District Commissioner Simpson, again upholding the Coryndon Definition, thought this inadvisable.

Commenting upon past arrangements made on an *ad hoc* basis by the Forest Department to prevent the right-holders from cultivating parts of the forest required for reafforestation, he noted that,

> . . . none of these arrangements has had the force of legal sanction behind them . . . Thus, to prosecute without being assured of a conviction and with acquittal resulting would, I am certain, open the eyes of the Tugen to the fact that they have been fooled for years, and that until rules are framed and made law, they can exercise their "rights" indiscriminately to the detriment of the forest and to the loss of power behind administrative order.[23]

The Tugen right-holders were already aware that the government were hamstrung by the terms of the Coryndon Definition. In August 1948, Simpson again wrote warningly to the Forest Department: 'In view of the fact that the Tugen in the Lembus Forest intend to make a political issue of its future, it is essential that we proceed now strictly according to the law'.[24] But which law, the Forest Ordinance or the Coryndon Definition?

At this time the Forest Department came to play a more prominent role in the planning of rural development in Kenya, but ironically, just as forest preservation became central to government development plans, events in Lembus conspired to make the implementation of forest policies virtually impossible. From 1948 to 1951 the Tugen campaign gained momentum, culminating in May 1951 with the Lembus locational Council sending a petition on the matter directly to the Secretary of State. By this time the Lembus question had become embroiled in nationalist politics, with representatives of the leading African political party, the Kenya African Union (KAU), visiting the forest to speak at public meetings. The prospect of action by the Forest Department to enforce husbandry rules in Lembus 'stirring up the KAU' and contributing to serious unrest now seemed very real.[25] A census of the forest, completed in August 1951, did nothing to alleviate the administration's sense of impending doom: there were now no fewer than 920 legitimate right-holders and their families, 300 of whom actually resided outside the forest but still determinedly exercised their rights in Lembus. These people therefore exploited their status as right-holders to cultivate lands within the forest as well as lands in the neighbouring Native Reserve.[26]

With political tension mounting in Kenya throughout 1952 as the Mau Mau crisis unfolded, the administration was pushed into direct consultations with the Lembus Tugen in the hope of finding a political settlement. A 'Working Committee', with four African members was set up in August 1952 to solve the problems of the right-holders, but this came too late to salvage the Forest Department's conservation strategy for Lembus. The following month J M ole Tameno, an African Member of the Kenya Legislative Council, wrote formally to the Chief Native Commissioner, declaring his fear of 'genuine unrest' in Lembus, and accusing the government of giving 'twisting answers' to the Tugen to avoid telling the truth about their rights and about plans for the future of the forest. On 5 November 1952, only a fortnight following the Declaration of

Emergency in Kenya, Tameno tabled a question in the Legislative Council on unrest in Lembus.[27] The political point was timely, for a recent ruling by the Solicitor-General, that the right-holders should be regarded as tenants-at-will, had given the Forest Department the necessary powers to override the Coryndon Definition and evict the Tugen from Lembus. However, the government was now very concerned to avoid confrontation in Lembus, and the legal victory of the Forest Department was lost amid the political realities of the situation. 'Whatever the legal position may now be', wrote the District Officer in 1952, 'even if we could evict these people, we have nowhere to put them'.[28]

With the legal solution they had so long struggled to achieve now denied them by the government's unwillingness to provoke further political unrest in Lembus, the Forest Department turned to ecological arguments. Backed by the African Land Development Board, the Forest Department insisted that to give way on husbandry rules in Lembus placed the whole development effort in the Perkerra catchment in jeopardy. By exercising their individual rights, the inhabitants of Lembus were accentuating erosion and degradation throughout the catchment.[29] This stand for strong conservation management in Lembus won the Forest Department many supporters, but only served to delay the now inevitable decision to find a practical and workable political solution to the Lembus problem. Although the forest was seriously overcrowded, and the lands to the east in Baringo were badly overgrazed, it was clear that none of the right-holders would meekly accept eviction from the forest. Political expediency prevailed over ecological considerations.

At a meeting of all sections of the administration involved in Lembus, held in March 1956, the Forest Department was finally forced to face the harsh political reality. The meeting accepted the Tugen's historical claim to Lembus, emphasizing that current political aspects were 'most important' in evaluating the situation. Thus, the Tugen had successfully 'tribalised' the Lembus Forest. It was therefore agreed that control of the forest should revert to the Baringo African District Council upon the termination of the Grogan concession in 1959, as 'any other course of action would meet with the bitterest opposition from the whole tribe and might well require a levy force to impose government orders upon the people'. In other words, it was felt that Forest Department husbandry rules could only be applied by coercion. More generally, it was recognized that to destock the forest glades and to control excessive cultivation and deforestation, would be 'well nigh impossible' without accommodating the wishes of the people. The victory of the right-holders was complete.[30]

Conclusion

Although the granting of the Grogan concession shaped the history of Lembus in a unique way, the subsequent experience of the Kenya Forest Department in their attempts to manage the forest offers a striking illustration of the conflicts between indigenous forest users, commercial forestry, and the aims of forest conservation. Lacking the necessary influence in political decision-making before the late 1940s, the requirements of the Forest Department were overlooked firstly in the granting of the forest lease, and then in the definition of native rights within the forest. While these decisions made the management of the forest difficult, it was the growth of the African political opposition to the implementation of broader colonial regulations governing land use that marked the final defeat for the Forest Department's policies in Lembus. The determination of the colonial government to transform African husbandry after 1945 was stimulated and justified by the aims of conservation (Cliffe, 1972; Anderson, 1984; Throup, 1985). With the deeply rooted history of conflict over the rights of Africans in the forest, it is hardly surprising that Lembus should have become the focal point of political opposition to colonial conservation policies. In many respects the political victory of the Tugen in defending their rights in Lembus should be applauded, but it should also be noted that conservation was, by default, a casualty of this struggle. Conservation measures advocated by colonial governments were often stigmatized in the eyes of Africans, creating a resistance to the enforcement of land husbandry rules that the governments of independent Africa have found it difficult to overcome. The failures of colonial conservation have sometimes left a powerful and undesirable legacy.

This chapter was taken from Andeson, D and Grove, R (1988) *Conservation in Africa: People, policies and practice,* Cambridge University Press, Cambridge.

NOTES

1. G S Cowley. 'Memorandum on Agricultural Aspects of the main Forest-free areas of the Lembus Forest'. Kenya National Archives (KNA) PC/NKU/2/1/31.
2. Eliot to Grant, 8 February 1904; Hill to Eliot, 25 April 1904; Linton (Conservator of Forests). 'Lease for Timber Cutting at Ravine'. May 1904; Grogan to Eliot, 2 June 1904; and 'Original Licence for Lembus Forest Concession', 15 July 1904, all in KNA AG 4/2313.
3. Crown Advocate, 'Memo, on the Lingham and Grogan Timber concession', 11 October 1910, and other correspondence in KNA AG 4/2313.
4. Acting Conservator of Forests to Attorney-General, 3 November 1920, 'Copy of Forest Lease, 1 March 1916', and related correspondence KNA AG 4/2313.
5. Conservator of Forests to Senior Commissioner/Kerio, 31 July 1925, KNA ARC(FOR) 7/2/128, for a summary of the Department's criticisms of ESM.
6. 'Report of the Select Committee of the Legislative Council appointed to consider Forest Royalties', 1919, KNA AG 4/919; Crown Advocate to Acting Chief Secretary, 23 April 1912 and Conservator of Forests to Attorney General, 3 November 1920, both KNA AG 4/2313; Nicholson to Colonial Secretary, 2 February 1928, KNA ARC(FOR) 7/2/128, Kenya's currency was counted in rupees until 1920, and in shillings and cents from 1921.
7. 'ESM Ltd Forest Concession', unsigned, 1923, KNA AG 4/2332. Tannahill to Commissioner of Lands, 20 January 1921, and Acting Chief Secretary to Commissioner of Lands, 10 March 1921, both KNA ARC(FOR) 7/2/128.
8. ESM to Conservator of Forests, 26 February 1924, and Conservator of Forests to Senior Commissioner/Kerio, 31 July 1925, both in KNA ARC(FOR) 7/2/128.
9. 'Grogan Forest Licence – Royalty Payments 1920–29', KNA ARC(FOR) 7/2/127; 'Grogan Forest Licence, 1921–28', KNA ARC(FOR) 7/2/128: 'Equator Saw Mills, 1923–9', KNA ARC(FOR) 7/1/41 II.
10. Grigg to Amery (Secretary of State), 12 November 1925, KNA ARC(FOR) 7/2/128;

Governor's Deputy to Secretary of State, May 1926: Conservator of Forests to Tannahill (ESM), 25 July 1924; and Monthly Returns, 1924–9, all KNA ARC(FOR) 7/1/41 II.
11. 'Grogan Forest Licence – Timber Payments, 1931–36', KNA ARC(FOR) 7/2/129.
12. E M Hyde-Clarke, 'Lembus Forest: Appreciation of the Position'. 28 July 1938, KNA PC/NKU/2/1/31.
13. 'Copy of Forest Lease, 1 March 1916', KNA AG 4/2313.
14. 'ESM Ltd Forest Concession', 1923, KNA AG 4/2332.
15. Governor Coryndon, 'Definition of Native Rights', 12 December 1923. KNA PC/NKU/2/1/31.
16. E M Hyde-Clarke, 'Lembus Forest', 28 July 1938, KNA PC/NKU/2/1/31.
17. H J A Rae (Forester), 'Lembus Right-Holders', 29 July 1938, and E M Hyde-Clarke, 'Lembus Forest', 28 July 1938, both KNA PC/NKU/2/1/31.
18. Londiani Forest Division Reports, 1931–6, KNA ARC(FOR) 7/1/1 II.
19. H J A Rae (Forester), 'Lembus Right-Holders', 29 July 1938, KNA PC/NKU/2/1/31.
20. Conservator of Forests of PC/Rift Valley. 11 April 1938, and PC/Rift Valley to DC/Baringo, 30 October 1940, both PC/NKU/2/1/31.
21. E M Hyde-Clarke, 'Lembus Forest'. 28 July 1938, and H J A Rae. 'Lembus Right-Holder', 29 July 1938, both KNA PC/NKU/2/1/31.
22. A B Simpson, 'Lembus Forest Area', 19 July 1949, KNA NKU/2/1/31; Denton (DO/Baringo) to Forest Officer/Londiani, 8 September 1952, KNA PC/NKU/2/13/3.
23. A B Simpson to Forester/Londiani, 19 May 1948, KNA PC/NKU/2/1/31.
24. A B Simpson to Forester/Londiani, 28 August 1948, KNA PC/NKU/2/1/31.
25. DC/Baringo to PC/Rift Valley, 13 May 1951, and DC/Baringo to Provincial Agricultural Officer/Rift Valley, 14 August 1951, both KNA PC/NKU/2/1/31.
26. PC/Rift Valley to Conservator of Forests, 22 May 1952, KNA PC/NKU/2/1/31.
27. J M ole Tameno to Chief Native Commissioner, 12 September 1952, and related correspondence, KNA PC/NKU/2/1/31.
28. DO/Eldama Ravine to DC/Baringo, 26 February 1952, and Solicitor General, 'Memo on Lembus', 15 November 1952, both KNA PC/NKU/2/1/31.
29. DO/Eldama Ravine to DC/Baringo, 26 February 1952; R O Hennings (ALDEV) to Colchester (Member for Forestry), 10 March 1956; PC/Rift Valley to Governor, 28 October 1955, all KNA PC/NKU/2/1/31. Graham (Acting Conservator of Forests) to Chief Native Commissioner, 17 September 1952, KNA PC/NKU/2/13/3.
30. 'Minutes of meeting to consider the Future of the Lembus Forest', 24 February 1956, KNA PC/NKU/2/1/31.

REFERENCES

Aldev (1962) *African Land Development in Kenya 1946–62*. Nairobi: English Press.

Anderson, D M (1982) 'Herder, settler, and Colonial Rule: a history of the peoples of the Baringo Plains, Kenya, *c* 1890–1940'. Unpublished PhD thesis, University of Cambridge.

Anderson, D M (1984) 'Depression, Dust Bowl, Demography and Drought: the Colonial State and Soil Conservation in East Africa during the 1930s' *African Affairs,* **83** (332), 321–43.

Anderson, D M and Millington, A C (1987) 'The political ecology of soil conservation in anglophone Africa' in *African Resources: Appraisal, Monitoring and Management* ed Millington, A C, Binns, A and Mutiso, S. Reading Geographical Papers Series, Reading University. (In press)

Beinart, W (1984) 'Soil erosion, conservationism, and ideas about development in Southern Africa' *Journal of Southern African Studies,* **11** (2), 52–83.

Blackburn, R H (1974) 'The Okiek and their history' *Azania,* **9**, 139–57.

Brasnett, N V (1942) 'Finance and the Colonial Forest Service' *Empire Forestry Journal,* **21**(1), 7–11.

Cliffe, L (1972) 'Nationalism and the reaction to enforced agricultural change in Tanganyika during the colonial period' in *Socialism in Tanzania: An Interdisciplinary Reader* ed Cliffe, L and Saul, J pp 17–24. Dar es Salaam: East African Publishing House.

Eliot, C (1905) *The East African Protectorate*. London: Edward Arnold.

Gardner, H M (1942) 'Kenya forests and the War' *Empire Forestry Journal* **21**(1), 45–7.

Graham, R M (1945) 'Forestry in Kenya' *Empire Forestry,* **24**, 156–75.

Helms, D and Flader, S L (ed) (1985) *The History of Soil and Water Conservation*. Washington DC: Agricultural History Society.

Hughes, J F (1949) 'Forest and water supplies in East Africa' *Empire Forestry,* **28**, 314–23.

Hutchins, D E (1907) *Report on the Forests of Kenya*. London: HMSO.

Hutchins, D E (1909) *Report on the Forests of British East Africa*. London: HMSO.

Johnston, H H (1902) *The Uganda Protectorate*, 2 vols. London: Edward Arnold.

Kenya, Colony and Protectorate (1950) *An Economic Survey of Forestry in Kenya and Recommendations regarding a Forest Commission*. Nairobi: English Press.

Kenya Forest Department (1904–63) *Annual Reports*. Nairobi: Govt Printer.

Kenya Land Commission (1934) *Kenya Land Commission (Carter): Evidence and Memoranda*, 3 vols. London: HMSO.

Logie, K P W (1962) *Forestry in Kenya. A Historical Account of the Development of Forest Management in the Colony*. Nairobi: Govt Printer.

McCracken, J (1982) Experts and expertise in colonial Malawi *African Affairs,* **81**(322), 101–16.

Matson, A T (1972) *Nandi Resistance to British Rule, 1890–1906*. Nairobi: East African Publishing House.

Murray, N U (1982) 'The other lost lands: the administration of Kenya's forests, 1900–52'. History Department Staff Seminar Paper, Kenyatta University College, Nairobi.

Nicholson, J H (1931) *The Future of Forestry in Kenya*. Nairobi: Govt Printer.

Ofcansky, T P (1984) 'Kenya forestry under British colonial administration, 1895–1963'. *Journal of Forest History,* **28**(3), 136–43.

Powell, J M (1976) *Environmental Management in Australia, 1788–1914: Guardians, Improvers and Profit: an introductory survey*. Melbourne: Oxford University Press.

Sorrenson, M P K (1968) *Origins of European Settlement in Kenya*. Nairobi: Oxford University Press.

Stebbing, E P (1922) *The Forests of India*, 3 vols. London: J Lane.

Stebbing, E P (1937) *The Forests of West Africa and the Sahara*. London: Chambers.

Stebbing, E P (1941) 'Forestry in Africa' *Empire Forestry Journal,* **20**(2), 126–44.

Throup, D W (1985) 'The origins of Mau Mau' *African Affairs,* **84**(336), 399–433.

Unwin, A H (1920) *West African Forests and Forestry*. London: Unwin.

Waller, R D (1985) 'Ecology, migration and expansion in East Africa' *African Affairs,* **84**(336), 347–70.

Ward, R (1937) *Deserts in the Making: A study in the Causes and Effects of Soil Erosion*. Nairobi: Kenya Arbor Society.

Chapter 4

Management of Tropical and Subtropical Dry Forests

Gill Shepherd, Edwin Shanks and Mary Hobley[1]

INTRODUCTION

Definition, extent and importance

While much of the attention of the North has been fixed on the results of the destruction of tropical rainforest, less attention has been focussed on the tropical and subtropical dry forests, which are at least as problematic, are more extensive, and are disappearing at a faster rate than rainforests. Because such forests/woodlands[2] occur in more densely populated regions than rainforests, their disappearance is likely to have a more severe impact on people living nearby the forests. Their location may mean that their disappearance will increase the prospects of desertification, and they are usually major suppliers of urban fuelwood to distant markets. Yet most of the interest and drama is currently attached to rainforests.

Tropical and subtropical dry forests are discussed here in the context of developing world management issues. For that reason, the extensive dry forests of Australia receive no mention, and studies are drawn only from Africa and Asia. Tropical and subtropical dry forests are treated as a single category from that point of view, and are therefore usually referred to in this chapter, for brevity, as 'tropical dry forests'.

The tropical dry forests are mainly to be found today in Africa, but there are also extensive tracts in Asia, Latin America and, of course, Australia. Forests cover roughly one-third of the world's surface. Worldwide, about 2.8 billion hectares are covered in closed forests, and 1.3 billion hectares in less densely wooded drier open forests. Forest regrowth on fallowed cropland covers an additional 406 million hectares and natural shrublands and degraded forests in developing countries an additional 675 million hectares. Adding all these categories together, the total – 5.2 billion hectares – covers about 40 per cent of the total world's land surface. Of the world's total forest cover, 54 per cent is closed and 46 per cent open forest, shrubland and fallowed land. However, if investigation is narrowed to developing countries only, open forests of various categories make up 58 per cent of the total, while closed forests make up only 42 per cent. Africa has two-thirds of the world's open tropical forests and two-thirds of its shrubland, while Latin America has the largest expanses of closed tropical forests. The tropical dry forest areas of the world are still vast, but disappearing rapidly. The most pressing concern is not their availability, but the identification of appropriate management regimes which might preserve them. It is to this task that this paper addresses itself.

Table 4.1 Distribution of the developing world's forest lands, 1985 (in millions of hectares)

Region	**Closed forest**							
	Broad-leaved	*Conif-erous*	**Open forest**	**Shrub-land**[1]	**Forest fallow**	**Total wooded area**	**Total land area**	**% of total land area wooded**
Africa	216	2	500	450	160	1328	2966	45
Latin America	666	26	250	150	170	1262	2054	61
Asia (exc China) and Oceania	317	30	83	45	76	551	1640	34
China	97	25	15	30	x	167	933	18
All developing countries	1296	83	848	675	406	3308	7593	44

Source: WRI/IIED, 1986, p 62)
[1]Shrublands, as defined in the source from which these figures are taken, mean areas with woody vegetation greater than 0.5 metres and less than 7 metres in height

The tropical and subtropical dry forests are important for a variety of reasons. First, like the rainforests, they have a protective function, not only by directly protecting and cooling soil (and in their case maintaining soil fertility) but also indirectly by locking up quantities of carbon and thereby providing the world with a 'breathing space' in which to adapt to the implications of global warming.

Second, while they contribute less moisture to the atmosphere than the rainforests on a per hectare basis, their size and their presence in lower rainfall areas makes their moisture contribution of very great importance. When they are removed, the increased albedo effect is marked.

Third, most tropical dry forests support larger numbers of people and domesticated animals than the rainforests, on a hectare for hectare basis – though the sustainable density is still relatively low. Humans have learned to live in symbiosis with these woodlands by relying both upon the milk and meat which animals produce from tree-browse, and upon the replenishment of soil fertility which the trees bring to agriculture. As a result, their disappearance may affect people living nearby at least as severely as those living in rainforests.

Fourth, the support offered by tropical dry forests may extend several hundred kilometres away to those living in towns or in higher rainfall agricultural areas, and who depend on them for fuelwood and poles, and for animal products obtained through trade. Recent work by the FAO has also highlighted the very important contribution of non-timber forest products in woodlands to the diets of a wide spectrum of individuals, who obtain them by gathering or by trade. Protein, minerals and vitamins, vital to complement the carbohydrate-rich farm diet are drawn from the woodlands.

Last, though, in most areas, tropical dry forests make no contribution to timber exports (a fact which may mean that their vital in-country contribution to biomass and timber needs is undervalued), they are nevertheless important hard currency earners. Not only are there modest earnings from non-timber forest products such as gum, but also the chief exports of a country like Somalia – hides, meat and live camels and cattle – are almost totally tree-generated. Dry forests can also save expenditure on imports: the commonest substitution being perhaps that of tree fodder – as manure – for imported fertilizer in countries such as Nepal, or leaf litter – as mulch – in the Sudanian zones of Africa.

The need for increased attention to natural forest management

Ever since the bench-mark of the Eighth World Forestry Congress in 1978, a slow process of change in the thinking of forestry professionals has been going on. Where once their primary task was seen as the protection of forests and maximisation of revenue for industry and the State, it is now becoming just as important to consider the dependence of rural people upon tree products and the effect upon them of forestry activities (Cortes, 1984).

Three factors lay behind the change. First, the oil price rises of the early 1970s made it clear that woodfuel would continue, indefinitely to be the primary energy source for billions of rural people, with the implication that rapid deforestation would follow. Second, it became apparent that the world's forests, which in earlier decades had been assumed to be an almost infinite resource, were rapidly disappearing. Third, development thinking itself was becoming increasingly concerned with the needs of the small producer. As a result tree planting projects with farmers and villagers were initiated in many parts of the world, largely in response to these concerns.

In the case of Africa, there was a further factor still. Following the severe Sahelian droughts of the early 1970s, greater recognition was given to the importance of tree growing on the desert margins. In response to this, governments and international donors began to channel resources into rural and peri-urban afforestation projects. Emphasis was initially on the establishment of large scale biomass and shelter plantations, usually of fast growing exotic tree species. By the mid-1970s efforts were also being made to encourage farm-level tree growing chiefly in the form of communally or individually managed woodlots. Almost without exception, these initiatives were driven by the perceived need to meet the demand for woodfuel and to help check desertification.

Tree planting or tree management?

Results were variable, however. It has gradually become clear that rural people usually only want to plant trees on permanently-owned lands, often only as a cash-crop, and only where land in general is so short in supply that there is no state or community-owned forest nearby from which tree products may be taken without the trouble of growing them. In Africa, in particular, but also in many parts of Asia, these prerequisites currently rule out the likelihood of successful villager tree planting projects in many areas. Forests and woodland, though dwindling, still cover millions of hectares and will continue to be used by the populations who live near them for most purposes, for the foreseeable future. At the same time, plantations of exotic species were less successful in dry areas than was expected.

As a result, the management of dry tropical forests – abandoned in many areas after the 1950s in favour of plantations – began to be tried again in the 1980s, with activity in African dry savanna woodland being a particular priority. However, much had changed since it was last tried. Africa's population had greatly increased, while rainfall levels had decreased, intensifying competition for resources; agriculture had also greatly expanded at the expense of woodland cover; above all, the beginnings of a cooperative relationship between foresters and villagers has evolved in the drier areas as a result of the village tree-planting projects of the 1980s: this in turn offers the opportunity of new working styles which are beginning to rule out coercive solutions to the management of natural woodland.

Although not apparent 25 years ago, it is now clear that plantations are unsuccessful in areas with less than 800 mm of rainfall, except where trees can be irrigated (Lamprey, 1986:125). Fast-growing exotics grow scarcely faster under these conditions than the pre-existing vegetation, and are outperformed by indigenous species in drought years. As a result, the low esteem in which dry tropical forest used to be held, is beginning to give way to the recognition that previous estimates of their productivity may have been too low (Jackson, 1983:12) and that, even at low increment levels, there are millions of hectares to be harvested. Most indigenous arid zone trees and shrubs continue to grow even when browsed, burned and lopped, and more protection leads to much improved productivity. However, good rains for Sahelian tree establishment only occur every decade or two, and woodlands tend to consist of large numbers of even-aged trees. The need is to ensure that a greater number of young, newly established seedlings and saplings survive into adulthood; such protection would probably be hundreds of times more effective in reafforesting dry areas than attempts to plant new trees (Lamprey, 1986:126).

Current research and knowledge

The increased interest in the management of dry tropical forest in recent years has been reflected in a variety of activities, though these have had a predominantly African focus. A benchmark paper was produced in 1983 for USAID and the Club Du Sahel under the title 'Management of Natural Forests in the Sahel Region' (Jackson, 1983). Literature searches were conducted and national forestry institutions visited in the Sahel. The report reviews the potential for natural woodland management in the Sahelian zone. It notes that productivity levels are 'admittedly low': figures averaging 0.5 cubic metres/

hectare/year in the Sahel zone rising to 1 cubic metre in the Sudan savanna zone are assumed. However, such figures are usually derived from unmanaged and often over-mature woodland, without fire or grazing protection. In places yields may be considerably higher than previously thought: for example in Bandin, Senegal (rainfall 600–700 mm/year) where the wood yield from *Acacia seyal* dominated woodland is in the order of 0.67 to 2.35 cubic metres/hectare/ year. This compared well with yields from a local *Eucalyptus camaldulensis* plantation (1.5 cubic metres).

An assessment was then made of the work done in eight countries: Cape Verde, Mauritania, Senegal, The Gambia, Mali, Burkina Faso, Niger and Chad. The picture which emerged was generally one of neglect; and of a large number of technically based management initiatives set in motion but rarely carried through or documented in such a way as to give guidance to future practitioners. It is noted that only in Senegal had natural woodland management been undertaken on any appreciable scale. A year later in 1984 the UK Overseas Development Administration (ODA), in an initiative intended to be complementary to the earlier survey, commissioned a similar study on semi-arid east and southeast Africa by T J Wormald. Neither study found a great deal of earlier experience of management on which to build. Though both make mention of trying to involve local people in management activities – Jackson because such involvement would help to eke out the meagre resources of forestry services, and Wormald because he could see that management ought to be more effective with such collaboration – neither found previous references to such styles of approach in the past.

The International Union of Forestry Research Organizations (IUFRO) held a research planning workshop for Sahelian and north Sudanian countries (from the Sahel in the North to the closed moist forests around the equator in the South) in 1986 entitled *Increasing the productivity of multipurpose lands* (Carlson and Shea, 1986). At the workshop, the two themes voted the most important were the genetic improvement of woody plants for the Sahel, and the need for increased attention to tropical dry forest management. It was nevertheless evident from the papers presented that any management role for the farmers and pastoralists who would be most affected by it was low on the list of topics worrying most delegates. Finally, a new generation of forest management projects have come into existence in the dry forests, and some of their experience will be reviewed in this chapter.

In Asia, the last decade of forestry practice has seen a shift from large-scale industrial plantation projects to small-scale village and individually based programmes. Although the types of forestry intervention diversified, the profession continued to embrace those traditional practices which propelled forestry further in its doctrines of 'timber primacy and sustained yield'. The profession was not ready to accept people first and trees second, and thus social forestry remained a top-down technical programme for reforestation.

The notion of community subsumed within the new doctrine of social forestry promoted policies which were directed towards the village as an undifferentiated entity, united for common action by its need for firewood and fodder. These policies ignored the differential access to both natural and political resources within a village and assumed all individuals would benefit equally from forestry programmes.

Current debates within and outside the forestry world have led to new initiatives focussed on a more active role for forest users in the management of their forest resources. However, the degree to which users actually take part in decision-making varies according to local power structures and also to the ideology of the organization. For some organizations social forestry means the empowerment of poor people and women through giving them control over access to forests and decision-making. For other organizations social forestry is a top-down intervention where local involvement is limited to the expenditure of labour in the creation of forests for the state. In most of India and Nepal, the forms of social forestry practised lie in between these two extremes – they are neither exclusively top-down nor exclusively bottom-up. Project initiatives are based on the active involvement of forest users in the management of forest resources but with continued technical support from the state forest department.

THE HISTORY OF MANAGEMENT OF TROPICAL AND SUBTROPICAL DRY FORESTS IN AFRICA AND ASIA

Communal management of forests in the semi-arid and sub-humid regions of Africa

This section is based on an extensive literature search and analysis, recently undertaken, of indigenous forest management practices in dryland Africa.[3] Management, in the sense in which it is used here, involves a series of mechanisms put into practice by rural people who are coordinating their actions with others, and at the command of some ideally local authority they regard as legitimate. In many cases, management is conducted almost entirely by people with their own rules and without coercion. In others, it is as important to know why people obey rules, as it is to know what the rules are.

Surprisingly, as the review was undertaken, it became clear that it is almost impossible to talk about the management of woodland as distinct from the management of trees on farmland.[4] First, in well over a third of the case examples identified, land alternates between woodland and farmland under swidden-fallowing[5] cycles, the former replenishing the fertility of the latter. Second, for the herder, the woodland is his farm in the sense that it is even more essential on a daily basis to his animals than it is for his crops. However, one can expect different and more wide-ranging methods of management of forest by herders than by agriculturalists, and must not extrapolate from the one to the other.

It also turns out that many management practices form a seamless continuum from management in the forest to management on the farm. It thus seemed a pity not to give some flavour of this process, since much of the information gleaned has useful implications for any proposed formal management of woodland.

Finally, it is now standard to deplore the inability of Third World governments, and colonial regimes before them, to distinguish between common property resources (CPRs) and open access land. It would be unwise to make the same mistake by lifting the woodland managed by local people from the matrix in which it exists – the primary economic activities of farming or herding – and treating it like a forest reserve. By and large, the tenure and authority regimes which once governed the successful use of forest are also those that govern the use of farmland and all other local resources. It is divorced management of the two resources – by different ministries, by local and non-local people – which has led to many of the present day problems. The starting point must be that ownership and management go together, and cannot be separated. Only by understanding tenure fully will we understand the conditions necessary for successful management.

Land ownership

Management of natural woodland is practised by those to whom it belongs, and as has been seen so clearly demonstrated in the case of tree planting, no serious investment of time and effort will be made unless the resource is owned. Thus the mechanism for ownership is this chapter's initial focus.

The herding lineage

In the case of many of Africa's herding groups, the genealogy of the lineage is the charter for access to land. The male descendants of one remote ancestor all share one large area, with subsets using subdivisions of it. Thus any particular area is strongly claimed by small numbers, but many people can assert secondary claims to it. Exclusive rights are most strongly asserted where a perennial asset such as water is at stake, and are vaguer where low and erratic rainfall makes it a gamble where the best grazing will be from year to year (Barrow, 1986; Behnke, 1980).

Sedentary kinship groups

For sedentary farmers, while a knowledge of genealogical links is important, the moral community which holds land is defined by both descent and residence, rather than descent alone. Very often this is expressed as the chief or the elders holding the land on behalf of the collectivity, or as the village holding corporate rights allocated by the village headman.

The household head and the household

Just as the lineage fits within the tribe, and the village is often a subset of the lineage defined by residence, so the household has, in the past, often been seen as the lineage writ small. This is clear from the fact that land would revert to the next collectivity upon the owner's death, rather than to his children. However, the household head's position is gaining in importance all the time, as other levels of the kinship system cease to have political meaning and as land registration becomes the norm. In the future, user groups for particular natural resource assets are more likely to be aggregated from otherwise unrelated households, rather than already being a subsection of a larger collectivity.

Typical management actions

For chiefs and lineage elders of herding lineages, management actions most commonly undertaken are likely to be the exclusion of outsiders, adjudication between insiders, and the promulgation of new rules. For instance, Turkana elders in Kenya in the 1980s issued new instructions about the lopping of *Acacia tortilis* so that only side branches were taken, and the main shoot stood a chance to grow rapidly above goat-browsing height (Kerkhof, 1990).

Many leaders have attempted to preserve the most valuable tree species in the area by linking them to chieftain status (Hammer, 1982; Norton, 1987). In southeastern Botswana, village chiefs would ban the felling of village amenity trees and arranged elaborate zoning for different categories of fuelwood collector (Shepherd *et al*, 1985). Around Mount Kenya (and widely elsewhere in Africa) sacred groves were used as the meeting places or burial grounds of chiefs or senior age-sets.

The creation of tenure through labour

Several of the documents surveyed make it clear that it is also the investment of labour which creates ownership. For the agriculturalist, this means being the first to clear and plant land once under forest or woodland. In all cases until present day land registration procedures, such cultivated land reverted, if it was abandoned, not to the wild but to the group to which the clearer belonged: so the individual had created rights for others as well as himself. Pastoralists maintain tenure by maintaining and defending the key dry season assets of grazing/browse and water (Barrow, 1986; Shepherd, 1989a). Tree planting, because it is more work than tree use, also creates tenure.

Individual tenure comes from the greatest and most constant labour investment. In Senegal (Postma, 1990), land was lost to the owner if it remained uncultivated for more than ten years, for instance. However, the investment of labour must come from the owner's own free will. Neither slaves nor tenants can easily create ownership for themselves by their labour, since that labour belongs to another within the terms of their status, or their contract with the landowner.

Indigenous woodland management methods

Management in the sense in which it is used here means some individual or group activity which organizes the utilization of tree resources in such a way that the resource is more equitably shared, more likely to remain into the future (ie more sustainably used), or will actually grow better if the management practice is not taking place.

Classical planned natural woodland management, of the kind practised in European forests in the past, is somewhat different. First, forest and farmland were early separated, and permitted grazing in forests has by no means been universal. Second, the formal drafting of management plans for rotations of particular lengths and specified products common in Europe is unknown in rural Africa, though planned actions which encourage some species and eliminate others or which encourage trees to produce different end products – thin poles, thick timber, for instance – are by no means uncommon. However, many of them take place in the context of farming, rather than dedicated forest management. What African managers are doing is devising rules for sparing certain species, or certain size classes, or simply saying who may and who may not use certain tracts of woodland. This may mean management of the individual tree, but more often it means management of the space which the trees occupy, along with the grass and water there too.

Within this context, any planned and deliberate activity which enhances the quantity/quality of woodland or makes its use more sustainable, is defined as management. There is not space here to set out all the management methods investigated. The full list includes the following: fallowing systems; the conservative use and selective maintenance of particular species; reservation and sacred groves; the opening and closing of areas by time and season; management by taboo and religious sanction; management by fire; animal grazing and browsing as management tools; and management of the individual tree. Just two methods are given as examples below.

Long and short fallow systems

Swidden-fallowing has been the most important method of woodland management in Africa for many hundreds of years. Patterns can be discerned in the large volumes of literature on the subject partly by focussing on the types of fallow associated with particular tree-species, and partly by attempting to identify the management procedures which governed the progression of the fallows.

In the Sahelian and Sudanian zones of Africa, for instance, where the inhabitants practise both agriculture and livestock raising, two particular fallowing systems predominate (Raison, 1988). Here, settlements create a variety of 'parklands' placed in constantly evolving concentric rings around the village. Moving out from the permanently cultivated village home-gardens, one passes through two zones before the unadapted bush is reached:

- the zone of permanent fields and short fallows set with *Faidherbia albida*;
- the zone of fields cleared in the bush, and of long fallows, where *Butyrospermum parkii, Parkia biglobosa* and *Ficus platyphylla* are found.

Butyrospermum parkii parkland – bush fallow (long fallow)

This formation is found throughout the Sudanian zone apart from Senegal. In the untouched wooded savannah, with initially perhaps 1000 trees per hectare, all but 100 or so are burned so that they die. *B parkii* are the main species preserved. Over the next four years, much of the dead wood is used up for fuelwood as the area is cultivated, and then the whole patch is rested for 20 years and the farmer moves on outwards from his village. If one area was to be re-used by one village for the duration of two to three successive cycles, ie

40–60 years, and with the *B parkii* protected all that time, the shade after 60 years would be so great that agriculture would suffer. So the whole village moves, using the earlier area as a fruit-gathering area, and the process start again. After two to three moves, ie 150–200 years, the *B parkii* will be very elderly, and the whole process of major land clearance will start again, with the preservation of younger trees.

While references to the political organization of fallow management in the literature surveyed are unfortunately scanty, what is impressive in the *Butyrospermum parkii* cycle is the lengths of time involved, the coordinated nature of village planting and protection, and the extent to which the landscape in these regions is man-made.

Faidherbia albida parkland – savannah fallow (short fallow)

This is the most interesting and developed of the types of parkland created by Sahelian farmers. The tree, because it is in leaf during the dry season and leafless during the rainy season, permits permanent cultivation beneath it. Also, at a density of 10–30 per hectare, fertilizes up to 50 per cent of the area. Crop rotation is practised under the trees.

However, it is clear that *F albida* alone cannot restore soil fertility – animal dung is needed too. One hectare of *F albida* parkland with 20 trees on it supports six cattle in the dry season – enough to keep the area fertile: but in the rainy season the tree is leafless, there are no agricultural residues available for fodder, and the bushland is then an essential complement. If it is nearby, livestock can browse there and be tethered on currently exploited fields at night to deposit their manure. If there is no bush nearby, more labour-intensive solutions must be adopted, such as sending animals north for part of the year, or growing forage for them. But more recently, with the departure of the young as labour migrants, animal herding has become too labour-intensive an activity. The result is often that *F albida* parkland fails and reverts to less intensively managed *Butyrospermum parkii* parkland (CTFT, 1988; Houérou, 1980; Kessler, 1990; Pélissier, 1966; Postma, 1990; and Raison, 1988).

Acacia senegal gum-gardens

It would seem that in the central Sudan each household was be allocated three or four plots of land, each of which would be farmed for three to six years on a rotational basis. *Acacia senegal* naturally regenerated in cleared plots and was then protected as it grew until it was ready to be tapped for gum by those who owned the fallows. Finally, when the trees were old, they were felled and sold to use as charcoal. This situation is now collapsing – because population densities are too high to sustain such large amounts of land per household (Hammer, 1982, 1988; Salem and van Nao, 1981; Seif el Din, 1987).

Reservation, sacred groves and religious sanction

Considering that trees are rarely set aside in indigenous management systems, but are protected, used and lived among, it is surprising that reservation exists at all. Yet it does. Some accounts make it very difficult to say whether trees are being reserved for soil conservation or watershed management reasons, or for religious and ritual purposes. But others make clear that trees are preserved in part to protect springs, to make rain, or to keep areas of original vegetation intact (Gerden and Mtallo, 1990). Some groves acquire their importance from the fact that ancestors' graves are beneath them. It is customary to put such clusters of graves on a hilltop or a ridge in some cultures, and so the trees may have an inadvertent conservation effect as well. Among the Kikuyu, such sites were commonly adorned with *Ficus natalensis* trees (Brokensha and Castro, 1987).

At the practical level there are a few instances of reservation for patently economic reasons. The grazing/browse reserves of the Pokot and Turkana are one such example from primarily pastoral areas (Barrow, 1986), while the hilltop grazing and woodfuel reserves of the Sukuma in Tanzania constitute an example in a more agricultural area (Shepherd, 1989b). Finally, in the past, some groups planted and guarded particular tree species as a reserve against famine (Jackson, 1983; Raison, 1988).

Religious sanctions, which are common, are a way of linking rulers with God and the ancestors, and enable such rulers to enhance their authority. Examples such as the attribution among the Kikuyu of calamities to the illicit felling of trees (Brokensha and Castro, 1987), and among the Tswana the creation of closed and open hunting or gathering seasons by the placing of religious taboos on infringements, make this clear (Schapera, 1943). While Europeans cannot always understand these taboos it is clear that they were usually obeyed and that they had meaning for those who obeyed them. One should assume, therefore, that such sanctions are rooted in long observation, and a good understanding of local institutions and the local ecosystem. For instance, it is not infrequently reported in some parts of Africa that villagers will say that anyone who plants trees will die, or that only God may plant trees and that it is impious for humans to do so. Closer investigation usually reveals that the proposed tree planting was on village Common Property Resources (CPR) land and that the trees would diminish the CPR rights of others: it is other people, not God, who would have been wronged in due course. The speakers are simply telling an ignorant outsider, in the strongest possible language, that elaborate arrangements are under threat.

In both these cases, trees are protected in ways superficially very different from formal forest management: in the first case, it is ultimately the agricultural cycle which protects certain species of trees; in the second, religious sanction provides intensive protection for small important pieces of reserved forest. Yet are these methods so strange? In the case of fallowed land, the boundary is secured against outsiders by inherited rights and by local residence; some trees are retained for many years, while others are felled and left to regrow on a 20-year cycle. In fact, the planning for short-, medium- and long-term gains is not so different from systems such as coppice with standards as practised, for instance, in Britain. In the case of sacred groves, we are merely seeing local political authority in action – backed by the supernatural rather than codified law.

The evolution of forest land tenure in India[6]

In the nineteenth century up to two-thirds of the land in India was under community control (Singh, 1986). Privatization and government appropriation have been the two main processes which have reduced this proportion. The settlement process of the British, as well as later revisions after independence, often missed the local distinctions in land classification so that many well-defined local resources such as pastures and forests were classified as state revenue land. Forest and revenue lands have historically been common property resources. In some cases these lands have become open access, where all may use the resource; in others access is controlled by a group, and excludes access to outsiders (Hobley, 1985).

For over 100 years a large proportion of India's land area has been under the formal custodianship of government forest departments. In 1975, state forest departments managed nearly 75 million hectares of land or 22 per cent of the nation's territory. Forest departments attempted to fulfil their dual mandate for revenue generation and environmental protection (Poffenberger, 1990). However, the main focus was placed on securing the nation's needs for timber and pulp. To meet this end much of the nation's forest land was protected by forest departments against local people who were using the forests for grazing and firewood.

Under the Indian Forest Act, state governments may assign to any village the rights of governance over forest land as a village forest. However, a December 1988 amendment to the Forest Conservation Act extended the requirement for central government approval to any state government action which assigned forest land to any private person or organization not owned, managed or controlled by the government. This has cast doubt on the validity of the large numbers of village forest agreements, and of the variety of other leasing and benefits-sharing arrangements that exist between state governments and local villages and individuals.

The amended Forest Conservation Act also places the same restriction of prior approval on the planting of forest lands with non-forest crops. As this term has been defined to include horticultural crops, oil-bearing plants, palms and medicinal herbs, concern has been expressed that growing certain tree and medicinal plant species, the produce from which figures prominently among that covered by usufructuary rights, may be discouraged.

Revenue lands comprise two categories: government wastelands which are owned by the government but used by the village; and grazing lands which are vested in village bodies. There is little *de facto* distinction between the two categories, as both are used for grazing, and are generally considered degraded. It has been estimated that on average there are about 20 hectares of such land per village, but there is much regional variation as well as variation between neighbouring villages (Chambers, Saxena and Shah, 1989).

Tenurial rights governing the use of land may be affected at one or more of five different levels:

1. Customary or traditional rights at the social custom level (eg village grazing rights);
2. Administrative orders regarding use of lands (eg Forest Department rules concerning collection of headload fees);
3. Court rulings regarding existing legislation;
4. State and national legislative statutes regarding rights over lands;
5. Constitutional law regarding citizens' rights in land (Singh, 1986).

In practice, the rights and practices which determine who has access to and can appropriate and use revenue lands are generally a matter of convention. The village panchayat may legally be in charge of these lands in the northern states, but as the panchayat consists of several villages, each having its own common lands, the authority of the panchayat over day-to-day control may be quite weak, and the elite of the village may exclude other villages of the same panchayat from using the commons.

In 1984 the National Remote Sensing Agency announced that between 1975 and 1982 India had lost 1.3 million hectares of forest annually. Sweeping restrictions were placed on commercial and local forest use by planners under the new draconian rules of the 1980 Forest Conservation Act (Poffenberger, 1990).

This marked the beginning of confrontation between environmentalists concerned about the rapid degradation of forests, individuals and organizations concerned about the local people's loss of access to forests, and the forest departments concerned that their lack of territorial staff prevented them from effectively policing the forest resource against incursions by rural people and others. Thus, local and national interests were directly conflicting over the management and protection of forest resources.

The solution to the rural fuelwood and fodder problem was social forestry programmes. However, as in Nepal, these programmes did not address the degrading of protected and reserved forest areas. Recent innovations look at methods by which local people in conjunction with forest departments can manage areas of degraded forest land. For example, in West Bengal, nearly 200,000 hectares are under joint protection, while in Orissa 3–10 per cent of all reserve and protected forest lands are estimated to be under informal local protection. Nationally, it is estimated that over 500,000 hectares of reserve and protected forest land are already under local protection through joint management agreements (Poffenberger, 1990).

Since many of India's dominant natural forest species have high regenerative capacity they are ideal candidates for protection under local management schemes. Simple silvicultural systems including coppicing and pollarding ensure both short and long-term benefits for local people. A recent analysis of Landsat and field data indicates that nearly 30 million hectares of degraded state forest land in India, or over 40 per cent of total forest area, has the potential to regenerate naturally with local protection (Poffenberger, 1990). The challenge now is to encourage the formation of mechanisms to allow effective interaction between local people and forest departments in the management of forest resources.

Recent developments in legislation facilitate participatory forest management in Nepal[7]

Nepal has furthered the ideas of local participation in management of forests through legislation and institutional change. The following description of the recent evolution in its forestry legislation shows the critical role legislation plays in the support of any type of participatory forestry.

The early 1960s ushered in a new period of local government within Nepal, known as 'panchayat polity'. The Forest Act of 1961, in conjunction with the institution of the panchayat system of government, had far-reaching consequences for local control of resources including a provision handing over 'protection' of forests to newly-formed panchayats (protection is used to designate control over, access to, and use of forests). Several categories of forest were delineated, and had different access rights assigned to them.[8]

Ownership of forest land remained with the government and control could be resumed whenever the government deemed it necessary. The panchayat had some powers to fine those who transgressed against the law. However, management decisions remained with the government forest service. Private forests that were considered to be poorly managed could be taken over by the government for a period of 30 years, and any income from the forest would be given to the owner with a sum deducted for management costs. The Forest Act legitimized panchayat control over local forests, but it did not have great impact in those areas distant from Kathmandu where local people continued to use forests for their subsistence regardless of legislation.

Important changes in forest legislation began as a result of the Ninth Forestry Conference held in Kathmandu in 1974. This conference convened forestry officers from all over Nepal. A radical group of foresters working in the districts promoted a new form of forestry, where local people were to be involved in management of forest resources, to be known as 'community forestry'.[9] The proceedings of this conference, in conjunction with the results of the findings of 'A Task Force on Land Use and Erosion Control' (National Planning Commission, 1974) formed the basis of the 1976 National Forestry Plan which reinforced the rulings of the 1961 Forest Act in allocating categories of forest land to the panchayats. However, wider powers were given to district forest officers under the 1976 Plan to formalise the transfer of nationalised deforested land to panchayat control.

In 1982, further legislation through the Decentralization Act formalized the duties and responsibilities of village panchayats and ward committees, and empowered them to form:

> 'People's consumer committees to use any specific forest area for the purpose of forest conservation and through it, conduct such tasks as afforestation, and forest conservation and management on a sustained basis' (Regmi, 1982:403).

Local-level management principles were incorporated into legal and project practice and propelled community forestry along the path of active involvement by local forest users. This produced the legal infrastructure necessary for the control of forest land to be handed down to the panchayats and to user committees. Recent proposals under a Master Plan for forestry in Nepal encompass people's participation in its short term objectives (Roche, 1990). To attain these objectives policy, legal and institutional reform are proposed, although under the current political climate it is difficult to predict the future of any such reforms.

PROBLEMS AND OPPORTUNITIES FOR IMPROVED MANAGEMENT

While some common themes have emerged, others have been seen to be specific to the countries and areas in which management is practised. For example, dry forest management is far more promising in much of Africa due to the relatively low population density (compared to Asia) and the so far almost complete lack of landlessness.

Dry forest management prospects in Africa

Woodland management and the state

One of the facts which has emerged from the research conducted by Shepherd is the tremendous paucity of formal forester knowledge about the management of tropical dry forests, and in spite of which lack of knowledge, enormous changes are being forced upon them in the name of better management and state control.

Key survey articles from the last few years highlight the first point only too clearly. According to Bonkoungou and Catinot (1986), there is only very slight experience of silvipastoral management based on natural regeneration techniques for mixed forest and grassland. Much indigenous knowledge waits to be collected. According to Jackson (1983), who undertook the biggest ever survey of Sahelian literature on woodland management, there were only two examples, up to 1980, of formal forest management in the Sahel: that of Bandin forest, Senegal (which failed in its aims), and the relatively successful management of *Acacia nilotica* on the Blue Nile in the Sudan. Other activities have apparently failed because they did not build on the much longer technical experience of local people.

Set against such thin knowledge is the imposition of European concepts of property and land tenure. This has had disastrous effects. The most important failure was the inability of the Europeans to understand the African fallowing systems, which had used the landscape sustainably for hundreds of years. Such systems were simply invisible to outsiders (Shepherd, 1986). Land being fallowed often looks abandoned and ownerless to the northern forester's eye. As a result, fallows, rested for years, well-wooded, and almost ready to be felled again, were gazetted by the state and turned into forest reserves. This turned intensively managed CPRs into open access land at a stroke, and as far as villagers were concerned, it was to be poached from rather than managed. As a consequence Africans became wary of fallowing their land and overworked it instead (Thomson, 1983).

Woodland management and other change

Tenure and land use changes

Villagers have also been turned from land owners into leaseholders in many countries eg Senegal (Postma, 1990), and the Sudan (Seif el Din, 1987); or indeed from land owners to landless wanderers. Barrow (1988) points out that customary tenure was taken into account in Kenya when agricultural land was demarcated, but neither investigated nor recorded in the dry areas. The best solution offered to Kenya's pastoralists has been group ranches, which lack the area flexibility of lineage-range management, and begin to look like some of Africa's other tenure ghettos, such as the communal lands in Zimbabwe.

Political changes affecting land tenure frequently destroy CPRs. For instance, during the Ujamaa period in Tanzania, the Sukuma were told to leave their dispersed granite outcrops and cluster in villages. Unoccupied, the small hilltop forests rapidly fell prey to urban charcoal burners (Shepherd, 1989b).

The diminished authority of local leaders

Much good local natural resource management has depended upon the understanding and political authority of locally born leaders. Central government has deliberately undermined their authority, but cannot replace them effectively. The substitutes are always centrally appointed officials who are not dependent on local economic realities for their livelihood, and who will shortly move on.

Population growth

Better health conditions and food security for humans, and the increased use of veterinary drugs, have increased both human and animal pressure on rangeland and farmland. More people together with their animals have been encouraged to live in settled rather than semi-mobile conditions, with a corresponding heavy local intensification of land-use and biomass loss, increases which have not been incorporated into land-management strategies. Many parts of Africa, too, have experienced bigger influxes of people than they can easily absorb – those moving away from drought areas; because of refugee problems, or because of the artificial withdrawal of large tracts of grazing land, as in the mechanized farming schemes in central Sudan (Wormald, 1984).

As populations rise and increased sedentarization occurs, the household grows in importance while clan and lineage structures – and the elders in charge of them – lose their importance. Grazing is privatized or shared by a much smaller subsection of a larger entity. All such actions tend to sound the death-knell of successful woodland management. It is a truism that rising population pressures on forests will cause deforestation. However, in areas with sufficient rainfall, the inevitable long-term result of population pressure actually creates a move to tree planting. In Rwanda (Gibson and Muller, 1987), a survey revealed that people plant trees in their greatest numbers where there is most pressure on land and much monocultural cash-cropping.

The shortening of fallows

Population pressure seems to have led to the shortening of fallow periods throughout semi-arid Africa (Seif el Din, 1987; Wormald, 1984). In western Sudan the whole cycle has halved in length, and the fallow period is less than a third of what it was (Hammer, 1988). The tendency is for the fallowing system to shrink to the point where it is replaced by a crop rotation of alternating millet and peanuts (Raison, 1988). Ironically, although the population has grown overall, more men and more young people have left the Sahel for ever and this out-migration of labour has prompted the extension of agricultural techniques. People are now trying to cultivate the most land they can in the least time, and at the very beginning of the rainy season. Shortage of labour has meant that older practices such as manuring, intensive sowing and weeding, planned fallowing and water conservation, have all had to be replaced by quick, easy farming (Thomson, 1983).

The growth of towns

The growth of towns is one of the largest threats to the dry tropical forests, because the clustering of large numbers of people who will need rural biomass for fuel, and the low annual increment of that rural biomass, mean that the fuelwood shadow from a large town or city spreads hundreds of kilometres out into already hard-pressed rural areas. Yet population growth in the Sahel is in the order of 2 per cent or less in rural areas and 6–10 per cent in urban areas. In addition, urban entrepreneurs are keeping cattle for cash near towns, whereas before they would have travelled widely with them (Bertrand, 1985; Hammer, 1982; Kerkhof, 1990).

Roads are built from towns to rural areas, and these then facilitate more commercial transactions, such as the sale of charcoal to towns, or wood-cured tobacco (Castro and Brokensha, 1987). Labour migration to towns hamstrings the effective farming of those left behind; low producer prices discourage complex husbandry. Thus, towns grow at the expense of rural people, undermining their resource base and extracting huge quantities of precious biomass from them (Shepherd, 1989a).

The state itself, as an urban-based institution, has exploited but failed to understand the dynamics of rural systems, has seen its own urban needs as paramount, and in sum has made it harder for rural people to build a surplus of animals, grain (because of labour shortages), or fallowing land. The result has been much increased vulnerability to cyclical droughts, at present in tandem with the prolonged dry phase currently being witnessed.

Conclusions: Key features of indigenous forest management systems

Although this paper has given a broad definition of management, certain themes stand out. These are relevant to any kind of effective management.

- In the past there have been strong capable managers in charge of woodlands and the exploitation of trees, managers with a lifetime of commitment since they, like the people they administrate, are making their living from the resource. Most management rules, as a result, are very well attuned to local needs and constraints, and have arisen in appropriate response to some perceived problem.
- Management has also been as simple as possible. Unless the resource has had some value or some scarcity, management is not undertaken. Rules are quite flexible, and can be modified as the need arises.
- Management has been geared to produce a set of interlocking benefits. It is quite hard to separate out woodland management from swidden-fallow management, herd management, and annual crop management. Moreover wood is far from being the only resource for which woodlands are managed.
- Rising population density is turning pastoralists into farmers, long swidden-fallow into short, the usufruct of clan land into individual title. So the management focus has narrowed and in many areas the numbers of locally born and locally significant decision-makers above the level of household head are dwindling.
- Formal political and economic authority has passed in most places to the State over the last 30–40 years. With local authority removed, unregulated exploitation increasingly takes over (CTFT, 1988). Indeed centralized political authorities still continue to deny, on the whole, the very ability of local decision-making bodies to manage their environment, though there is ample evidence to the contrary. Yet, if local management structures are not identified, forestry becomes no more than forest reserves and village forestry schemes, neither of which replicate the integration with trees practised in the past.

The prognosis for adaptive change looks poor. Many previous woodland management practices can only work if managers are also owners, yet it is rare for control over reserved forest land to be returned to the people who are being asked to cooperate in its management.

There is also the question of size. It would seem that rural people only want to manage a resource if all can benefit to some extent. Small patches of hilltop forest can be managed by small numbers of people, or be designated a local no-go area (along the lines of earlier sacred groves) for a larger number if there is a political institution which can guard it. Otherwise, rural people often prefer such small patches to be looked after by the respective country's forest department, so that they are not involved in high social costs for a low return.

Ironically, of course, tree management increases the less forest there is. Under certain special circumstances trees are managed as forest, but increasingly the preferred management is on farms. A publicly owned forest is an anachronism in a tightly farmed landscape, and in many cases represents the superimposition of the will of outsiders on the local population.

Lessons for the future

The most fundamental management technique which emerges from research on indigenous management systems is so obvious that it gets overlooked: some recognition of ownership of the resource – by descent, residence, or any other principle – so long as all agree to it. No cooperation by local people in management will take place without it. And the recognition of these owners in a public way, by formal tenure registration as well as by allowing their views on management to be heard, is essential.

Another lesson to emerge from the study of indigenous management systems is that the area to be managed must be consonant with the scale of the management entities available. Pastoralists have always had reason either to be mobile themselves, or to make sure that their animals can have access to a wide and diverse area through carefully maintained kinship ties. They can manage quite large areas under certain circumstances. Villagers can most happily manage the fallowed and protected land which radiates out from their village. Neighbours can manage the valley grazing resources which run between their privately farmed fields, or the small wooded hills which rise among them. It is very difficult for any of these categories to manage large state forest reserves, for the same reasons that the state cannot manage them either – the resource is too diffuse and guarding it is too costly.

The implications of this are that more successful management demands the breaking up of many large reserved areas into smaller and more easily protected segments. The more that umbrella structures such as clans and lineages disappear, the smaller must be the entities which it is feasible to try to protect. In nearly all other cases, the most promising focus for local people is the creation of tree resources on farms, leaving patches of environmental reserve to the state – the successor of the leaders who managed the sacred groves. Only in rarer cases can management be passed to small local user groups whose internal composition and resources are clearly defined and felt by them to be manageable.

Forest management prospects in India

Plans for improved forest management in the context of a new role for forester–villager relations

Despite an apparently only moderately successful social forestry era, the last decade in India has seen a tremendous amount of experimentation with tree planting and management arrangements between villagers and officialdom. The success of the various innovations in West Bengal, Orissa and Haryana, among others, has led to the passing of an important resolution by the central government Ministry of Environment and Forestry, encouraging cooperation between state departments, non-governmental organizations (NGOs) and local people in forest management. It outlines guidelines for the development of legally binding working arrangements between the various parties, so that a more positive era of forest management can begin. The main recommendations of the resolution are as follows:

- developing partnerships between villages and forest departments, facilitated by NGOs when helpful.
- an access and benefits system, though only for organized villages undertaking regeneration, with equal opportunity based on willing participation.
- rights to usufruct ie all non-wood forest products and percentage share of final tree harvest to villages, subject to successful protection and conditions approved by the state.
- a 10 year working scheme with microplans detailing forest management; institutional and technical operations should be developed by village management organizations and local foresters.
- funding from the Forest Department social forestry programmes for nursery-raising with encouragement for villages to seek additional funds from other agencies.
- the use of rules with strict adherence to no grazing, agriculture or cutting trees before maturity except as outlined in the working scheme. (From Poffenberger, 1990)

Guidelines for forest management policy and programmes

At a recent workshop held in Delhi with senior Indian policy makers, foresters, NGO leaders and social scientists, the following steps were suggested to provide an appropriate framework for local involvement in forest management. The following points summarize the interesting direction in which forest management thinking is now proceeding in India, (Poffenberger, 1990):

- State-level resolutions: The issuance of a state-level resolution encouraging the Forest Department to work with villages in the management of forest lands through the formation and/or recognition and empowerment of communities protecting forest lands. It was felt this should be an enabling provision rather than a directive. State forest departments and rural communities should be allowed to move at their own speed in establishing decentralized management systems. Joint management programmes should build on existing groups and local management systems if these organizations are functioning effectively and allowing broad-based participation.
- Terms of management partnership: The resolution should provide a framework for joint management agreements, specifying resource-sharing rights and protection responsibilities, so that forestry field staff and rural inhabitants have a clear understanding regarding the terms of the management partnership. At the same time, the resolution should allow community organizations flexibility to determine operational procedures for managing production and protection activities.
- Communication systems: Experience made it clear that good communications between foresters and villagers lead to more effective joint management systems. Some departments find this dialogue to be effectively initiated through a micro-level planning process for each joint

management area, where officials and villagers formulate a local management plan together. The need for some type of jointly prepared management plan should be indicated in the resolution.

- Benefits for rural poor: For low income rural families to take part in joint management programmes it is essential that material benefits begin flowing as soon as possible. Harvesting of regenerating grasses and leaves can begin during the first year. Enrichment planting activities should emphasize fast growing species, especially those providing raw materials needed for village industries.
- Budget provision: Some budget provisions may also be necessary to support forest management groups who are losing income during the early phases of forest regeneration. These allocations should support employment which will speed ecological recovery and enhance forest productivity. Funding for improving coppice growth, enrichment planting, and nurseries are particularly important in the first years of protection. Micro-level plans should chart employment requirements over a minimum period of five years, but avoid creating heavy dependencies based on employment subsidies.
- Forest Department – NGO coordinating committees: To help sustain the community forest management groups, it may be useful for the resolution to encourage forest departments and NGOs to help management groups establish coordinating committees with representation from their members. These 'federations' could be endorsed by appropriate local government institutions.
- Formal recognition of a role for NGOs: Government orders should allow NGO participation in assisting forest departments and villages to establish joint protection and management systems.
- Local-level grazing controls: Uncontrolled grazing, mostly by scrub cattle and goats, suppresses the regeneration of India's forests. Resolutions should require villages taking part in joint management activities to develop methods to control grazing through social fencing and stall feeding. Village micro-plans should assess fodder requirements and identify ways to meet fodder needs through regeneration and enrichment grass planting, while developing strategies to reduce low quality livestock.
- Budgetary flexibility: In order to enlist village involvement in joint forest management programmes, forest departments need greater budgetary flexibility both within routine budgets and special allocations to respond to village priorities and emergency situations such as acute water shortage, irrigation system problems and the disruption of infrastructure and communications.
- Reorientation and training: As departments expand joint forest management systems and activities, extensive staff reorientation and training is necessary to develop attitudes and approaches which engender supportive interactions with community management groups. A national committee should be formed to identify and develop useful training

materials and case examples from forest departments and NGOs in India and other developing countries. Emphasis should be placed on helping foresters develop creative problem solving skills with villages, rather than rigid manuals and procedures for project implementation. Existing training programmes should be reviewed to determine how they could be improved to better prepare staff.

Forest management prospects in Nepal

Nepal has advanced further than India in the process of finding new forest management styles, having accepted that local people should be involved in the management of forest resources for their own needs. The pioneering work of the Nepal–Australia Forestry Project and the Koshi Hills Development Programme has focussed on reorienting field-level forestry staff and villagers to bring them together in the management of forest resources.

Reorientation and its central role in forest management for local users[10]

Early experiences in establishing forest committees and identifying forest users led both projects, along with the Forest Department, to realize that they did not have the necessary skills to facilitate widespread forestry management with local users. To institutionalize the changes necessary within villages and the Forest Department it was necessary to set up a systematic programme of reorientation and training.

Legally, before a user group can use a community forest, it has to submit an Operational Plan for that forest. The plan is prepared by the users of the forest, not by professional foresters or natural resource planners. Sufficient time has to be allowed for all members of the user group, weak and strong, to reach a consensus on the management of the forests. As has been seen this process rarely takes less than three months. The users regard their plan 'as rules for our forest' detailing, for example, access to the forest and forest products as well as protection and decision-making mechanisms. The plan is sanctioned by the district forester and, until the recent political change, by the local pradhan pancha (chairman of the village or town panchayat). An executive forest user group committee is then elected by the user group members to oversee the implementation of their plan. (Gronow and Shrestha, 1990)

Mechanisms for facilitating involvement at the local level have been implemented in Nepal through Forest Department field staff. However, prior to involving local people in forest management a major reorientation of field staff was required. To ensure changes promoted in Forest Department implementation are sustained it is necessary to:

- change government forest policy away from policing and towards collaborative forestry;
- change the value systems and hierarchies government officials and project advisers impose on the field staff;
- establish relationships of respect and trust between policy-makers and field staff and devolve more decision-making responsibility to the field staff;

- promote experience-recounting, reflection and confidence-building among field staff;
- help field staff to identify problems and define new approaches;
- support field staff and applaud their efforts. (Gronow and Shrestha, 1990)

Field staff are expected to adopt a new people-based approach to forestry which is alien to their previous training. To help field staff acquire the necessary skills several fundamental changes need to occur. The following discussion draws on the work of two facilitators of reorientation programmes, Gronow and Shrestha (1990). The practical steps followed included workshops and field support techniques.

Workshops
Workshops encourage a democratic, two-way learning approach. Each person participating in the workshop is encouraged to share their experience and knowledge with others, including the facilitators. Workshop location is of central importance to this process of change within the Forest Department, hence all workshops are held in district forest offices and involve representatives from all levels of the district hierarchy from Ranger to District Forest Officer. If field-level staff are to have the support of officers higher in the hierarchy it is essential that these officers are part of the training process, and fully understand the need for local management of forests. (King *et al*, 1990a and 1990b)

Field support
Field support is essential if the reorientation process begun in the workshop is to be sustained. Gronow and Shrestha (1990) stress the importance field staff place on this continued support:

> The field staff have repeatedly said that working on their own presents difficulties with regard to security, credibility, confidence and political pressure. Their youth in relation to the villagers, low official status and the negative reputation of the department of forests make them feel insecure. The villagers' lack of faith in the Department works not only against their participating in community forestry but also against the field staff's attempts to adopt the new role of facilitators.

Institutional change for forest management

Institutional change is another key factor in sustaining these new forms of forestry. As has already been mentioned it is essential for all levels of the forestry hierarchy to support these new initiatives. The target-oriented philosophy of forest departments does not provide the supportive framework necessary for the more service-oriented role to be played by field level staff. If staff are only rewarded on the basis of the number of nurseries built or seedlings distributed, there are no incentives for staff to pursue the more difficult role as village-level facilitators. Local level management of forest resources cannot happen without the institutional capacity to support it.

Although institutional development is a prerequisite for change, this must be supported by policy and legislation. As has been seen from the brief description of Nepal's forest legislation, considerable power has been handed over to the user group level. If user group rights are embodied in law, members of user groups are more able to defend their rights against outsiders, and therefore there will be a greater incentive to protect resources for the long term.

The Nepal experience reinforces the understanding gained in India. The major constraints to successful local management of forest resources lie at several different levels. The first level of constraint begins with the resolution of conflict between different forest user groups. Social forestry projects have had a large degree of success when they have worked with homogenous groups ie one caste or one tribal group. In Nepal, it was necessary to work below the level of the village to ensure that low castes and women were also able to represent their views at the village level.

In all projects it was essential to define local people's rights with respect to the forest resource, so that they could defend their rights against other local people, and also uphold their rights against the Forest Department.

Without an institutional structure to support the legislative and local changes, such local forest management initiatives cannot be sustained. In Nepal, reorientation and appropriate training of the Forest Department has begun to secure the future of local forest management. Skilled Forest Department facilitators are beginning to catalyse change at the local level, as well as reinforcing institutional change within the Forest Department itself.

Non-governmental organizations obviously have a role to play as intermediaries between local people and forest departments in certain circumstances. In Orissa, the experience of Gram Vikas has shown how effective an NGO can be in negotiating agreements on behalf of local people with government departments. However, in Nepal, where there is no history of non-governmental organizations, it appears that Forest Department staff themselves are able to acquire the necessary skills to facilitate effective local management systems. Above all, flexibility in approach must be built into any such local management initiative (Arnold, 1989). National legislation needs to be supportive of such approaches, but should, in the case of India, also allow for regional differences.

CONCLUSIONS

As this paper has demonstrated, alongside the gloomy deforestation figures obtained from Land Satellite imagery and the loss of tree cover which is going on all over the world, it is just as important to note some of the interesting changes going on in forest management approaches in the tropical dry forest areas.

While resources were ample, dry forests were only marginally regarded, especially in comparison with the riches of the tropical moist forest hardwoods. However, a variety of initiatives now suggest that dry forest is being more highly valued as it becomes scarcer, and therefore more hotly competed for.

When forests were first reserved, there was no question but that they were being reserved from local people, for the needs of the state. Models for forest management were taken from European models for forests. Such models assumed private land-tenure, the exclusion of those who lived nearby for all but minor usufruct access, and the management of the forest for timber – and perhaps some fuelwood.

This model has never reflected the reality of the developing world, though when there were fewer people and more trees the contradictions were less acute. Now the question constantly raised is, 'what is a forest to be managed for, and for whom?' There is debate in all three of the areas reported on about the devolution of forest management to local people, and more interest in management for locally important products, many of them in fact non-timber products.

Many countries which would once have unflinchingly asserted state rights to forest in perpetuity are now looking for arrangements which relieve them of some of the expense and management, particularly in drier, lower value areas. Foresters' experience of working with farmers in farm and social forestry programmes has undoubtedly paved the way to more friendly and informal working styles in the area of forest management as well.

As a result there is now a situation where local people are asking for stronger tenure rights in forest resources, where in some places there is already a sufficient shortage to make planting worthwhile, and where the forestry profession is in many cases more prepared to entertain the idea of local people's forest management than it has ever done before.

Acting to maintain and increase tree cover

In the circumstances, we should be acting while there is still time. It is possible to propose plans for environmental management which address protection and the enhancement of tree cover through one of the following mechanisms:

- Farm Forestry programmes on clearly owned and usually private land;
- management of land to which local people have clear, locally devised and legally recognized common property rights, and where such land – be it forest or watershed – has sufficient importance for local people and their needs to be worth protecting from their point of view;
- government protection where neither of the two above situations apply.

Government land and the environment

Often, government finds itself in a situation where it cannot manage effectively all the land it owns, yet will not relinquish it either. Such contradictions should be tackled. Where individuals or groups are keen to own and manage natural resources (and this will by no means happen everywhere) governments would often be better off by giving up some of their sovereignty, and concentrating state resources on lands which, for whatever reasons, must be protected but will never attract more specific ownership.

All too often, however, governments seriously expect that while they continue to own the land, local people should manage it voluntarily. Needless to say, such hybrid arrangements, in which the party which owns the resource experiences no expense while the party which does not faces costs, can never work.

Strengthening the rights of local people to practice communal management
For the best environmental management, land rights for local people are probably the best solution – and this is now an area of great experimentation. Where the rights offered are too limited, or bring no obvious benefits, local people decline them, yet governments in many countries are plainly too weak or too corrupt to have much success either. More documentation is gradually becoming available which points to the elaborate traditional woodland management practices of local people in many parts of the world. Not many projects are as yet building on this experience, though under the right conditions this approach holds promise, and is likely to be much experimented with over the next few years. Land rights, as always, will be the key.

Adapting resources to the capabilities of new managers
Several cases show that communities manage small resources best. This means that well-intentioned forestry departments cannot hand over tracts of forest which they have been managing without working through a process of defining, with the would-be managers, the area they in turn feel capable of managing.

Training foresters
The top priority for forest management by the many is the training required in extension methods and more informal work styles for professional foresters. This, in conjunction with improved land tenure, is essential if more effective styles of forest protection and management are to become widespread. Foresters working in the tropical dry forests of Africa and Asia now have a several-year lead in these newer working styles, but more work is needed to develop norms for action which will in time spread to the tropical moist forest areas too.

TAILPIECE

Paradoxically the deforestation which, in the dry forests, led to a panic about fuelwood shortages and to the social forestry programmes of the 1980s has taken us in the direction of new solutions. Social forestry's contribution has not only ended up being the provision of more fuelwood, but has also provided a strenuous training ground for foresters in more participatory approaches to rural people and management; approaches which are essential now if sustainable forest management activities are to be the order of the day, and for which the developing countries were, on the whole, ill-equipped at the start of the 1980s.

What began as a large-scale rural tree planting programme based on plantation models popular since the 1960s, has thus, over the past decade, set in motion the makings of an inevitable and major paradigmatic shift in the discipline of tropical forestry, as it struggles to respond to dwindling forests, rising populations, and growing numbers of securely owned and protected trees.

This chapter first appeared as 'National Experiences in Managing Tropical and Subtropical Dry Forests' in Sargent, C and Howlett, D (1991) *Technical Workshop to Explore Options for Global Forestry Management*, ITTO, ONEB and IIED, London.

NOTES

1. Social Forestry Network, Overseas Development Institute, London, United Kingdom.
2. The term 'forest' generally suggests a closed canopy, while 'woodland' suggests a more open scattered formation. Much dry tropical forest is open and could probably more accurately be called woodland. Furthermore, there has been a tendency for dry tropical forest to be called 'woodland' in Africa and 'forest' in Asia. The terms may be read interchangeably in this chapter.
3. This article arises from research on the management of natural woodland in Africa, which is to be published in full both as an ODI Occasional paper, *Communal Management of forests in the semi-arid and sub-humid regions of Africa*, Gill Shepherd, ODI, 1991 and as part of an FAO book on tree-management by rural people.
4. Many of the articles used here are summarized in the ethnographic present, in line with the article itself. It is vital to check the date of the item. In this analysis, the ethnographic present is similarly used for the sections on management practices which were still extant at the time they were described. However, as the succeeding sections on change and the future make clear, we should not assume in too sanguine a way that all management practices are still functioning.
5. The hybrid term swidden-fallow has been used for two reasons. Firstly, it links two terms which, rather arbitrarily, are normally used respectively for non-African and African situations. Secondly, it stresses (in a way terms such as 'shifting cultivation' and 'slash-and-burn agriculture' do not), the fact that farmers do not abandon cleared land when they leave it to restore its fertility and plant crops on new land. They obtain other products from it, may retain continuing control over trees growing up on it, and will return to it when their own particular cycle is complete.
6. This section draws on the work carried out by Arnold and Stewart in 1990. The authors wish to thank them for their permission to reproduce their findings in this study.
7. This section draws on an unpublished PhD (Hobley, 1990a) and on an article, (Hobley, 1990b).
8. **Panchayat Forests:** Any government forest or any part of it, which has been kept barren or contained only stump, may be handed over by HMG (His Majesty's Government) to the village panchayat for plantation for the welfare of the village community on the prescribed terms and conditions; **Panchayat Protected Forests:** Government forest of any area or any part of it may be handed over to any local panchayat for protection and management purposes; **Religious Forests:** Government forest located in any religious spot or any part of it, may be handed over to any religious institution for protection and management purposes; **Contract Forests:** Any government forest area having no trees or sporadic trees may be handed over by HMG in contract to any individual or institution for the production of forest products and their consumption.
9. From a taped interview with T B S Mahat (1986) former Divisional Forest Officer for Chautara. T B S Mahat was a member of this radical group of foresters, he based his views on experience gained as the Divisional Forest Officer in the Chautara Forest Division where he had already supported local initiatives including establishment of forest nurseries in two panchayats. Chautara was to become the focus of Australian forestry aid activities. The same group of foresters were also involved in the 1978 FAO/World Bank mission to identify a community forestry project in the Middle Hills of Nepal. These foresters were sent to FAO in

Rome to write up part of the project preparation mission's report. Arising out of this mission was the Community Forestry Development Programme which operates in the hill districts of Nepal.

10. This section draws heavily on the pioneering work of two social forestry projects: the Nepal-Australia Forestry Project, a bilateral project between the Government of Nepal and the Government of Australia and the Koshi Hills Development Programme, a bilateral programme between the Government of Nepal and the Government of the UK.

REFERENCES

Arnold, J E M (1989), People's participation in forest and tree resource management: a review of ten years of community forestry, a ms prepared for the Policy and Planning Service, Forestry Dept, FAO, Rome (ODI library).

Barrow, E G C (1986), 'Value of traditional knowledge in present day soil conservation practice, the example of the Pokot and the Turkana', paper presented to the Third National Workshop on Soil and Water Conservation, Kenya, Example 1 The Pokot. Example 2 Turkana (ODI library).

Barrow, E (1988), Trees and Pastoralists: the case of the Pokot and Turkana, *ODI Social Forestry Network Paper 6b*.

Barrow, E, Kabelele, M, Kikula, I and Brandstrom, P (1988), Soil conservation and afforestation in Shinyanga Region: potentials and constraints. Mission Report to NORAD (ODI library).

Behnke, R H (1980), *The Herders of Cyrenaica*, University of Illinois Press, Urbana, Illinois.

Bertrand, A (1985), 'Les nouvelles politiques de foresterie en milieu rural au Sahel. Réglementations foncières et forestières et gestion des ressources ligneuses naturelles dans les pays de la zone soudano-sahélienne.' *Revue Bois et Forêts des Tropiques*, No 207, 1er trimestre, pp 23–40.

Bonkoungou, E G and Catinot, R (1986), 'Research on and development of natural regeneration techniques for the silvopastoral management of existing forest resources', in Carlson and Shea (eds), pp 89–124.

Brokensha, D and Castro, A H P (1987), 'Common Property Resources'. Background paper presented February 1988, Bangalore for the Expert Consultation on Forestry and Food Production/Security, FAO, Rome (ODI).

Carlson, L W and Shea, K R (eds) (1986), *Increasing the productivity of multipurpose lands*, IUFRO Research Planning Workshop for Sahelian and Northern Sudanian Zones, Nairobi, Kenya, 9–15 Jan 86, IUFRO: Vienna.

CTFT – Centre Technique Forestier Tropical (1988), *Faidherbia albida* (Del) A Chev (Synonyme: *Acacia albida*) Monographie, CIRAD, pp 29–36/50–61.

Chambers, R, Saxena, N C and Shah, T (1989), *To the hands of the poor: water and trees*, Oxford University Press, New Delhi and IBH Publishing.

Cortes, E V (1984), 'Management of natural forests', in *Tropical Forests*, Vol 1, No 1, Jan–Mar, pp 8–11.

Gerden, C A and Mtallo S, (1990), Traditional Forest Reserves in Babati District, Tanzania. A study in Human Ecology. Swedish University of Agricultural Sciences, IRDC, Working Paper 128, Uppsala 1990.

Gibson, D C and Muller, E V (1987), Diagnostic Surveys and management information systems in agroforestry implementation: A case study from Rwanda, Working Paper No 49, ICRAF and CARE.

Gronow, J and Shrestha, N K (1990), From policing to participation: reorientation of Forest Department field staff in Nepal, HMG Ministry of Agriculture–Winrock International Research

Report Series No 11, Kathmandu, Nepal (ODI library).

Hammer, T (1982), 'Reforestation and community development in the Sudan', *Energy in Developing Countries Series*, Discussion Paper D-73M, Resources for the Future, Washington (ODI library).

Hammer, T (1988), 'Wood for Fuel – Energy Crisis Implying Desertification: The case of Bara, Sudan', in *Whose Trees? Proprietary Dimensions in Forestry*, Fortmann and Bruce (eds), US: Westview Press, pp 176–181.

Hobley, M (1985), 'Common Property does not cause deforestation', *Journal of Forestry*, Vol 83.

Hobley, M (1990a), Social Reality, Social Forestry: the case of two Nepalese panchayats. PhD thesis, ANU, Canberra, Australia (ODI library).

Hobley, M (1990b), 'From Passive to Active Participatory Forestry: Nepal', in Oakley, P *Projects with People: The Practice of Participation in Development*, Geneva: ILO.

Houérou, H N le (1980), 'Agroforestry techniques for the conservation and improvement of soil fertility in arid and semi-arid zones', in le Houérou, H N (ed), *Browse in Africa: the current state of knowledge*, International Livestock Centre for Africa, Addis Ababa, pp 433–435.

Jackson, J K (1983), Management of the Natural Forest in the Sahel Region, Technical Report prepared for USDA Forestry Support Programme (Technical Assistance Programme for AID's Forestry Development Activities) AID/USDA/USFS PO 40-319R-3-00273, 2/9/83 Washington DC (ODI lib.)

Kerkhof, P (1990), 'Turkana Rural Development Project, Kenya', in *Agroforestry in Africa: a survey of project experience*, G Foley and G Barnard (eds), Panos Institute, London, pp 161–170.

Kessler, J J (1990), 'Agroforestry in the Sahel and Sudan zones of West Africa', in *BOS Nieuwsletter* No 20, Vol 9(1) 1990, People and Trees in the Sahel, pp 27–33.

King, G C Gilmour, D A and Hobley, M (1990a), 'Management of Forests for local use in the hills of Nepal. 1. Changing forest management paradigms'. *Journal of World Forest Resource Management*, Vol 4:93–110.

King, G C Hobley, M and Gilmour, D A (1990b) 'Management of forests for local use in the hills of Nepal. 2. Towards the development of participatory forest management', *Journal of World Forest Resource Management*, Vol 5, pp 1–13.

Lamprey, H F (1986), 'The Management of indigenous trees and shrubs for the rehabilitation of degraded rangelands in the arid zone of northern Kenya', in Carlson and Shea (eds), pp 125–138.

Norton, A (1987), 'The socio-economic background to community forestry in the Northern Region of Ghana', ODA Community Forestry Project in the Northern Region of Ghana (ODI library).

National Planning Commission Nepal (1974), Draft Proposals of the Task Force on Land Use and Erosion Control, Kathmandu (ODI library).

Pélissier, P (1966), 'Les paysans du Senegal', *La civilisations agraires du Cayor à la Casamance*, pp 252–273.

Poffenberger, M (ed) (1990), 'Forest management partnerships: regenerating India's forests', Executive summary of the Workshop on Sustainable Forestry, New Delhi, 10–12 September 1990, Ford Foundation.

Postma, M (1990), Land and tree tenure in the Wolof village M'borine, Senegal. Doctoral report for the Section of Forest Management, Department of Forestry, Wageningen Agricultural University, Netherlands.

Raison, J P (1988), Les Parcs en Afrique: état des connaissances et perspectives de recherches, Document de Travail, Centre d'études africaines, EHESS, Paris.

Regmi, M C (1982), 'Decentralization Act, 1982', Regmi Research, Nepal Miscellaneous Series, Vol 15/84 (ODI library).

Roche, N (1990), Community Forestry in Nepal: a case study of the Integrated Hill Development Project, Dolakha District, Oxford Forestry Institute, MSc paper (ODI library).

Salem, B Ben and van Nao, T (1981), 'Fuelwood Production in Traditional Farming Systems', *Unasylva*, Vol 33, No 131, pp 13–18.

Schapera, I (1943), *Native Land in the Bechuana Protectorate*, Cape Town, Lovedale Press.

Seif el Din, A G (1987), 'Gum Hashab and Land Tenure in Western Sudan', in *Land, Trees and Tenure*, J B Raintree (ed), ICRAF Nairobi, pp 217–224.

Shepherd, G *et al* (1985), for Energy Resources Limited and International Forest Science Consultancy, 'A study of energy utilization and requirements in the rural sector of Botswana', consultancy report prepared for ODA, UK and the Ministry of Mineral Resources and Water Affairs, Botswana, Vol 1 Report, Vol 2 Appendices.

Shepherd, G (1986), Forest Policies, Forest Politics, *ODI Social Forestry Network Paper 3a*.

Shepherd, G (1989a), 'The reality of the Commons: answering Hardin from Somalia' in *Development Policy Review*, London: Sage, Vol 7, No 1, March 1989, pp 51–63.

Shepherd, G (1989b), An evaluation of the village afforestation project, Mwanza, Western Tanzania, ODI for IIZ Austria (ODI library).

Shepherd, G (forthcoming), Communal Management of Forests in the Semi-arid and Sub-humid Regions of Africa. A literature search and analysis prepared for the FAO Forestry Department, with the assistance of Joe Watts, Althea Ifeka and Daniel Blais (ODI library).

Singh, C (1986), *Common Property, and Common Poverty. India's forests, forest dwellers and the law*, New Delhi, India: Oxford University Press.

Thomson, J T (1983), Deforestation and desertification in twentieth century arid Sahelian Africa, paper prepared for the conference on 'The World Economy and World Forests in the Twentieth Century', University of North Carolina (ODI library).

WRI and IIED (1986), *World Resources 1986*, Basic Books, Inc, New York.

Wormald, T J (1984), The management of the natural forests in the arid and semi-arid zones of East and Southern Africa, report for ODA (ODI library).

PART II
LAND USE AND LAND DEGRADATION

Chapter 5

Defining and Debating the Problem

Piers Blaikie and Harold Brookfield

LAND DEGRADATION AND SOCIETY: INITIAL STATEMENTS

Land degradation as a social problem

Land degradation should by definition be a social problem. Purely environmental processes such as leaching and erosion occur with or without human interference, but for these processes to be described as 'degradation' implies social criteria which relate land to its actual or possible uses. Other processes, such as acidification and salinization, are only rarely recognized under natural conditions, at least in an acute form, and have a more directly human origin. The word 'degradation', from its Latin derivation, implies 'reduction to a lower rank'. The 'rank' is in relation to actual or possible uses, and reduction implies a problem for those who use the land. When land becomes degraded, its productivity declines unless steps are taken to restore that productivity and check further losses. In either case, the yield of labour in terms of production is adversely affected. Land degradation, therefore, directly consumes the product of labour, and also consumes capital inputs into production. Other things being equal, the product of work on degraded land is less than that on the same land without degradation.

However, it may be argued that, if there is abundant land or if losses in productivity can be made up by the provision of chemical fertilizers, degradation is neither an economic nor a social problem. This argument can be turned around: without degradation it would not be necessary to move to new land with the attendant costs; without degradation, such large inputs of chemical fertilizers would not be necessary in order to sustain production at constant levels, and efficiency of their use by plants would be greater. Either way, there are both economic and social costs. There are also secondary costs, such as the nitrification of water supplies, which are purely social in nature in that they affect people and ecological conditions away from the site.

The social significance of degradation has been the subject of a wide variety of views rather than of engaged debate for reasons which are outlined by Blaikie (1985a: 12). Under defined conditions it is a problem of a major order. Decline in the productivity of the land and of labour can be viewed as the 'quiet crisis' which nevertheless erodes the basis of civilization – to adapt two phrases of Lester Brown (1981). This view claims that the problem is pervasive, often insidious but crucial to the future of humankind. There are elements of environmental fundamentalism in claims of this type, but they underline the essentially social nature of the problem. Also, there is an important link between the chronic, slow-moving phases of the problem and the acute. When production conditions are adverse, as in a drought, the margin of productivity or of survival for a producer on degraded land is smaller than that of a producer on better managed land. When, as in large parts of Africa in recent years, climatic conditions have remained adverse over a long period, farmers on badly degraded land suffer a particularly severe penalty. Land degradation, as well as drought, has been partly responsible for the severity of famine in agricultural areas of Ethiopia and the Sudan (Eckholm 1976).

These simple considerations should alone be sufficient to establish land degradation as a problem of social significance. But it is also necessary to this argument to show that land degradation has social causes as well as consequences. While the physical reasons why land becomes degraded belong mainly in the realm of natural science, the reasons why adequate steps are not taken to counter the effects of degradation lie squarely within the realm of social science. Yet the problem of resource deterioration has been curiously neglected by the latter. There have been a few classic texts warning of the problems, such as Malcolm (1938), Jacks and Whyte (1939), Glover (1946), Rounce (1949) and Hyams (1952), but they are rarely cited in recent work. Also, neither classical nor Maxian economics have satisfactorily attacked the methodological problems of studying land degradation, thus depriving social scientists of a developed theoretical base. For a variety of reasons there has been remarkably little in the way of either empirical or methodological work on the economics of land and water conservation, by contrast with the economics of pollution which has a large literature. The *Journal of Soil and Water Conservation* and the output of some Departments of Agricultural Economics in the United States Midwest are perhaps honourable exceptions.

Blaikie (1985a) has recently sought to open the issue of degradation of land as a social problem. Essentially, *The Political Economy of Soil Erosion in Developing Countries* built a number of theories to explain different aspects of degradation and conservation, drawn mostly from the standpoint of political economy. *Land Degradation and Society* (Blaikie and Brookfield, 1987) from which this chapter is drawn offers a greater diversity of approach. A number of central social issues in land degradation which received only thematic treatment in the earlier book are discussed in Blaikie and Brookfield (1987) in detail. These include the problems of measuring and economically appraising losses, and different institutional arrangements for land management, including common property and private property institutions and the state.

More particularly, it also draws on a long and varied historical perspective in order to focus on the reasons why land management fails to be effective.

Issues of significance

Central to the issues discussed in Blaikie and Brookfield (1987) is the role of the 'land manager'. Land managers may find themselves responding to changes in their social, political and economic circumstances quite independently of changes in the intrinsic properties of the land which they employ. They may be denied access to common resources, or be forced to grow crops by landlords, market or social demand, or by the state. They have to find a strategy with which to meet such pressures, and do this on land which itself changes in nature. The intersection of circumstances and strategies forms the subject matter here.

Any interference by humans with the natural processes of soil formation, evolution and erosion has an effect upon these processes, often unforeseen. Leaching, compaction and erosion of the soil, changes in plant cover and hydrological regime, changes in soil and water chemistry all take place naturally in the absence of any human intervention or even presence; in some environments these processes take place quite rapidly under natural conditions. Violent atmospheric events can cause rapid changes in environments empty of people. In some islands of recent geological origin it can be shown that the soil had been eroded and/or become able to support only a limited biota long before the arrival of people. Yet human interference has modified and usually accelerated all these processes and has created the conditions under which new sets of processes, previously absent or insignificant, come into play. With the exception of the work of bulldozers, explosives, trail-bikes and other tools of malice, all the processes of land degradation occur in nature, but human activity on the land changes the conditions of their operation. The task of land management is to recognize these changes and find some means of bringing them under control.

Of course, the effect of human interference is not the same at all times and in all places. Human management of the land without leading to degradation is not only possible in a great majority of environments, but has been frequently accomplished in human history. However, the same human skills are not useful and effective in all places; under similar systems of management the productivity of some land is well sustained, while that of other land deteriorates rapidly. The problem is further compounded by the fact that degradation has occurred at one period but not at another on the same land. Agro-technology has not only changed through time, but has also been applied with differing degrees of care and perception.

Human-induced degradation occurs when land is poorly managed, or where natural forces are so powerful that there is no means of management that can check its progress. Some degradation is caused when land that should never have been interfered with is brought into use, but most land now subject to accelerated degradation is capable of more effective management than it receives. Our basic question is why these failures have occurred, and whether or not the problem has been perceived as such by those responsible at the place and time.

Since land degradation has occurred in such a wide variety of social and ecological circumstances, it is clearly futile to search for a uni-causal model of explanation. Equally, there are a number of hypotheses which have useful explanatory power, such as 'population pressure' or the exploitation by people of people, and these are examined in Blaikie and Brookfield (1987). However, it is apparent that while there are many causes where population pressure has contributed to land degradation, in others a marked *decrease* in population densities has led to the same result. Likewise, an onerous burden of taxes, inequitable distribution of landholdings, corvée labour systems and the like have probably led to declining management on the part of the exploited, but not invariably. On the other hand, there are many examples where very favourable prices for agricultural commodities or for timber have led to accumulation of profits, but also to land degradation. The complexity leads us away from any single theory of land degradation, since there are so many conjunctural factors operating at one place and time. Rather, case-study material and discussion of methodological issues together suggest a general approach to the problem of land degradation to provide an illustrated manual with which readers can approach their own empirical evidence.

DEFINITIONS OF VALUE, CAPABILITY AND DEGRADATION

Choices in defining degradation

As the opening paragraph of this chapter states, the definition of degradation is 'reduction to a lower rank'. The term is therefore perceptual and implies at least a 'rank' scale of relative measurement. As a perceptual term, however, it is open to multiple interpretations. To a hunter or herder, the replacement of forest by savanna with a greater capacity to carry ruminants would not be perceived as degradation. Nor would forest replacement by agricultural land be seen as degradation by a colonizing farmer. Usually there are a number of perceptions of physical changes of the biome on the part of actual or potential land-users. Usually, too, there is conflict over the use of land – whether it be between farmers and conservationists, pastoralists and peasants, small farmers and the state, developers and concerned landholders. Since degradation is a perceptual term, it must be expected that there will be a number of definitions in any situation. It is, therefore, essential that the researcher recognizes any such conflict over the use of land and, therefore, the definition of degradation. Sometimes the definition is given as the 'ruling' one or the state-supported one, in the sense that land should be used in a certain way and degradation is, therefore, defined as reduction in capability to fulfil this demand. However, it is important to take other criteria into consideration, often derived from one's own political and technical viewpoint.

It is of course more usual to employ the language of natural science to describe degradation, from the perspective of the soil scientist or agronomist. However, the processes are varied and, from a social point of view, their impact may be felt in very different ways. Erosion, especially gully erosion and massive sheet or rill erosion, is very obvious, although the role of a human agency may not be. Modification of horizon structure, partial removal of fine particles, pan formation, podsolization, compaction and similar changes are less obvious and have only a more gradual effect on the productivity of the land. Changes in hydrology affect the flow of streams and ground water, affecting storage and the supply of water to livestock and people as well as to the soil. Impoverishment of vegetation, the invasion of weeds and the selective elimination of soil fauna and the larger fauna which live on them affect the whole quality of environment as well as of the land; new environments, such as the Mediterranean *maquis*, may be created and come to be regarded as natural. Among more insidious processes, salinization becomes persistently severe in dry areas and periodically severe where drought is of irregular incidence, where it is seen as a problem mainly at such times. Acidification, on the other hand, affects the rooting depth of plants in a more lasting manner, but its build-up is very slow and it is not at once perceived as a problem.

These physical changes have also to be evaluated in social terms. The first step is to estimate reductions in crop yield, livestock or useful vegetation resulting from these changes. A useful review can be found in Stocking and Peake (1985). This is a relationship which those in the field are only beginning to be able to quantify, and there are many crucial gaps in both our basic understanding and in orders of magnitude under different conditions. The second step is the evaluation of degradation in economic terms. As chapter 5 in Blaikie and Brookfield (1987) indicates, there are on-site and external dis-benefits of degradation, now and in the future; however, these are generalized income benefits expressed in money terms. Although these are of obvious and overriding importance in assessing the impact of degradation, they leave unanswered the problem of varying and competing perceptions of degradation. For example, a reduction in income for agriculturalists may result in an increase for herders. There are also issues concerning the distribution of losses from degradation between different groups, and access to alternative means of livelihood (eg new land) or to new technologies which can limit the effects of degradation or reverse them. All such issues affect the boundary conditions for accounting the social impact of land degradation.

The 'value' of land

There is also another issue which should be discussed before proceeding to a definition of land degradation, and this concerns the 'value' of land, which in some way is reduced for the user by degradation. It raises a number of theoretical problems. In none of its forms does the theory of value take adequate account of the 'value' contained in the natural source of all energy in the ecosystem, the sun's energy and of the stored products of that energy, which include the weathered material and nutrients which constitute the soil. Such value cannot be said to be created by labour, does not have a cost of production, and is priced by the market according to a mixed set of utilities, including location, which often ranks higher than quality. Insightful comment on the failure of economics, specifically but not only Marxian economics, to take account of the physical processes underlying production is provided by Alier and Naredo (1982), Alier (1984) and Gutman (1985). While these authors, and the nineteenth-century socialist, Podolinsky, also fail to consider land degradation, they call attention to the failure of economics to consider energy flows or to come to terms with the notions of energy, except in a very imperfect manner.

Marx did, in fact, come somewhat closer to an appreciation of the role of land in production than did most other classical writers. He recognized that:

> Man . . . can work only as nature does, that is by changing the form of matter. Nay more, in this work of changing the form he is constantly helped by natural forces . . . labour is not the only source of material wealth, of use-values produced by labour (Marx 1887/1954: 50)

But while there is a recognition of land as the product of natural forces, land – and other natural resources – were considered 'free' inputs into production and did not produce value since it was only labour that was considered to perform this function. On the contrary, it is clear that land may need to be 'paid' a great deal in order to continue to 'exist' at the same quality, as Blaikie and Brookfield (1987) seek to demonstrate. Even modern resource-depletion models fail almost entirely to consider the environment itself as a degradable resource (Hufschmidt *et al* 1983: 57). It is difficult, therefore, to use the term 'value' in relation to land, and even Robinson's (1963: 46) cop-out in regarding value as a metaphysical concept without empirical meaning does not help; the term is therefore avoided.

Capability of land

The term used instead is *capability*. When land is degraded, it suffers a loss of intrinsic qualities or a decline in capability. This term is not one used within economic literature. It is, however, used in modern agronomic literature with something like the sense which is required. As a first step towards clarification, degradation is defined as a reduction in the capability of land to satisfy a particular use. If land is transferred from one system of production or use to another, say, from hunter-gathering to agriculture, or from agricultural to urban use, a different set of its intrinsic qualities become relevant and provide the physical basis for capability. Land may be more or less capable in the new context. This is important, because it must not be supposed that deforestation, for example, necessarily constitutes degradation in a social sense, even though it certainly leads to changes in micro-climate, hydrology and soil. Socially, degradation must relate to capability, and it is only if the degradation process under one system of production has reduced the initial capability of land in a successor system, actual or potential, that degradation is, as it were, carried across the allocation change. In actual practice, this is often the case, since more serious degradation reduces capability for most, if not all, future possible land uses.

A definition of degradation

It has been noted that the effect of human interference need not always be deleterious. It is also possible to restore and improve land, and to create new productive ecosystems of which the outstanding example is the irrigated rice-terrace. The land itself also has its own means of repair: new soil is formed, gullies grass over and become graded; nutrient status is restored under rest. Just as it is necessary to take account both of the interaction between natural processes and human interference in degrading land, it is also important to recognize both natural reproduction of capability and of human artifice in assisting this reproduction. Bidwell and Hole (1965) made a useful distinction between 'beneficial' and 'detrimental' effects of human works on the soil. So also should the beneficial and detrimental processes in nature be distinguished.

Degradation is, therefore, best viewed not as a one-way street, but as a result of forces, or the product of an equation, in which both human and natural forces find a place. We could say that:

Net degradation = (natural degrading processes + human interference) − (natural reproduction + restorative management)

A neat example of the variation of natural reproduction and its impact upon net degradation is provided by a comparative study of Hurni (1983) in which he compares the soil-loss tolerance in the mountains of Ethiopia and the hills of northern Thailand. In the former case, cultivation has been going on for 2000 years with a fairly low rate of soil loss. However, the cumulative loss and slow rates of natural soil formation have both served to produce very serious land degradation. In northern Thailand, however, with higher rates of soil loss, the local land-management system has 'compensated' for this and the capability of the land, in which soil formation is more rapid than in Ethiopia, is maintained.

THE ROLE OF LAND MANAGEMENT

Ways of managing land

With a definition of degradation as a reduction of capability, the role of land management becomes clear. Land management consists of applying known or discovered skills to land use in such a way as to minimize or repair degradation, and ensures that the capability of the land is continued beyond the present crop or other activity, so as to be available for the next. There is no system of land use, anywhere in the world, that does not have agro-technical means with which to achieve or at least approach these ends, provided they are practised in natural environments suitable for their employment.

At the simplest level, rotational grazing and shifting cultivation are effective strategies if well managed, with sufficient land over which they can be applied.

Both are 'avoidance' rather than 'control' strategies (Kellman 1974) in that they leave reproduction of capability to natural repair, and avoid the need for intensive inputs on site. Many control strategies are, however, incorporated into modifications of these simple and ancient methods of management; slope control and water control are both employed in association with shifting cultivation, and so also is the addition of fertilizer. Rotational grazing is more easily managed with the addition of fencing and tethering. The major step forward from these strategies in temperate lands was mixed farming, in which both cultivation and grazing were rotated in relation to one another. Thus, perhaps as early as from the eighth century onwards, the two-, three-, and four-field systems of Europe emerged from an essentially shifting-cultivation base. In the humid tropics, mixed rotational farming was less widely suitable and wholly arable technologies evolved, generally involving massive inputs of human labour aided by livestock and their manure to make possible the permanent cultivation of land. Modern technology has added a range of artificial fertilizers, leguminous crops employed in rotation, and the ability to undertake much larger site-management works.

Fundamentally, the land manager's job is to manage natural processes by limiting their degrading consequences, both on-site and downstream. By downstream is meant external effects away from the site, whether actually downstream, downslope or downwind, or effects which undermine the efforts or exacerbate the problems of neighbours, wherever located. The natural processes involved fall into two main groups, the mainly biological/biochemical and the mainly physical. They have a different range of impact, and present different, though related, management problems. The main problems of biological/biochemical management are on-site, though they have important downstream consequences through the movement of mobile ions which can lead to salinization. The basic problem is to cope with the fact that purposive plant growth and removal for use tends to extract mineral and organic elements from the soil faster than they can be reproduced. Natural replacement requires a rest period or the planting of crops and trees which often have a low value in use. Reproduction of the capability of the land itself is usually the secondary objective of farming systems, but it is a vital objective and one that can absorb a great deal of labour.

The natural rate of soil formation varies enormously over the world, from close to zero in a thousand years in parts of Africa and much of Australia, to the formation of a capable solum in as little as ten years on some volcanic ashfalls under humid tropical climates. The impact of the cumulative loss of soil upon crop yields is also probably extremely variable. It has been estimated that a 15 mm loss from an Oxisol in an experiment in Indonesia (Suwardjo and Abyamia 1983, reported in Stocking and Peake 1985) caused a 40 per cent yield reduction, while a mere 2 mm loss from an Ultisol caused a 15 per cent yield reduction. Also, if these results are compared with data from the United States, it appears that the tropical Oxisol suffered a yield reduction ten times that of temperate soils and the tropical Ultisol twenty times that of temperate Ultisols, with similar soil loss. Even if these preliminary data are only approximately correct, they indicate great variation in the manager's task of maintaining land capability.

Landesque capital

It is important to distinguish between land management in relation to the current crop, the object of which is the production of that crop and the consequences of which are incidental, and purposive land management designed to secure future production. In the nature of things, most of the latter is in the physical area, though if a clearing for shifting cultivation is to last two or three crops, then a part of the labour put into initial clearing creates 'capital' for the second and third crops. The institutional costs of the reorganization of land tenure to make the installation of the three-field system possible in ancient Europe constituted capital which endured for centuries. However, there is a class of works, including stone walls, terraces and such improvements as field drains, water meadows, irrigation systems and regional drainage and reclamation systems which is much more purposive in intent, the specific object of which is to create capital for the future maintenance of land capability. Investments of this nature have a long life and are sometimes described as landesque capital, which refers to any investment in land with an anticipated life well beyond that of the present crop, or crop cycle. The creation of landesque capital involves substantial saving of labour and other inputs for future production. There is very little literature on this subject, and what there is suggests that the private benefits to land managers of costly landesque investments are seldom enough over the term of typically perceived discounting rates. It is therefore necessary to supplement these (rather sparse) economic explanations with others to explain why landesque capital is (and was) created at all. Blaikie and Brookfield (1987) make clear that sheer necessity created by a lack of other options (in order to ensure the survival of the land managers themselves) and particular coercive relations of production are two of the most common explanations, amongst other social and political reasons.

There is a need to be aware that conservation decisions, including the investment in landesque capital, are not often made by individual decision-makers, who will bear all the costs and reap all the benefits. Therefore, one must be able to identify clearly the land manager(s) or hierarchy of land managers, whoever they may be – farmer, developer, landlord, agri-business, manager, government official or whoever. This issue of identification is discussed fully in chapter 4 of Blaikie and Brookfield (1987), but it is enough to say here that managers may have different decision-making environments and different claims or demands upon the same tract of land.

CONCEPTUALIZING THE ROLE OF LAND MANAGEMENT

There remains the need to define briefly the task of land management in relation to the natural processes which require to be managed. These are two-fold and concern the role of land management respectively in checking the natural processes of degradation and in aiding the natural processes of repair. The characteristics of the land that is being managed must therefore be defined simply and unambiguously, in such a way that will specify the nature of the land-management task.

Sensitivity and resilience of land

There are two qualitative terms which are useful in describing the quality of land systems (soil, water, vegetation) and these are *'sensitivity'* and *'resilience'*. A number of other terms have also been used including 'susceptibility' and 'fragility' (Winiger 1983; Glaser 1983), but some of these are loaded terms. The first term chosen here is sensitivity and it refers to the degree to which a given land system undergoes changes due to natural forces, following human interference. The term used here refers to sensitivity to erosion as well as to other forms of damage, such as the accumulation of mobile ions (which can give rise to salinization).

The second group of land characteristics of importance in land management concern the ability of land to reproduce its capability after interference, and the measure of need for human artifice toward that end. This restoration of capital in the form of organic matter, nutrients and soil structure occurs naturally under forest or grassland fallow, as Nye and Greenland (1960) demonstrated in a manner that is still relevant. It occurs, however, at very different rates in different situations, while the depletion under cultivation which creates the need for restoration also takes place at very different rates. Certain ecosystems offer high initial productivity but this is rapidly depleted; in others, productivity is better sustained under repeated use. This property of standing up to, or absorbing the effects of, interference, is only partly correlated with what is loosely termed 'fertility' of the soil.

Broadly following Holling, the authors propose to term this property '*resilience*'. Holling wrote of the resilience of a natural system where 'resilience is a property that allows a system to absorb and *utilize* (or even benefit from) change' (1978: 11). Where resilience is high, it requires a major disturbance to overcome the limits to qualitative change in a system and allow it to be transformed rapidly into another condition. Also, resilience is independent of the quantitative primary productivity of the site, be it small or great.

It will be apparent that, where a site is highly resilient and also insensitive to the forces of damage, the task of land management is relatively easy. Many wetlands, even though they require some initial drainage and may be liable to occasional flood, have both these properties, as do alluvial plains in humid climates. It may be for this reason that, as recent research has established, most early agriculture in southern Europe and the Middle East, and perhaps elsewhere also, was on moist land; it was fixed-plot cultivation on land easy to manage, from which there has been subsequent differentiation into various forms of wetland and dryland farming (Sherratt 1980, 1981). Even shifting cultivation, adapted to land of low resilience, is seen as a subsequent development in this argument.

Usually the resilience of land has limits, and the task of land management becomes one of supplementing natural resilience with devices such as land- and crop-rotation, manuring and fertilization, the planting of legumes and a range of tillage and land-preparation methods, many of which are also linked in part to the control of sensitivity to damage. It is a part of our argument to show that almost all land other than the most infertile or least capable, least resilient and most sensitive, can be managed at some level of production wherever there is water and a sufficient growing period. Recent research even in the Amazon basin has shown that only about 3 per cent of its soils are incapable of management in some form, despite the acidity and low fertility of 75 per cent of the remainder (Sanchez *et al* 1982; Wade and Sanchez 1983). The cost of management may, however, be very high whether in terms of labour or material inputs.

Characteristics of land and the implications of land use and management can be summarized as follows:

- a land system of low sensitivity and high resilience only suffers degradation under conditions of very poor land management and persistent practices which remove soil, increase compaction, cause salinity and so on;
- a land system of high sensitivity and high resilience suffers degradation easily but responds well to land management designed to aid reproduction of capability;
- a land system of low sensitivity and low resilience is initially resistant to degradation but, once thresholds are passed, it is very difficult for any system of land management to restore capability;
- a land system of high sensitivity and low resilience easily degrades, does not respond to land management, and should not be interfered with in any major way by human agency, except (paradoxically) where major works create the landesque capital of a wholly new agro-ecosystem. The comparison between the impact of soil loss on productivity in temperate and tropical soils indicates that the latter tend, as a class, to have a relatively high sensitivity and low resilience and, hence, present more difficult management problems.

Two examples may serve to illustrate further the implications of different degrees of sensitivity and resilience for land. The first concerns the middle hills of the Nepal Himalaya, where some of the world's worst induced erosion is said to be taking place (eg Eckholm 1976). It is now established that the Tibetan plateau has been uplifted some 1000 m over the past 100,000 years (Ives 1981). Over the whole period, this is a mean rate of 1 cm/year. The Himalayan face has been uplifted at a lower rate, creating high natural erodibility as the slope becomes steeper, but an estimate of current uplift in the middle hills is 1 mm/year (Iwata, Sharma and Yamanaka 1984). In a small catchment in central Nepal, Caine and Mool (1982) calculate an annual lowering rate from mass wasting of 1.2 cm/year, while Williams (1977, cited in Carson 1985) calculates total denudation rates in four large catchments ranging from 0.51 to 2.56 mm/ year. Regional uplift and regional degradation are natural processes, and the effect of terracing for agriculture has often been to check natural surface erosion rates, though with no significant effect on the more sporadic and localized mass wasting processes (Carson 1985). The management of such terrain presents enormous problems. This is an example of land with high sensitivity and of variable resilience.

The second example is from the lowlands of western and central Europe, which would seem on *prima facie* grounds to present a much less sensitive environment, with geological stability, a climate of low erosivity, and low relief. However, the whole region is mantled by a loess-type periglacial *limon*, of low permeability and, in the absence of management designed to ensure such permeability, has been subject to substantial erosion, leading to the redeposition of colluvial material. Discussed further in chapter 7 of Blaikie and Brookfield (1987), this region has been shown to have quite high sensitivity and to be subject to episodic damage. Sensitivity is not always readily explained and the less obvious it is the greater perhaps the danger that a relaxation of management might lead to damage. However, under better management, the land system was able to reproduce its capability and even to increase it as a result of the degree of its resilience.

RELATIONSHIPS BETWEEN SOCIETY AND LAND DEGRADATION

Having defined the key terms, the next task is to outline the main characteristics of the relationship between land degradation and society, and then to draw conclusions about an appropriate method of analysis. The authors identify three main characteristics: the interactive effects of degradation and society through time; the crucial considerations of geographical scale and the scale of social and economic organization; and the contradictions between social and environmental changes through time.

Interactive effects

As in many complex issues of social or physical change, there is a reflexive and two-way relationship between land degradation and society. To take the similar case of population growth and development, for example, rapid population growth can, under certain conditions, adversely affect economic development and the living standards of the majority of the population unless the economy can be expanded at a comparable or greater rate. Conversely, however, many aspects of poverty lead couples to have large families, and thus encourage a high population growth rate. In the same way, land degradation can undermine and frustrate economic development, while low levels of economic development can in turn have a strong causal impact on the incidence of land degradation. Blaikie (1985a: 117) offers examples of 'desperate ecocide' by peasants and pastoralists under extreme pressures to survive, and chapter 2, section 4 in Blaikie and Brookfield (1987) gives a further illustration.

These interactive effects also take place through time. A period of rapid degradation may reduce the range of options over the possible uses to which land can be put in the future, unless there is effective repair. The future history of the affected region therefore takes a different course. This simple observation is somewhat complicated when establishing the impact of such land degradation upon the future history of the relevant people who use, or would have used, the land. The problem revolves around the convenient word 'relevant'. First of all, land degradation can affect, presumably adversely, the options of people living in the afflicted area, and future generations. However, if these future generations have the option of migrating elsewhere the issue becomes hypothetical. If, on the other hand, they do not have this option – perhaps because of national barriers as in the case of the Sotho of Lesotho, if the option of working in the South African gold mines is closed in the future – then the impact of degradation of a region on the present population becomes a very real question for analysis. This issue is one of 'option values' which is discussed in chapter 5 of Blaikie and Brookfield (1987).

Interaction and scale

The scale issue is crucial to the definition of land management because it focuses on the boundary problem of decision-making and of allocating costs and benefits. One person's degradation is another's accumulation, and this is equally true of uphill and downhill positions of a slope, regions, nations and even continents. For example, the 'hollow frontier' of Brazil in the early twentieth century, and that of the United States in the nineteenth, might be said to have contributed to the process of accumulation and the development of infrastructure on a national scale in the form of railways, roads and services. The fact of degradation on the settlement frontier had its effect on future options there, but the immediate effect of extracting short-term profits from the land was beneficial in the national context.

On a smaller scale, the physical transfer of fertility via riverborne silt and dissolved minerals, or by deliberate transportation of organic or mineral fertilizer from one place to another, makes it necessary to develop a more sophisticated set of criteria with which to analyse the impact of land degradation in one area upon the wider society. The exceptional case of Nauru has particular point here. The removal of rock phosphate from Nauru since 1900 has destroyed the agricultural capability of the island, which was never high, in the interests of overcoming phosphate deficiency in the soils of Australia and New Zealand. Latterly, the Nauruans have received good compensation for this loss, which they have invested mainly in the Australian economy, and on the proceeds of which they now largely live.

Contradictions between social and environmental change

The third aspect for debate concerns the possible contradictions between the criteria used for land degradation, and those for beneficial social change, or development, through time. An increase in cash incomes through commercial cropping and ranching can yield a temporary increase in rural incomes, maybe even over several generations, but can lead to degradation through lack of attention to management of the land, and hence to subsequent income reduction. Examples of this contradiction are legion. With the development of synthetic fertilizers, and their manufacture in larger and larger quantities, it can be argued that those pioneers who put profit first and good land management second made the right decisions, since the deleterious consequences of their actions are now masked by inputs of industrial origin. Moreover, while it may be that the modern oil-based fertilizers will not always be available, and more certainly will not be available so cheaply as oil resources finally approach exhaustion, the optimists would maintain that substitutes will be invented as the need arises. It is impossible to refute this argument, other than by pointing to the lower long-term cost of adopting management strategies which rely more upon natural processes of regeneration and repair.

THE APPROACH ADOPTED IN *LAND DEGRADATION AND SOCIETY*

Demands made by the society/land degradation relationship upon the method of analysis

Three characteristics of the relationship between land degradation and society have been identified: the importance of interactive and feedback effects through time, the importance of scale considerations, and the contradictions between social and environmental changes through time. These have to be recognized as placing difficult demands upon the way in which land degradation and society is studied.

One of the chief demands is a great deal of data, and there immediately arise technical problems of definition, measurement and availability. The second set of data problems involves the relationship between physical changes in soil and vegetation and declines in the productivity of the land (eg crop yields, livestock production). Again, this is partly a technical exercise, and much of the biophysical modelling of these relationships is beset by enormous uncertainties and errors (Amos 1982), but it is also an exercise which must try to distinguish the impact of physical changes in soil and vegetation from the impact of other purely socioeconomic changes in the circumstances of the land manager. Thirdly, there are difficult problems in the quantification of flows between people and regions. These derive from several distinct sources: the problem of conversion of flows of qualitatively different types to a common measure where energy, nutrients, available calories for human consumption, and market or shadow prices are only sometimes interchangeable; more abstract theoretical problems of incorporating the 'value' of resources found in nature; and lastly the 'unit of account' problem.

Wide degrees of error can therefore be made in the assessment of the important causes and rate of degradation and the reduction in capability of land. The ambiguity is compounded by the scantiness of data on farming and pastoral practices. Over long periods particularly, the causes of degradation usually involve social and economic changes which are difficult to measure, even if it is possible to reconstruct qualitative processes. If, for example, it is suggested that onerous rates of taxation and rents were responsible for heavy-handed and exploitative management of the soil, the challenge is to prove it. A rigorous explanation linking the cause and effect would also have to predict that a reduction in rates of taxation and rents would reduce exploitation of the soil. This account of the problems should not be a charter for sloppy reasoning and inadequate empirical verification, but it does indicate that the extent of rigour in any analysis is as much a matter of circumstance as it is of necessity.

What then is our response to these demands for data which probably cannot be met? Presented with these problems it looks as if the task of explanation outruns the prospect of empirical verification. Part of the response is an adaptation and development of the ideas of Thompson and Warburton (1985a, b) who suggest ways of 'getting to grips with uncertainty'. The first element in our approach is to accept 'plural perceptions, plural problem definitions, plural expectations and plural rationalities' (Thompson and Warburton 1985a: 123). There *are* competing social definitions of land degradation, and therefore the challenge of moving away from a single scientific definition and measurement must be taken up. This means we must put the land manager centre stage in the explanation, and learn from the land managers' perceptions of their problems. Thus land becomes a 'resource-in-use', inextricably related to the people and society that uses it. It also means that we avoid single hypothesis explanations of degradation. Degradation at one place and time will be conjunctural and complex. There are patterns that repeat themselves in human–environment relations, but their modelling can only be partial at best. Case-study material therefore becomes crucial, and is a dominant feature in Blaikie and Brookfield (1987). But it is easy to lapse into a mere recording of unique events full of 'emic' data, which are difficult to relate to each other. Therefore an approach is suggested which allows for complexity, uncertainty and great variety, and one which takes as its point of entry those data which are beset with least uncertainty – the direct relationship between the land-user and manager and the land itself.

The other response to uncertainty leads us in a different direction, but one which is not contradictory. This is to try and improve the means of measuring and evaluating land degradation. If outside institutions are to make any contribution to the reduction of land degradation and of the incomes of people who rely on the land for their livelihoods, they will have to know if there is a problem and how great it is. Therefore, reliable methods of measurement of land degradation are crucial. Of course data are not reliable, they are constructed, and considerable attention in the book under consideration in this chapter is devoted to their ideological nature, but this does not detract from the necessity to improve techniques of measurement. To this end chapter 3 of Blaikie and Brookfield (1987) explores the problems and prospects. Also, we need a methodology to evaluate the importance of land degradation in economic terms and a contribution to this is offered in chapter 5. First of all, the theoretical basis of the approach to land degradation and society is outlined in the next section.

The approach of 'regional political ecology'

The complexity of these relationships demands an approach which can encompass interactive effects, the contribution of different geographical scales and hierarchies of socioeconomic organizations (eg person, household, village, region, state, world) and the contradictions between social and environmental changes through time. The authors' approach can be described as 'regional political ecology'. The adjective 'regional' is important because it is necessary to take account of environmental variability and the spatial variations in resilience and sensitivity of the land, as different demands are put on the land through time. The word regional also implies the incorporation of environmental considerations into theories of regional growth and decline.

The circumstances in which land managers operate in their decision-making over land use and management can be considered in the context of core–periphery relations. Location-specific studies of the settlement frontiers of Brazil, the United States and South East Asia, as well as of agricultural decision-making in economically declining areas, provide considerable evidence for suggesting that declining regional economies provide an important context for lack of initiative and investment of labour and capital in managing land. Chapter 6 in Blaikie and Brookfield (1987) gives examples from hill and mountain areas of this link between the status of regional decline and the circumstances of decision-making in land management. Chapter 7 on the other hand provides evidence from eighteenth-century France to show that both the downswing and the upswing in a rural economy can almost equally press on the welfare and freedom of those who occupy the most vulnerable position in the social order.

The phrase 'political ecology' combines the concerns of ecology and a broadly defined political economy. Together this encompasses the constantly shifting dialectic between society and land-based resources, and also within classes and groups within society itself.

We also derive from political economy a concern with the role of the state. The state commonly tends to lend its power to dominant groups and classes, and thus may reinforce the tendency for accumulation by these dominant groups and marginalization of the losers, through such actions as taxation, food policy, land tenure policy and the allocation of resources. The agrarian history of Europe provides abundant examples (Abel 1980; Kriedte 1983). Very recent work on the relationship between cumulative soil losses and crop and livestock yields has shown a negative exponential relationship (Stocking and Peake 1985; Hufschmidt *et al* 1983: 146) which strongly encourages the state to allocate resources to protect productive and still capable land, rather than to repair already degraded land which has fallen to a low level of productivity. Such a trend may be accentuated by the need of dominant groups to protect the source of major commercial crops. The allocation of state-controlled resources in rural development therefore usually disfavours the physical and social margin. This is shown for Latin America by Posner and MacPherson (1982) and for Nepal by Blaikie, Cameron and Seddon (1980). It may be added that the efforts of international agencies have hitherto tended to concentrate in the same direction, notwithstanding contrary statements of policy. These ideas are developed in an introductory fashion later in this chapter.

Extended examples of regional political ecology which consciously use theoretical material from the core–periphery model, applied theories of the state, and the ecology of agricultural systems, are offered in chapter 2, section 4 and chapter 6, section 6 of Blaikie and Brookfield (1987). In the latter, it is hypothesized that many areas of the Third World suffer from a set of related symptoms which combine the results of land degradation, political and economic peripheralization, stagnant production, out-migration and poverty. However, there are clearly important variations in the politico-economic and physical histories of peripheral areas. Some areas, especially in hills and mountains, have avoided colonization and have preserved elements of ancient culture and social structure, such as segmented tribal organization and unformalized rules of land tenure. Other areas and their people have been intensively colonized and have attracted metropolitan capital into plantations, large farms and ranches, but are limited by sensitive and unresilient environments of a a different type altogether. *Land Degradation and Society* clearly draws distinction between these two in chapter 6 and again in chapter 10.

However, there were and still are political economies which predate the world capitalist system, or remain only loosely articulated with it in modern times. Today, post-1945 Albania is an example and historically the Asian and tropical-American empires grew and differentiated on the basis mainly of internal division of labour and trade, with only peripheral dependence on external exchange. Such writers as Chevalier (1963) and Borah and Cook (1963) have shown how a class structure had evolved in central Mexico under the Aztec empire, how this was reflected in the management of land and the exaction of tribute, and how remoter groups brought under Aztec rule were incorporated into this system in a peripheral relationship. Degradation and erosion were substantial (Cook 1949). In Blaikie and Brookfield (1987) the more remarkable – because little stratified – case of the highlands of Papua New Guinea is analyzed in chapter 8; here a political economy based on surplus production for competitive prestation evolved in the 300 or so years before there was any direct contact with the world political economic system, and a significant degree of land degradation was brought about under that isolated system.

In chapter 7 of Blaikie and Brookfield (1987) a more specific historical inquiry into the conditions of degradation in the past is undertaken. The authors seek to explain how and why erosion of a type generally associated with sub-humid areas of southern Europe came to prevail in quite large parts of central and western Europe in the past, reaching a peak, at least in France, in the eighteenth century. Finding the evidence to favour a preponderantly human causation, it is hypothesized that pressures on the peasantry came to be translated into inadequate management of the land. Landlords, the emergent bourgeoisie and the state all contributed to these pressures. This historical example, and other historical material in this book, are introduced for a very specific set of reasons. Not only was the early-modern condition of the peasant and working classes in the west comparable with, or worse than, that of their modern counterparts in the Third World, but the pressures on them assumed a severity rarely encountered today. The historical examples thus provide something of an extreme case of our thesis that damage to the land and damage to certain classes in society are interrelated. Moreover, they also provide long-term depth of material that is not generally available to use in the Third World or in countries of recent European settlements, and hence provide both an illustration of political ecology in time depth, and also a corrective to facile conclusions that might otherwise be drawn from the examination only of contemporary problems.

The economic concept of the margin

The concept of the marginal unit of a factor of production, that last unit which when brought into use yields exactly its own cost and no more, is implicit in the classical theory of rent. Ricardo (1951) developed the theory of rent in regard to qualities of land; when all land of the first, and by definition uniform, quality has been brought into production, and land of the second quality is then employed, the cost of production on the latter will be higher than on the first. For this to be possible the price must rise, and so all land of the first quality will receive an unearned income in consequence of the incorporation of the second; the unearned income of labour inputs on the land is rent. If land is more intensively cultivated, the law of diminishing returns will apply. Hence the schedule of production will form a parabola, so that at the optimum ratio land and labour will both be utilized fully, and beyond this point there is a shortage of the forces of natural growth relative to the input of labour. Further increases in demand will therefore make it necessary to bring in new and inferior land, and the last land to be brought into use, or to be intensified, will just repay the cost of production and no more; this is the margin.

Von Thunen (Hall 1966) noted that beyond the optimum point of intensification it is a combination of constant land and increasing labour that becomes less productive, so that it is the additional units of labour that will in fact earn less. Gossen (as cited in Heimann 1945) noted that the value of any given unit of a quantity, wherever produced, is appraised like the marginal or last unit and thus has the same utility, and showed that the value of any individual unit produced must be equal to the marginal utility. The marginal unit is therefore that whose marginal cost is equal to the marginal utility, and if we are writing of land qualities, then this unit is the marginal land (Heimann 1945: 186–7). Add to this Von Thunen's arguments about the effects of intensification as the margin is approached on the distribution of returns to the factors of production, and we also have a link with the political-economy view of the margin which is developed below.

The ecological concept of the margin

In principle, at least, the ecological concept of the margin is comparable with the neo-classical one. For a given plant, or association of plants such as a forest, the marginal unit of land is that where natural conditions will just permit the plant to survive. However, an ecological view cannot avoid the question of environmental variability, so that we have to define the margin in terms of expected adverse conditions, recognizing that in some years plants can grow well beyond their secure domain. This being so, a marginal environment for plants is better interpreted as the area or zone within which there is expected killing stress, but over which a plant or plant association can expand when that stress is absent. The same concept applies to marginal habitats for wildlife, and by extension also to crops and livestock.

Discussion of the ecological margin does not always follow this logical approach. Perhaps it is better to be more restrictive and to define the term by extrapolation of the neo-classical definition to take account of environmental variability. The Sahel, for example, is thus defined as a marginal zone within which droughts of great severity and length can be expected. Discussion of the advance of the desert margin into this zone (Stebbing 1935; Rapp 1976) means essentially that its marginality is becoming accentuated as human interference assists natural forces in the elimination or pauperization of plant communities, and makes their re-establishment in good years less likely.

Ecological marginality need not relate only to natural conditions. Agro-ecosystems created by people immediately acquire a new set of relevant environmental variables. In all irrigated land, the availability and the quality of water becomes paramount. A clear example of ecological marginality in the context of created agro-ecosystems is provided by the annually reconstructed fields made in the gravelled beds of rivers in parts of the Mountain Province of the Philippines, while another is the gardens fed with human manure that were until recently encountered on embanked portions of the sea beach around the inlets which penetrate the New Territories of Hong Kong. Both were economically better than marginal land, otherwise they would not have been constructed, but both were ecologically marginal and at grave risk from storm and flood.

The political-economic concept of marginality

The political economy approach concerns the effect on people as well as on their productive activities of on-going changes within society at local and global levels. Use of the term in this context has arisen in the Latin American literature, where it was used to describe the sort of process described by many writers from Mariategui (1971) onward, and pithily summed up by Stavenhagen:

> The channeling of capital, raw materials, abundant foods, and manual labour coming from the backward zones permits the rapid development of these poles or focal points of growth, and condemns the supplying areas to an increasing stagnation and underdevelopment. (1969: 108)

At about the same time, Casanova (1970: 123) wrote of the 'marginal masses' who are outside the political system of Mexico, and of the 'marginal population' which is disorganized, uninformed and which can make demands only 'in the traditional forms of supplication, petition and complaint'. The term was quickly adopted (Parra 1972) to refer to a whole class of people who are excluded from employment, services, participation in decision-making, opportunity and secure housing (Brett 1973). Gaining wider currency, 'marginalization' has been used in the continent's feminist literature to describe the exclusion of women from productive employment (Hartmann 1976; Young and Moser 1981), and in being widened to this and other contexts has perhaps lost something of the force contained in the original Latin American formulation.

The relation between three concepts of the margin

Writing of Kenya, Wisner (1976) wrote of marginals created by colonialism and capitalism who, in the process of social allocation of space, were quite literally pushed into marginal places. However, socio-political and ecological or economic marginality are not necessarily correlated in this way. Marginalized peasants can, and do, occupy smallholdings on highly fertile land, while ecologically marginal land that is also near marginal in the neo-classical sense can, if a holder has enough of it, offer the basis for a highly profitable commercial operation. Much of northern Australia is ecologically marginal, but while most of it would be sub-marginal for commercial agriculture as has repeatedly been shown, it can support very profitable pastoral operations when coarsely divided into properties and chains of properties the size of small European countries. However, the Aboriginal people dispossessed of their land and now working on these estates share none of this affluence, and have been marginalized within the new relations of production.

If we control the comparison within a single mode or system of production, however, a relationship can more readily be established. An Asian rice-growing community has land sharply differentiated by fertility and hydrology, and its upland areas are sensitive under interference. When *sawah*-rice terraces are created, these new agro-ecosystems differ greatly in their ecological security. If they are on unstable slopes, the terrace walls may collapse. Some are difficult to supply with water in dry years, while others lose water readily by seepage. Under a high population density all the land capable of *sawah*-rice production has been taken up and converted; some of this is ecologically marginal even though economically secure in most years. Great differences in rent are yielded by the *sawah*-rice parcels. Some farmers without or with insufficient *sawah*, take up dry land for swidden cultivation on the ecologically marginal slopes, where they get good short-term returns from dry crops, but are at risk from erosion and loss of fertility. Those who are most marginalized in the socioeconomic sense have no land, and are forced to seek casual work from others. This is a hypothetical example, but is not unlike an upland West Java village (*kampung*) studies by members of the International Rice Research Institute (IRI). They conclude:

> As growth of population presses hard on limited land resources under constant technology, cultivation frontiers are expanded to more marginal land and greater amounts of labor applied per unit of cultivated land; the cost of food production increases and food prices rise; in the long end (*sic*), laborers' income will decrease to a subsistence minimum barely sufficient to maintain stationary population and all the surplus will be captured by landlords as increased land rent. This is exactly what has occurred in the *Kampung*. (Kikuchi *et al* 1980: 15)

It will be useful to summarize some of the postulated and demonstrated relationships. To clarify, we identify the three concepts of marginality as economic (EN), ecological (EC) and politicoeconomic (PE) in what follows.

Land managers can become marginalized (PE) through the imposition of taxes, corvée labour and other relations of surplus extraction. The responses they make may be reflected in land use and in investment decisions over the preservation of productivity of their land. Adversity of this sort can produce innovations which raise productivity – to pay for the extraction of surpluses – as well as safeguard future productivity. However, more extreme marginalization (PE), often involving a whole number of readjustments, particularly a loss of labour power (through war, conscription or emigration), has frequently led to changes in land use and the inability to keep up longer-term investments in soil and water conservation (eg repair of terraces and cleaning or irrigation and drainage ditches). The land then becomes economically marginal (EN) and the result is a decline in capability and marginality (EC) of the agro-ecosystem.

Spatial marginalization (PE) may also accompany these changes. Dominant classes may gain control and use more fertile land and force others to use more marginal land (EN). The attempts of the latter to make a living with reduced resources have often led to land degradation. Marginal land (EC) which has a high sensitivity and low resilience to even skilful or light interference by land managers can attract land uses, for this reason, which permanently damage the capability of the land. Here the emphasis rests not only upon the socially imposed marginality (PE) of the land manager, but also upon the intrinsic marginality (EC) of the land itself. Commercial ranching in the Australian interior is a prime example. If land degradation comes about as a result of either commercial exploitation or socially induced marginalization (PE) of land managers, a vicious circle of increasing impoverishment and further marginalization (EC) of land and land managers (EN) can sometimes result. Hence land degradation is both a result of *and* a cause of social marginalization (PE). It can accentuate the physical marginality (EC) of land by reducing its present capability, and marginalize (EN) it for present alternative uses. Much of Ethiopia, the Sahel region as a whole, and other areas of low resilience find themselves in this position.

DEGRADATION, HAZARDS AND THE ENVIRONMENTAL PARADIGM

The approach of regional political ecology taken in Blaikie and Brookfield (1987) is compatible with the new directions in hazards and disaster research. Both approaches share an historical and a dynamic approach to human–environment relations. Nature is seen to be in constant flux, and measurement must constantly be updated. Also, nature is not universally or statically defined; resources become resources when people define them as such (Blaikie 1985c). The multiple definitions of natural resources and degradation by three groups of land-users and three government departments in the Indonesian case study are a good illustration. Both approaches emphasize underlying social order rather than capricious nature in the explanation of calamitous events:

> Causes, internal features and consequences (of natural disaster) are *not* explained by conditions or behaviour peculiar to calamitous events. Rather they are seen to depend upon the ongoing social order, its everyday relations to the habitat and the larger historical circumstances that shape or frustrate these matters. (Hewitt 1983: 25)

The three concluding chapters in Hewitt's book provide the basis of this alternative approach, linking the ongoing social order to hazardous events. Susman, O'Keefe and Wisner (1983) build on the work of O'Keefe (1975) and of Wisner (1976) who for a decade have linked disasters to processes of marginalization and proletarianization. The trigger events which start disasters or catastrophes have explanatory linkages with land degradation because both arise from the conjunction of physical and social processes. Sayer urged that we must start with the essential and necessary unity of society and nature, and that 'to start in the conventional manner with . . . a separation followed by a listing of interactions would be to prejudice every other aspect of the exposition' (1980: 22). Approving this view, Watts goes on to argue that 'the subject matter of human ecology is accordingly *inner*-actions with nature' (1983: 234). This formulation is close to the idea of a resource-in-use used earlier in this chapter. Also shared with the alternative approach in hazards research is an avoidance of relegating natural processes to a mere context or backdrop to hazards or degradation. Some radical literature has tended to do this and to imply that studies of climatic change in the Sahel, for example, are no more than a smokescreen and decoy to cover the tracks of the 'real' culprit – capitalism. It is vital to understand (as accurately as data, measurement and modelling will allow) the natural forces which create a variable management task to which decision-making, subject to political economic conditions of choice, has to respond.

THE SOCIAL SCIENTIST'S CONTRIBUTION: THE NEED FOR OPEN MINDS

Blaikie and Brookfield set out initially to write *Land Degradation and Society* from position papers which adopted respectively Marxist and behavioural approaches, in each case with qualifications. What happened instead was something unforeseen: large areas of agreement emerged between the two authors, and several of the contributors also. While a more abstract (and no doubt rigorous) analysis of the two positions would undoubtedly expose fundamental contradictions, there is a broad area within which the explanation of land degradation can draw upon similar themes. There is something to be said for declaring a truce on the more abstract structural differences in the interpretation of social change, however important these differences may be, if it allows cross-fertilization of approaches. There are certainly fundamental contradictions between the 'human adaptation', neo-classical and various Marxist approaches, to take these three only. However, they share the objectives of understanding and problem solving, and of bringing about change in the situation, albeit in different degrees and in different ways. While there are epistemological reasons why Marxists have not been too interested in decision-making models, there is nothing inherently revisionist in building them. Likewise, there is no betrayal of the profession of neo-classical economics in trying to pursue the quantification of costs and benefits of degradation and conservation into the realms of politics and unquantifiable conjecture. Nor is there any reason why the study of human behaviour should fail to take advantage of the insights of theory about economic rationality or disregard the contradictions inherent in all social change and social formation.

There is a need for open minds, too, in the use of quantification and model building. There is an extraordinary schism between two self-perceived epistomological camps, the one which measures, creates its own data and uses others' in model building, and the other which calls itself 'radical' and eschews analysis of this sort as positivist, and the data as ideologically tainted and reductionist. Whilst Blaikie and Brookfield (1987) amply show that data do not simply exist but rather is constructed, they also argue strongly for technically better *and* more ideologically aware measurement of process, costs and benefits. Quantitative modelling of resources-in-use and land managers themselves need not be mindless number crunching. Nor need a central concern for the social meaning of degradation and for conscious ideological choice in explanation be dismissed as biased and not 'real' science.

Open minds assist in clarifying and sharing objectives. There are many blocks to open minds: the criteria for excellence and promotion differ between various practitioners (academics of different disciplines, consultants, administrators, politicians); there can be interdisciplinary rivalry between different academic departments (particularly between natural and social science); and more specific epistomological differences, mainly about the domain and status of proof in discourse and research. Land degradation and society, because of its complex and multidisciplinary nature, and its theoretical and practical elements, encounters most of these blocks.

If these blocks are not removed, the issue of land degradation will remain shrouded in controversy, uncertainty and incomprehension. What people cannot understand, they tend to avoid; what is unclear, people cannot decide upon. So it is with policy-makers and land degradation. While solutions will be as multiple as the causes of land degradation, the general approach outlined here aims to unify, but through an appreciation of plurality of purpose and flexibility in explanation. For the discipline of geography at least, Carl Sauer put the problem and challenge perfectly more than 45 years ago:

> Surely nothing could be more geographic than critical studies of the wastage of surface and soil as expressions of abusive land occupation. On the one hand are the pathological physical processes; on the other, the cultural causes are to be studied. Next come the effects of continued wastage on survival of population and economy, with increasing tendency to degenerative alterations or replacement. Finally, there is the question of recovery or rehabilitation . . . Geographers have given strangely little attention to man as a geomorphologic agent . . . The theme was clearly indicated as a formal problem of geography three-quarters of a century ago by Marsh. Geographers have long given lecture courses on conservation of natural resources and considered the evils of soil erosion. But what have they done as investigators in the field, which may actually lie at the doorsteps of their classrooms? Is the answer that soil students should study sheet wastage, geomorphologists gullies, agricultural economists failing agriculture, rural sociologists failing populations, and the geographer prepare lectures on what others investigate? (Sauer 18–19)

SUMMARY AND CONCLUSION

All aspects of the relationship between land degradation and society are both social and physical – a commonplace statement that is self-evidently true, but not trivial. It means that degradation is perceptual and socially defined. There may well be competing perceptions and these can be put into the context of the political economy as a whole, in which different classes and groups perceive and use land and its resources in different ways. Our four central terms – land management, land degradation, resilience and sensitivity – are all defined in a social context, and with explicit reference to ongoing processes of social change. There are extremely severe problems of data availability and of verification and proof. The approach taken in Blaikie and Brookfield (1987) must respond to this problem of uncertainty and does so by seeking a point of entry where uncertainty is least, at the point of the land manager. The land manager is then contented, and her or his actions explained within a set of dynamic human–environment relationships which we call regional political ecology. The various definitions of the margin and marginality are central to this approach.

It will be obvious that the book avoids an ethical and fundamentalist approach to land degradation. The definition of degradation and whether it is bad or not are both related to the people who use the land. The field of interest in this book does not include difficult environmental-ethical questions such as the extinction of endangered species, or conflicts between national parks and other human uses of the biome, where ethical judgements assume greater importance. The approach taken in Blaikie and Brookfield (1987) is that land degradation is judged in terms of the altered benefits and costs that accrue to people at the time and in the future.

This chapter is taken from Blaikie, P and Brookfield, H (1987) *Land Degradation and Society,* Methuen Press, London.

REFERENCES

Abel, W (1980) *Agricultural Fluctuations in Europe from the Thirteenth to the Twentieth Centuries*, translated by O Ordish from *Agrarkrisen und Agrarkonjunktur* (1978, Paul Parey, Hamburg) Methuen, London.

Alier, J M (1984) *A History of Ecological Economics*, Facultat de Ciencies Economiques, (English version of part of a book in Spanish published by la Fundacion Juan March, Madrid).

— and Naredo, J M (1982) 'A Marxist precursor of energy economics: Podolinsky', *Journal of Peasant Studies*, 9, pp 207–23.

Amos, M J (1982) 'Economics of soil conservation', unpublished MSc dissertation, National College of Agricultural Engineering, Silsoe, Bedfordshire.

Bidwell, O W and Hole, F D (1965) 'Man as a factor of soil formation', *Soil Science* 99, 65–72.

Blaikie, P M (1985a) *The Political Economy of Soil Erosion in Developing Countries*, Longman, London.

— (1985b) 'Natural resources and social change', Unit 7 II (Analysis: aspects of the geography of society) in *Changing Britain, Changing World: Geographical Perspectives* (course D 205), The Open University, Milton Keynes.

— and Brookfield, H (1987) *Land Degradation and Society,* Methuen Press, London.

—, Cameron, J and Seddon, D (1980) *Nepal in Crisis: Growth and Stagnation of the Periphery*, Oxford University Press, New Dehli and Oxford.

Borah, W and Cook, S F (1963) *The Aboriginal Population of Central Mexico on the Eve of the Spanish Conquest*, Ibero-Americana 34, University of California Press, Berkeley.

Brett, E A (1973) *Colonialism and Underdevelopment in East Africa: the Politics of Economic Change 1919–1939*, Heinemann, London.

Brown, L R (1981) 'Eroding the base of civilisation', *Journal of Soil and Water Conservation*, October, 36, pp 255–60.

Caine, N and Mool P K (1982) 'Landslides in the Kolpu Khola drainage, middle mountains, Nepal', *Mountain Research and Development*, 2, pp 157–73.

Carson, B (1985) *Erosion and Sedimentation Processes in the Nepalese Himalaya*, ICIMOD occasional paper no 1, International Centre for Integrated Mountain Development, Kathmandu.

Casanova, P G (1970) *Democracy in Mexico* (translated by Salti, D from *La Democracia en Mexico* 1965, Ediciones Era, Mexico City) Oxford University Press, New York.

Chevalier, F (1963) *Land and Society in Colonial Mexico: the Great Hacienda* (translated by Eustis, A, from *La Formation des Grands Domaines au Mexique*, 1952, Institut d'Ethnologie, Paris) University of California Press, Berkeley.

Cook, S F (1949) *Soil Erosion and Population in Central Mexico*, Ibero-Americana: 34, University of California Press, Berkeley.

Eckholm, E P (1976) *Losing Ground: Environmental Stress and World Food Prospects*, Norton, New York.

Glaser, G (1983) 'Unstable and vulnerable ecosystems: a comment based on MAB research in island ecosystems', *Mountain Research and Development,* 3, pp 121–23.

Glover, Sir H (1946) *Erosion in the Punjab: its Causes and Cure,* Feroz printing works, Lahore.

Gutman, P (1985) 'Teona economica y problematica ambrento: un dialogo difical', *Desarrollo Economico: Revista de Ciencias Sociales*, April–June, 25 (97), pp 47–70.

Hall, P (1966) *Von Thunen's Isolated State*, an English edition of *Der Isolierte Staat* by Johann Heinrich von Thunen, translated by Warenberg, C M, edited with an introduction by Peter Hall, Pergamon, London.

Hartmann, H (1976) 'Capitalism, patriatchy and job segregation by sex', in Blaxall, M and Reagan, B (eds) *Women and the Workplace*, University of Chicago Press, Chicago, 137–69.

Heimann, E (1945) *History of Economic Doctrines: An Introduction to Economic Theory*, Oxford University Press, New York.

Hewitt, K (1983) 'The idea of calamity in a technocratic age', in Hewitt, K (ed) *Interpretations of Calamity, from the Viewpoint of Human Ecology*, Allen and Unwin, Boston, pp 3–32.

Holling, C S (1978) *Adaptive Environmental Assessment and Management*, John Wiley, Chichester.

Hufschmidt, M M, James, D E, Meister, A D, Bower, B T and Dixon, J A (1983) *Environment, Natural Systems and Development: An Economic Valuation Guide*, Johns Hopkins University Press, Baltimore.

Hyams, E (1952) *Soul and Civilisation* (second edition 1976) John Murray, London.

Ives, J D (1981) 'The Heavenly Mountains: an excursion to the Tien Shan, Peoples' Republic of China, 1–24 June 1981' *Mountain Research and Development*, 1, pp 293–98.

Iwata, S T, Sharma, K and Yamanaka, H (1984) 'A preliminary report of central Nepal and Himalayan uplift', *Journal of the Nepal Geography Society*, 4, pp 141–49.

Jacks, G V and Whyte, R O (1939) *The Rape of the Earth: A World Survey of Soil Erosion*, Faber and Faber, London.

Kellman, M (1974) 'Some implications of biotic interactions for sustained tropical agriculture', Proceedings of the Association of American Geographers, 6 pp 142–45.

Kikuchi, M, Hadif, A, Selah, C, Hartoyo, S and Hayami, Y (1980) *Changes in Community Institutions and Income Distribution in a West Java Village*, IRRI Research Papers Series 50, Manila, International Rice Research Institute.

Kriedte, P (1983) *Peasants, Landlords and Merchant Capitalists: Europe and the World Economy, 1500–1800* (translated by Berghahn, V R from *Spatfeudalismus und Handelskapital: Grundlinien der Europaischen Wirtschaftsgeschichte vom 16, bis zum Ausgand des 18, Jahrhunderts*, (1980) Gottingen, Vandenhoeck und Raprecht) Berg Publishers, Leamington Spa.

Malcolm, D W (1938) *Sukumuland – An African People and their Country* International African Institute and Oxford University Press, Oxford.

Mariategui, J C (1971) *Seven Interpretive Essays on Peruvian Reality* (translated by Urquidi, M from *Siete Ensayos sobre la Realidad Peruana*, 1928, NS, Lima) with an introduction by Basadre, J, University of Texas Press, Austin, Texas.

Marx, K (1887/1954) *Capital: a Critique of Political Economy*, vol 1, Progress Publishers, Moscow, from the English edition of 1887, Engels, F (ed).

Nye, P H and Greenland, D J (1960) *The Soil under Shifting Cultivation*, Technical Communication 51, Commonwealth Agricultural Bureaux, Farnham Royal, Bucks.

O'Keefe, P (1975) *African Drought: A Review*, occasional paper no 8, Disaster Research Unit, University of Bradford.

Parra, V R (197) 'Marginalidad y Sudesarrollo', in Cardone, R (ed) *Las Migraciones Internas,* Editorial Andes, Bogota, chapter 5.

Posner, J L and MacPherson, M F (1981) 'Agriculture on the steep slopes of tropical America: the current situation and prospects', *World Development*, May, pp 341–54.

Rapp, A (1976) 'The Sudan' in Rapp, A, Le Houerou, H N and Lundholm, B (eds) *Can Desert Encroachment be Stopped?* Swedish Natural Science Research Council Ecological Bulletin 24, National Science Research Council, pp 155–64.

Ricardo, D (1951) *On the Principles of Political Economy and Taxation*, 3rd edition, John Murray, London, reprinted with an Introduction by Straffa, P, with the collaboration of Dobb, M H as *The Works and Correspondence of David Ricardo*, vol 1, Cambridge University Press, Cambridge.

Robinson, J (1963) *Economic Philosophy*, Aldine, Chicago.

Rounce, N V (1949) *The Agriculture of the Cultivation Steppe of the Lake, Western and Central Provinces (Uganda)*, Longman, Cape Town.

Sanchez, P A, Bandy, D E, Villachica, J H and Nicholaides, J J (1982) 'Amazon basin soils: management for continuous crop production', *Science*, 216, pp 821–7.

Sauer, C O (1941) 'Foreword to historical geography', Annals of the Association of American Geographers, 31, pp 1–24.

Sayer, A (1980) *Epistemology and Regional Science*, School of Social Science, University of Sussex, Brighton.

Sherratt, A (1980) 'Water, soil and seasonality in early cereal cultivation', *World Archeology*, 11, pp 313–30.

Sherratt, A (1981) 'Plough and pastoralism: aspects of the secondary products revolution', in Hodder, I, Isaac, G and Hammond, N (eds) *Patterns of the Past: Studies in Honour of David Clarke*, Cambridge University Press, Cambridge pp 261–305.

Stavenhagen, R (1969) 'Seven erroneous theses about Latin America', translated and revised by Stavenhagen, R, from an original article in Spanish, published in *El Dia* (newspaper, Mexico City) (1965) in Horowitz, Il, de Castro, J and Gerassi, J (eds) *Latin American Radicalism: A Documentary Report on Left and Nationalist Movements*, Random House, New York, pp 102–17.

Stebbing, E P (1935) *The Creeping Desert in Sudan and Elsewhere in Africa*, Ministry of Agriculture, Khartoum.

Stocking, M A and Peake, L (1985) *Erosion-induced Loss in Soil Productivity: Trends in Research and International Cooperation*, paper presented to the IVth International Conference on Soil Conservation, Maracay, Venezuela, 3–9 November, undertaken under Soil Conservation Programme, Soil Resources, Management and Conservation Science, Land and Water Development, FAO, Rome.

Susman, P, O'Keefe, P and Wisner, B (1983) 'Global disasters, a radical interpretation', in Hewitt, K (ed) *Interpretations of Calamity from the Viewpoint of Human Ecology*, Allen and Unwin, Boston, pp 263–83.

Suwardjo and Aibyamia, S (1983) 'Crop residue mulch for conserving soil in upland agriculture', paper presented to Malama Ana 83, Honolulu, Hawaii.

Thompson, M and Warburton, M (1985) 'Uncertainty on a Himalayan Scale', *Mountain Research and Development*, 5, pp 115–35.

Wade, M K and Sanchez, P A (1983) 'Mulching and green manure applications for continuous crop production in the Amazon basin', *Agronomy Journal*, 75, pp 39–45.

Watts, M (1983) 'On the poverty of theory: natural hazards research in context' in Hewitt, K (ed) *International of Calamity from the Viewpoint of Human Ecology,* Allen and Unwin, Boston, pp 231–62.

Williams, V (1977) 'Neotectonic implications of the alluvial record in the Sapta Kosi Drainage Basin, Nepalese Himalayas', PhD thesis in Geology, University of Washington, Seattle.

Winiger, M (1983) 'Stability and instability of mountains ecosystems: definitions for evaluation of human systems', *Mountain Research and Development*, 3, pp 103–11.

Wisner, B (1976) 'Man-made famine in Eastern Kenya: the interrelationship of environment and development', Institute of Development Studies discussion paper no 96, IDS, Brighton.

Young, K and Moser, M (eds) (1981) *Women and the Informal Sector*, IDS Bulletin 12, p 3.

Chapter 6

Deforestation in Amazonia: Dynamics, Causes and Alternatives

Anthony B Anderson

Tropical deforestation is one of the major environmental crises of our time (Gomez-Pompa *et al* 1972; Myers 1979, 1980; WRI 1985). In drier tropical zones, where human population densities are frequently high, deforestation has often led to irreversible changes in vegetation, soil, and possibly even climate, undermining the carrying capacity of ecosystems as well as their capacity to recover. The results are erosion, devastating floods, chronic fuelwood shortages, and desertification. In moister tropical zones,[1] where human populations are generally low, large-scale deforestation is a recent phenomenon, and its environmental effects are currently far less apparent than in drier zones. But an increasing array of evidence suggests that the long term effects could be equally severe.

The dangers of deforestation in the moist tropics are largely due to the nature of ecosystems that predominate in this zone. Despite their luxuriant growth, moist tropical forests generally occur on soils of low fertility. Most of the nutrients are stored in the biomass, and cycling of nutrients takes place via litter fall and root uptake. Prolonged removal of this biomass – as is currently occurring over increasing areas of the moist tropics – effectively depletes essential nutrients from the ecosystem while promoting erosion, soil compaction, and weed infestation. Although the possibility of converting moist tropical forests into a 'red desert' (Goodland and Irwin 1975) may be nil, over-intensive use of deforested sites in this zone could lead to permanent degradation of regional ecosystems (Uhl 1983; Uhl, Buschbacher, and Serrão 1988). Furthermore, destruction of moist tropical forests leads to the demise of indigenous cultures that are best adapted to these ecosystems (Davis 1977; Hemming 1978; Posey 1983), and to the increased marginalization of newly arrived settlers as well (Wood and Schmink 1978; Barbira-Scazzocchio 1980; Hecht, Anderson, and May 1988).

In moist as well as dry zones, tropical deforestation can have a potentially disruptive effect on global geochemical cycles and, ultimately, climate. The popular conception of the Amazon forest as the world's 'lungs' that generate the oxygen we breathe has been debunked (Fearnside 1985). But at least half of the rainfall in the Amazon Basin comes from water evapotranspired from the forest itself (Salati 1985), and widespread deforestation is likely to result in a pronounced reduction in regional rainfall. Moreover, moist tropical forests contain approximately 35 per cent of the world's living terrestrial carbon pool (Brown and Lugo 1982); release of this pool into the atmosphere through felling and burning could contribute substantially to global warming.

Perhaps the greatest danger of tropical deforestation is the loss of genetic diversity (Buschbacher 1986). Widespread destruction of tropical moist forests, which contain 40–50 per cent of the earth's species, will inevitably result in large scale extinctions. If allowed to occur on this scale, such extinctions will eliminate a genetic heritage not only of inestimable aesthetic value, but of current use in essential ecological services (ie pollination, seed dispersal, etc), and of future use in development of agricultural crops, pharmaceutical products, and industrial materials (Myers 1979, 1984).

The scale of deforestation in the moist tropics has only recently been quantified on a worldwide basis. Data compiled by the FAO show that during 1980–1985, loss of tropical closed forests[2] averaged 75,000 km^2 per year,[3] or an area roughly equivalent to the country of Panama (WRI 1986). Rates of deforestation during this period varied considerably from country to country, ranging from 0.1 per cent in Papua New Guinea to 5.9 per cent in the Ivory Coast. On a regional basis, deforestation rates during 1981–1985 were highest in Latin America, both in absolute area (43,000 km^2 per year) and in percentage of total forest area (0.64 per cent per year).[4]

DEFORESTATION IN AMAZONIA

Dynamics

The Amazon Basin, drained by the largest river system on earth, covers a total land area of 7.05 million km^2, of which approximately five million km^2 are covered by rain forest (Sioli 1984). This forest varies considerably in structure and composition (Ducke and Black 1954; Pires and Prance 1985); on a regional basis, its richness of bird, fish, and insect species is unmatched (eg Haffer 1969; Goulding 1980; Brown 1982). This extraordinary diversity may in large part be due to repeated contractions and expansions of the Amazon rain forest during the Pleistocene period (Prance 1985).

More recently, the rain forest has been subjected to human-caused disturbances following the arrival of indigenous groups approximately 12,000 years ago (Meggers 1985). Evidence suggests that population densities along the major rivers may actually have been higher prior to European contact, when the indigenous population of Amazonia was perhaps six million people (Denevan 1976). Even today, most of the rural population continues to live in close proximity to rivers and subsists by shifting cultivation and extraction of forest products.

Until recently, deforestation seemed to be a remote problem in this region.[5] Available data indicated that both the absolute area and rate of deforestation were exceptionally low in Amazonia. For example, in 1980 Brazil's Institute of Space Research (INPE) published results obtained from satellite imagery of the Brazilian Amazon during 1978 (Tardin *et al* 1980). The study concluded that only 1.55 per cent of the region's forest cover had been removed, and that the rate of deforestation was a mere 0.33 per cent per year (Table 6.1). Although the extent and rate may have been underestimated (Fearnside 1982), this study reflected the fact that the impact of deforestation was, at that time, practically insignificant given the immensity of the region's rain forest.

Since 1980, however, the situation has changed dramatically. The 1980s have been a decade of ambitious, government-supported development projects in the Amazon. A network of highways begun in the 1960s now crisscrosses the region, hydroelectric dams are being planned or constructed in many of the principal river systems, and large scale extraction of minerals and fossil fuels is now underway.

In recent years the development of Amazonia has moved increasingly beyond the control of the public sector. Whereas in the past government incentives and infrastructure were a *sine qua non* for private sector involvement in regional development, today groups such as ranchers, farmers, miners, loggers, and charcoal producers operate independently. Up until a decade ago, conversion of rain forest to cattle pastures in the Brazilian Amazon required massive government incentives and costly infrastructure such as highway construction and maintenance. But today, ranchers near the town of Paragominas in the state of Pará are currently constructing their own highways to the neighboring state of Maranhão, for extraction of timber and establishment of new ranches – largely on public lands (Oren 1988). In the vicinity of the government-controlled Carajás mining project in southern Pará, private processing plants have begun producing pig iron using minerals from the mine and fuel from the native forest. By the 1990s, the area of forest that will have to be cleared to support this activity is estimated at between 900 and 2000 km^2 per year (CODEBAR/SUDAM 1986; Mahar 1988). Privately held gold mines are springing up throughout the region and polluting major river systems. An estimated 250 tons of mercury from gold mines have entered the Tapajós River basin in Pará, and government attempts at regulation have thus far failed.

Whether spontaneous or planned, colonization in the Amazon Basin is now largely under private initiative, and settlements are spreading along the southern flank of the region and wherever new roads are built. The influx of new settlers often overwhelms the capacity of frontier communities to absorb them.

The convergence of these factors is producing a marked increase in the rate and scale of deforestation in the Amazon. Recent estimates based on LANDSAT data indicate that deforestation has accelerated sharply in the 1980s.[6] As shown in Table 6.1, the total area deforested increased to almost 600,000 km^2 by the end of 1987. This area represents 12 per cent of Legal Amazonia[7] and is larger than France. Most deforestation is occurring in an arc along the western, southern, and eastern edges of the region, which coincides with current zones of frontier expansion (Figure 6.1). On a percentage basis, deforestation has been highest in the states of Rondônia and Mato Grosso,

Table 6.1 LANDSAT surveys of forest clearings in the Brazilian Amazon (after Mahar 1988)

		AREA CLEARED (km^2)				PER CENT OF STATE OR TERRITORY CLASSIFIED AS CLEARED			
State or Territory	*Area in Legal Amazonia (km^2)*	*Through 1975*	*Through 1978*	*Through 1980*	*Through 1988*	*Through 1975*	*Through 1978*	*Through 1980*	*Through 1988*
Amapá	140,276	152.5	170.5	183.7	571.5	0.1	0.1	0.1	0.4
Pará	1,248,042	8,654.0	22,445.3	33,913.8	120,000.0	0.7	0.8	2.7	9.6
Roraima	230,104	55.0	143.8	273.1	3,270.0	0.0	0.1	0.1	1.4
Maranhão	257,451	2,940.8	7,334.0	10,671.1	50,670.0	1.1	2.8	4.1	19.7
Tocantins (Goiás)	285,793	3,507.3	10,288.5	11,458.5	33,120.0	1.2	3.6	4.0	11.6
Acre	152,589	1,165.5	2,464.5	4,626.8	19,500.0	0.8	1.6	3.0	12.8
Rondônia	243,044	1,216.5	4,184.5	7,579.3	58,000.0	0.3	1.7	3.1	23.7
Mato Grosso	881,001	10,124.3	28,355.0	53,299.3	208,000.0	1.1	3.2	6.1	23.6
Amazonas	1,567,125	779.5	1,785.8	3,102.2	105,790.0	0.1	0.1	0.2	6.8
Legal Amazonia **(total)**	5,005,425	28,595.3	77,171.8	125,107.8	598,921.5	0.6	1.5	2.5	12.0

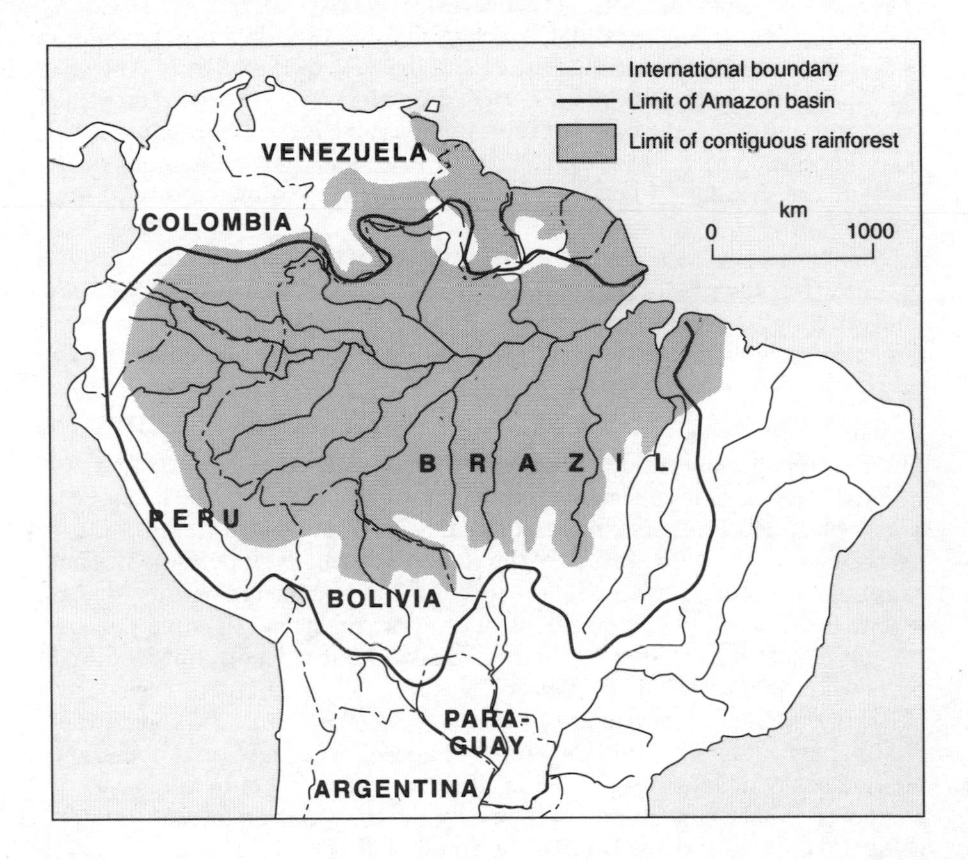

Figure 6.1 The Amazon Basin

followed by western Maranhão, Acre, and northern Goias.[8] On this basis, deforestation has been less in the more centrally located and larger states of Pará and Amazonas. But in absolute terms, the impact has been immense: in both states, over 100,000 km^2 have been cleared, mostly in their southerly portions (the Tocantins basin in Pará and the Madeira basin in Amazonas). The northern states of Roraima and Amapá,[9] as well as the northerly portions of Pará and Amazonas, have thus far suffered little deforestation in both relative and absolute terms.

The data in Table 6.1 indicate that not only the scale but the rate of deforestation is rapidly increasing. From 1976 through 1978, the average annual rate was 0.3 per cent of Legal Amazonia. From 1979 through 1980, the rate had increased to 0.5 per cent, and from 1981 through 1988, to nearly 1.2 per cent.[10] This trend suggests that the rate of deforestation in Amazonia is increasing exponentially, which could have ominous implications for the future of the region's remaining forests (cf Fearnside 1982).

Recent evidence suggests that deforestation in Amazonia occurred on an unprecedented scale during the particularly dry year of 1987. On 24 August of that year, INPE detected 6800 fires just in the states of Mato Grosso and a small portion of southern Pará and eastern Rondônia (Silveira and Nestlehner 1987). It was the largest scale burn recorded in the history of the region. Smoke from the Amazon fires lasted until December and resulted in the closure of most regional airports.

Once essentially local in scale, the destruction of the Amazon rain forest has become, for the first time, a regionwide phenomenon. And given the immensity of the region and its recent propensity for growth, deforestation in Amazonia is soon likely to ocur on a scale unprecedented in human history and exert impacts on a global level.

Causes

The underlying causes of deforestation are extremely complex and often defy rational analysis. One persistent myth is that tropical deforestation is caused by overpopulation. Excessive population densities may indeed contribute to deforestation in certain areas of Asia and Africa, but even here the ultimate cause often boils down to unequal land distribution. Excluding the Amazon forest, Brazil has a population density of approximately 23 people per km^2, roughly equal to the United States, the world's greatest exporter of food. If all potential farmland outside of Amazonia was equally distributed, each person in Brazil could have four hectares. Instead, 4.5 per cent of Brazil's landowners hold 81 per cent of the country's farmland, and 70 per cent of rural households are landless (Caufield 1984). With insufficient land, rural poor are constantly compelled to seek new frontiers where they can clear the rain forest for shifting cultivation. Often maligned as a cause of deforestation, shifting cultivators are more frequently victims of an unequal distribution of resources (Wood and Schmink 1978).

Government incentives, often for establishment of land uses that would otherwise be economically as well as ecologically inviable, have played an enormous role in promoting deforestation in the region (Maher 1988). For example, during the past three decades over ten million hectares of Amazon rain forest were converted to cattle pastures, largely through government policies. A recent calculation by Uhl and Parker (1986) reveals just how unproductive Amazon pastures are: one-half ton of rain forest is required to produce a single quarter-pound hamburger.[11] Moreover, on government-subsidized ranches, for every quarter-pound hamburger that cost an average of $0.26 to produce, Brazil expended $0.22 in the form of subsidies (Browder 1988).

Even given current reductions of government incentives in debt strapped Amazonian countries, combined with the economic inviability of this form of land use, cattle pastures continue to spread in frontier areas throughout the region. This apparently irrational activity seems to be motivated by (1) the value of land as a speculative investment in inflationary economies (such as those of Brazil, Peru, and, until recently, Bolivia) which often outweighs returns from the actual *use* of the land (Hecht 1985; Hecht, Norgaard, and Possio 1988); and (2) the facility of establishing claim to large areas of land once they have been converted to pasture, which is especially important in a region of chronic land conflicts (Branford and Glock 1985). The long term investment required by most potentially sustainable forms of land use is unattractive when inflation is high and tenure insecure. As a result, most regional investment is geared towards shorter-term profits that can be generated either by simply clearing the forest or selectively extracting its components. In the current social and economic environment, the Amazon rain forest is usually treated as an obstacle to development rather than as a foundation for sustainable forms of land use.

The irrationality of Amazon development is not only socioeconomic but political in nature. Development policies in the region are frequently incoherent because they are determined by a plethora of conflicting government agencies (Mahar 1979). Long-range planning is rarely a component of development projects in Amazonia. A classic example is the implantation of the Transamazon Highway and its associated colonization project in the Brazilian Amazon during the 1970s. This project was initially conceived to settle 70,000 families between 1972 and 1974, but largely due to lack of planning and inadequate infrastructure, a mere 5700 families were effectively settled by the end of 1974 (Maher 1979; Moran 1981; Smith 1982). The opposite occurred during the more recent colonization of the Brazilian state of Rondônia during the late 1970s and 1980s. Here plans were carefully made to integrate construction of a new highway with a World Bank-supported settlement programme designed to receive small scale producers from other regions of Brazil. But massive displacement of small producers due to lack of land reform in other parts of the country, relatively fertile soils along extensive portions of the highway in Rondônia, and an enormously effective government campaign to attract settlers resulted in explosive demographic growth – at an annual rate of 28 per cent during most of the 1980s, Rondônia's growth has been over ten times the national average and well beyond the capacity of the state's meagre infrastructure.

In short, deforestation in Amazonia results, to a large degree, from socioeconomic and political processes that originate outside of the region. These processes are frequently complex and elude simple analysis, as well as simple solutions.

ALTERNATIVES TO DEFORESTATION IN AMAZONIA

Given the complex situation described, it would appear that the complete destruction of the Amazon rain forest is only a matter of time. Yet this text is based on the firm conviction that the current scenario *is* reversible, and that viable alternatives to deforestation – both within Amazonia and elsewhere in the moist tropics – do exist.

The myriad problems associated with tropical deforestation have been widely trumpeted in the popular and scientific media, while potential solutions have gone relatively unnoticed. One reason for this discrepancy is that the problems are far more dramatic – and, hence, more newsworthy – than the solutions, many of which have been quietly occurring in rural communities or in the minds of a small contingent of researchers. Another reason is that the people involved in the search for, or implementation of, alternatives to tropical deforestation represent a heterogeneous group comprised of ecologists, foresters, agronomists, anthropologists, geographers, extension agents, policymakers, and – most significantly – the actual inhabitants of tropical landscapes. Such people rarely have an opportunity to communicate among themselves, much less in concert to a wider audience. The few attempts to promote such communication to date (eg Gradwohl and Greenberg 1988), although important for promoting public awareness, have invariably occurred in temperate countries, far removed from the landscapes and inhabitants that have the ultimate stake in tropical deforestation and its alternatives.

The book from which this chapter is taken, *Alternatives to deforestation – steps towards sustainable use of the Amazon rainforest*, presents a selection of papers presented at an international conference held in Belém, Brazil, from 27–30 January 1988. The location of this conference in the Amazon region of Brazil was appropriate for three reasons. First, the region contains the world's largest expanse of intact tropical forests. Second, current development policies and programmes in Brazil are aimed at removing these forests on an unprecedented scale. And third, the destruction of the Amazon rain forest has recently aroused the concern of the general public and policymakers, not only within Brazil but worldwide. For the first time, public officials at the highest levels have expressed resolve to change the destructive policies of the past and seek more sustainable alternatives. This book is about such alternatives.

Conservation is frequently cited as one alternative to tropical deforestation. But *Alternatives to deforestation* is not a book about conservation as most people in developed countries understand the term – that is, protection of natural ecosystems from human intervention. Rather, this book is about how tropical forests can be used sustainably – which, in the final analysis, must be a part of any long-term effort to promote conservation. In Amazonia, sustainable forms of land use are currently limited to a small minority of the rural population under highly specific conditions. Developing such practices for the population as a whole, including small farmers and ranchers as well as private and public companies, has only just begun.

The subtitle of this volume expresses its underlying theme: *steps toward sustainable use of the Amazon rain forest*. These steps consist of innovative technologies or approaches that permit simultaneous use and conservation of the rain forest, and that have the potential to provide a better way of life for people than do the land uses that currently predominate in the region. Although some of these steps may be promising, many are merely tentative and most seem insignificant when compared with the scale and velocity of deforestation in the region. Thus, although this text offers hopeful alternatives to the spectre of tropical deforestation, one should not forget that these alternatives are currently remote.

Natural and human disturbances

Developing viable land use alternatives to deforestation in Amazonia requires an understanding of how the Amazon rain forest responds to disturbances. A popular misconception is that, prior to recent deforestation, Amazonia was undisturbed. Yet an increasing array of evidence suggests that rain forests in Amazonia and elsewhere in the tropics have suffered repeated contractions and expansions in the geologic past (Prance 1985). In addition, natural disturbances on both a small and large scale have been common components of the Amazon rain forest throughout its history. Clearing of small patches of forest for agriculture has been a widespread practice since the arrival of humans approximately 12,000 years before present, and it continues to be widely practised today. It is only in the past three decades that large-scale, prolonged deforestation has assumed major importance in Amazonia. Yet, albeit vastly greater in scale and duration, today's deforestation is part of a continuum of disturbance that has long been a part of the Amazonian landscape.

Studying how the rain forest responds to different kinds of disturbances offers insights for the design of more sustainable forms of land use. In *Alternatives to deforestation*, ecologist Chris Uhl and co-authors examine a spectrum of disturbance types (ranging from small scale, natural disturbances to large scale, anthropogenic disturbances), and use their findings to suggest ecologically sound land use alternatives. For example, their studies of small scale disturbances indicate that land use systems such as agroforestry, which incorporate many of the structural and functional aspects of natural forest ecosystems, can protect fragile soils and and provide steady yields. Likewise, their research on abandoned, degraded cattle pastures is helping to develop appropriate technologies for restoring the native forest cover on these derelict lands. Overall, they have found that management practices that mimic natural disturbances in size, duration, and frequency can protect the functional integrity of forest ecosystems while providing modest economic returns. Their paper provides a theoretical foundation for many of the land use alternatives discussed in subsequent essays.

Natural forest management

Natural forest management has been maligned due to its low yields, slow economic returns, difficulty of implementation, and vulnerability to disruptive land uses such as shifting cultivation (Spears 1984). From this perspective, silvicultural plantations are seen as the only viable source of wood products in the tropics. Yet factors such as changing market demands and pest outbreaks make plantations a risky alternative, and they cannot replace all the ecological functions of natural forest. In short, plantations and natural forest management are not competitive but complementary forms of land use that provide different types of products and are suited to different terrains (Schmidt 1987).

The search for successful forms of tropical forest management should begin in traditional communities where such land uses have been carried out for centuries or even millennia. Indigenous peoples throughout the tropics frequently manage natural forests for a wide variety of so-called minor forest products, including fibres, fuelwood, fruits, latex, resins, gums, medicines, and game. Some of the best research on tropical forest management by indigenous peoples is being conducted in Mexico. Ecologist Arturo Gomez-Pompa and anthropologist Andrea Kaus describe the major indigenous forms of forest management in this country. They show that many so-called natural forests have in fact been managed by aboriginal groups in the past, and that this past management accounts, at least in part, for the current distribution of many economic tree species. Among present-day traditional farmers, agriculture and forest management are part of a spatial and temporal continuum, as evident in such practices as the maintenance of kitchen gardens and the enrichment of forest fallows. In contrast to the patterns of land occupation and degradation predominant today, these practices have supported high population densities over long periods of time. They thus provide important clues in the search for sustainable land uses in the humid tropics.

The forms of land use utilized by indigenous peoples often require a profound knowledge of highly complex ecosystems. Yet many forested ecosystems in Amazonia and elsewhere in the moist tropics are dominated by one or a few economic species (Anderson 1987); such ecosystems can be managed with relative ease. In *Alternatives to deforestation* paper 4 describes one such management system utilized by nonindigenous river dwellers in floodplain forests of the Amazon estuary. The simplicity of this management, its low input requirements, significant economic returns, and general acceptance by the rural population make it an attractive land use model in flood-plain forests of the Amazon.

In the following paper, ecologist Chuck Peters examines the viability of harvesting native fruits in a similar ecosystem in the Peruvian Amazon. Peters demonstrates that natural populations can produce astonishingly high fruit yields that can be harvested without undermining the regenerative capacity of these populations. His paper provides a scientific basis for sustainable (as well as potentially profitable) utilization of Amazonian forests.

The technical success of natural forest management ultimately depends on the regeneration of desirable species. In paper 6, forester Virgilio Viana examines the ecological factors that govern the availability of seeds and seedlings, using a case study of a promising timber species in Central Amazonia. Viana illustrates how theoretical principles derived from ecology can have practical application for the management of tropical forests.

Although its potential is high, implementing forest management on a commercial scale has been an elusive goal in the Amazon Basin (Rankin 1985). Part of the reason for this is the lack of long-term research efforts aimed at testing management alternatives. The longest continuous research on sustained timber production in Amazonian forests has been carried out by Dutch foresters in Surinam. N R de Graaf and R L H Poels describe a management system based on selective harvesting of high-quality timbers and periodic, low-level silvicultural treatments over a 20-year cycle. Its minimal ecological impacts and relatively high economic returns make this system especially attractive as an extensive land use alternative in Amazonia.

Implementation of natural forest management in the moist tropics is often undermined due to a lack of support and involvement of local residents. Ecologist Gary Hartshorn closes the section on natural forest management by returning to a rural community in the Peruvian Amazon, where Yanesha Indians have established a forestry cooperative. He describes the cooperative's forest management system, which appears to be ecologically sustainable as well as economically profitable. Hartshorn's paper demonstrates how programmes for managing tropical forests can be designed so that local inhabitants can make a living from renewable natural resources, which gives them a stake in making these programmes work.

Agroforestry

Agroforestry has recently gained considerable attention as a promising alternative to tropical deforestation that is both potentially sustainable and highly adaptable to the needs of smallholders (eg Weaver 1979; Hecht 1982). Yet this form of land use often eludes precise definition. At one extreme, which can be referred to as extensive agroforestry systems, the degree of management intervention is low, and the conservation of native forest structure and function is high. Agroforestry systems of this type are extremely common in the Amazon Basin and throughout the American tropics. In paper 9, ethnobotanist Janis Alcorn compares such systems between indigenous groups in Mexico and the Peruvian Amazon. The land uses of both groups are characterized by the inclusion of native trees and natural successional processes, the multiple use of a given species, and the spreading of risk through maintenance of high biotic diversity.

Extensive forms of agroforestry such as those described generally occur in regions characterized by low land use pressures and loosely defined tenure relationships. In contrast, 'intensive' agroforestry systems, which substitute native forest vegetation with plantations that must be maintained against weed and pest invasions, are associated with relatively high land use pressures and more rigidly defined tenure relationships. Although intensive forms of agroforestry are exceptionally common in Asia and Africa, they are relatively rare in tropical America (Nair 1985) and extremely so in traditional communities of Amazonia.

It is probably no accident that the classic example of intensive agroforestry systems in Amazonia is practised by immigrants to the region. In paper 10, ecologists Scott Subler and Chris Uhl describe the Japanese agroforestry systems in Tomé-Açu, a small farming community in eastern Amazonia. These systems are characterized by intensive cultivation of relatively small areas and require high inputs of capital, labour, and materials. The emphasis is on high-value cash crops, and these systems are both profitable and apparently sustainable. The intensity of land use characteristic of Japanese farms is part of a complex legacy that will be difficult for local farmers to adopt. In fact, the few cases in which Japanese farming techniques have been adopted involve relatively wealthy Brazilians from other regions who are, in a sense, immigrants themselves.

Although agroforestry systems are an integral part of indigenous and Japanese land-use strategies, they are surprisingly rare among newly arrived settlers in frontier zones of Amazonia. This rarity is largely due to the nature of frontier expansion in the region, where large-scale land uses such as cattle pastures frequently encroach on small-scale production systems. Faced with uncertain land tenure, distant markets, and lack of technical support, the small holder is frequently compelled to practise degenerate forms of shifting cultivation for subsistence.

Promoting sustainable land uses such as agroforestry in frontier zones requires effective extension services aimed at small-scale producers. Agronomist Robert Peck describes a project in Amazonian Ecuador that addresses many of the problems characteristic of extension through the use of on-farm demonstrations. These demonstrations effectively promote agroforestry practices in isolated areas, encourage participation and experimentation by resident farmers, and can be periodically modified to suit local needs. As a result, agroforestry systems have become an integral part of the local landscape. Through practical extension efforts such as these, settlers in frontier zones of Amazonia can be induced to adopt more sustainable forms of land use.

Landscape recovery

Despite efforts to promote agroforestry, degraded landscapes are an increasingly familiar sight in frontier zones throughout Amazonia. Degradation here refers to long term modifications of natural ecosystems that undermine their capacity to recover through natural successional processes. Nature, especially in the moist tropics, abhors a vacuum, and natural forest ecosystems in this zone exhibit an extraordinary (and frequently underestimated) capacity to recover from human-induced disturbances. Yet this capacity can be seriously or even permanently undermined in areas where such disturbances are excessively widespread, prolonged, or intensive. And it is precisely in such areas that settlement in Amazonia tends to be concentrated.

In paper 12 of *Alternatives to deforestation*, agronomist Jean Dubois reviews case studies showing how rural inhabitants in the South American tropics are actively engaged in restoring forest cover on sites previously cleared for agriculture or other activities. Although secondary forests are frequently indicators of land abandonment following shifting cultivation, Dubois shows that they are in fact frequently utilized and managed. Management techniques include selective weeding and enrichment planting, which result in high-density stands of economically important forest resources on sites that were previously marginal. Dubois illustrates the significant role of rural inhabitants in the process of landscape recovery and suggests that intensified use of secondary forests in already settled areas can reduce land use pressures at the frontier.

Peck demonstrates that, even in highly altered ecosystems, people can continue to utilize forest resources and manipulate natural processes on a sustainable basis. On sites where these processes have been seriously impaired, however, land restoration requires other approaches.

With a current coverage of over 100,000 km^2, abandoned pastures constitute the most extensive form of land degradation in Amazonia. Two alternatives exist for recovering these lands: restoration of the pasture or regeneration of the natural forest. In paper 13 of *Alternatives to deforestation*, agronomists Adilson Serrão and José Toledo evaluate prospects for the first alternative. Restoring the productivity of pastures that are already severely degraded requires energy- and capital-intensive measures such as mechanized soil turnover and application of fertilizers and pesticides. They conclude that the most rational strategy is to reduce degradation and extend the useful life of still productive pastures through introduction of better adapted forage germplasm, together with appropriate techniques for pasture establishment and grazing management.

If the land use goal on highly degraded pastures is to re-establish the natural forest cover, ecologist Dan Nepstad and co-authors recommend a high information, low energy approach in paper 14. Their research indicates that an economic way to accelerate natural forest regeneration is through dissemination of seeds from selected, large-seeded, drought-tolerant tree species. Rather than blanketing the landscape, reforestation should focus on 'high resource' tree islands, from which regeneration can naturally spread. By investigating the barriers to forest regeneration on highly degraded sites, this last paper illustrates how Amazonian ecoystems recover following severe disturbance. Comparison with similar research on landscape restoration in Costa Rica (Janzen 1986) provides interesting insights concerning forest regeneration processes in moist and dry tropical zones (see also Ewel 1977). A key underlying difference involves the microclimatic gradient between forested and nonforested sites. In the moist tropics, natural forests are relatively closed, which means that their microclimate is considerably less harsh than on nonforested sites; in contrast, the relative openness of natural forests in the dry tropics makes this difference less pronounced. As a result, plants that grow in dry tropical forests can often grow on nonforested sites as well. The propagules of many tree species are dispersed by wind, which is typically high in the dry tropics and continues to blow even when other dispersal agents, such as mammals and birds, have been reduced by hunting or habitat destruction. In contrast, the steep environmental gradient between forested and nonforested sites may be beyond the physiological tolerance of many forest trees in the moist tropics. As Nepstad and colleagues show, forest species that do exhibit such tolerance generally possess large propagules, which contain sufficient metabolic reserves to permit rapid development of deep root systems; the latter, in turn, permits tapping of deep soil moisture reserves. Such species, however, rarely arrive on nonforested sites, as dispersal of large propagules from forested sites appears to be minimal. Wind is generally less pronounced in the moist tropics, and few of the dispersal agents that remain surmount the steep environmental gradient surrounding forest remnants.

As a result, reforestation of highly degraded sites is a far more difficult task in the moist tropics than in the dry tropics. Technical research in this field must focus on species that can tolerate the harsh conditions of deforested sites. Establishment of such species can help restore essential ecological functions and may eventually serve to promote subsequent establishment of such species can help restore essential ecological functions and may eventually serve to promote subsequent establishment of a wider range of native species.

Although complex, the technical issues involved in landscape recovery in the moist tropics are probably resolvable. The success or failure of landscape recovery, as well as of other alternatives to deforestation described in this text, ultimately requires the participation of the people who inhabit Amazonia. And in a region of constantly beckoning frontiers, a key challenge is to engage rural inhabitants in restoring the land in areas where the frontier has already passed.

Implications for regional development

The papers summarized in the preceding pages indicate that a wide variety of alternatives to tropical deforestation exist and are currently practised by rural inhabitants. In Amazonia, these alternatives are generally at an early stage of development. All of the alternatives described – natural forest management, agroforestry systems, and landscape recovery – appear to be technically feasible, but data are generally lacking on their ecological and economic sustainability. Because these alternatives promote maintenance or restoration of a diversified forest cover, one can safely assume that they are more ecologically sustainable than land uses currently in vogue; the long term practice of many such alternatives by rural inhabitants in Amazonia and elsewhere in the tropics lends credence to this assumption. A more crucial gap in our knowledge is the current lack of data on the economic costs and benefits of these alternatives. The few economic data that are presented (eg papers 3, 4, 7 and 8) suggest, however, that they are highly competitive.

Although the alternatives presented in *Alternatives to Deforestation* (from which this chapter is taken) would thus appear to be promising, one must not forget that they comprise only a small fraction of the land uses that currently predominate in Amazonia. Ecologist Phil Fearnside puts these alternatives in perspective by analyzing land uses practised on a large scale, such as logging, shifting cultivation, and cattle ranching. In frontier zones of Amazonia, none of these land uses utilize native forest ecosystems in a sustainable way. Even when carried out selectively, logging causes severe damage and leaves forests highly susceptible to subsequent destruction by fire (Uhl and Buschbacher 1985). Other major land uses require wholesale removal of the rain forest.

Although shifting cultivation incorporates forest regeneration during the fallow period, this period is often truncated as populations increase and/or available land decreases (Sioli 1973; Hecht, Anderson, and May 1988). With a useful life that typically ranges from four to ten years before abandonment, Amazonian pastures have been aptly characterized as a prolonged form of shifting cultivation (Buschbacher 1986). But when pastures are abandoned after intensive use, a scrubby, fire-resistant vegetation often takes over, and forest regeneration is brought to a standstill (Uhl, Buschbacher, and Serrão 1988). The ecological and economic unsustainability of these land uses may ultimately lead to increased acceptance of more rational alternatives, provided that current policies that promote landscape degradation are changed.

As emphasized earlier in this essay, development policies in Amazonia have largely been determined outside of the region and, until recently, were rarely questioned by the people who stood most to lose by their implementation. But whereas development was frequently viewed as a panacea in the past, today it is increasingly being questioned by the regional press, politicians, and popular groups in both urban and rural areas of Amazonia. One of the most significant results of such questioning has been the birth of a social movement comprised of rubber tappers engaged in non-violent resistance of deforestation in the Brazilian state of Acre. This movement, described by anthropologist Mary Allegretti, is now supported by a broad coalition of indigenous and environmental groups, as well as by policy-makers and government officials in and out of Brazil. The rubber tappers have called for the establishment of so-called 'extractive reserves', comprised of public lands designated for the specific purpose of exploiting forest products such as rubber and Brazil nuts on a sustainable basis. By late 1988, three such reserves had been established in Acre, and others were being planned elsewhere in the Brazilian Amazon. In effect, establishment of extractive reserves represents an important step toward implementing simultaneous use and conservation of the Amazon rain forest.

As the rubber tappers' movement clearly illustrates, development of the Amazon need not imply its destruction. In the final paper of *Alternatives to deforestation*, sociologist Donald Sawyer argues that the current environmental destruction taking place in the region is not an inevitable consequence of development, but rather the result of highly artificial – as well as socially and economically questionable – policies that can be reversed. Sawyer examines policy initiatives that could radically change current scenarios of land use in Amazonia, such as (1) cutting economic incentives to deforestation within the region; (2) relieving pressures for settlement by strengthening land, health, and welfare reform programmes outside of Amazonia; (3) consolidating existing settlement by improving infrastructure in already occupied areas of Amazonia; and (4) promoting ecologically and economically sustainable forms of land use. If adopted in concert, such policies could contribute substantially toward reversing current deforestation trends in Amazonia.

Implementing alternatives to tropical deforestation is a process that has just begun. *Alternatives to Deforestation: steps toward sustainable use of the Amazon rain forest* illustrates how – with technical knowledge, political will, and sensitivity to the needs of local inhabitants – the Amazon region can be developed in ways that assure its preservation.

This chapter was taken from Anderson, A (1990) *Alternatives to Deforestation*, Columbia University Press.

ACKNOWLEDGEMENTS

Many of the ideas in this paper were conceived while preparing a report for the Ford Foundation on forest management issues in the Brazilian Amazon. I thank Steve Sanderson and Peter May of the Ford Foundation for their encouragement, and Chris Uhl and Marianne Schmink for their critical review of the manuscript.

NOTES

1. 'Dry' and 'moist' are used here to refer to different climatic zones and their associated forest vegetation in the tropics. The 'dry' tropics refer to regions with a pronounced dry season (ie monthly precipitation under 100 mm for at least four months); the native forests in these regions are distinctly deciduous. In contrast, rainfall in the 'moist' tropics is less seasonal and native forests are generally evergreen. Of the world's currently existing 9 million km^2 of tropical forests, approximately one-third are of the former type and two-thirds of the latter (Postel and Heise 1988).
2. Closed forest has a cover of trees sufficiently dense to prevent the development of a grass understory. By contrast, open forest has at least 10 per cent tree cover but enough light penetration to permit the growth of grass under the trees (WRI 1986).
3. The overall rate of deforestation in both the moist and dry tropics has been estimated at approximately 110,000 km^2 per year (cf WRI 1986). According to Myers (1986), the total rate of tropical forest destruction is closer to 200,000 km^2 per year, if one includes forest depletion through selective harvesting of wood by commercial loggers and fuelwood gatherers.
4. This rate is expressed as percentage of total forest area in existence in 1980.
5. The following discussion on deforestation in Amazonia focuses on Brazil, which encompasses 80 per cent of the region and where deforestation rates appear to be highest.
6. As *Alternatives to deforestation* was in press, Brazil's Institute of Space Research (INPE) published an updated report on forest clearing in Amazonia (INPE 1989) that differs considerably from the World Bank estimates (Maher 1988) reproduced in Table 6.1. According to the INPE report, total forest clearing for the region through 1988 was approximately 250,000 km^2, or a mere 5.1 per cent of the total area of Legal Amazonia. This report, however, has been received with considerable scepticism within Brazil (Antônio 1989). For example, prominent members of the national scientific community charge that the figures were underestimated to still international criticism of Brazil's environmental policies in the Amazon. Likewise, technicians at INPE complain that the report was prepared too hastily (less than one month), and that the scientific staff was excluded from the final data analysis and synthesis. Finally, 'old' clearings made before 1970 were excluded from the estimates, including approximately 87,000 km^2 in the states of Pará and Maranhão alone (INPE 1989; Marcos Pereira, personal communication). Due to the questionable manner in which this report was carried out, I have decided to exclude it from current analysis of the scale of forest clearing in Amazonia, which is based exclusively on the World Bank figures.
7. 'Legal Amazonia' is a concept used in Brazil to define seven states (Amazonas, Pará, Mato Grosso, Rondônia, Roraima, Acre, and Amapá) and parts of two others (Goiás and Maranhão). This definition, which is used for regional planning and political purposes, includes extensive areas outside of the Amazon Basin as well as over one million km^2 of nonforest savanna and scrub vegetation.

8. In 1988, northern Goiás was transformed into the state of Tocantins.
9. In 1988, the territories of Roraima and Amapá became states.
10. An annual deforestation rate of 1.2 per cent for Amazonia translates into an area of almost 60,000 km^2 per year, or nearly 80 per cent of the recent estimate of 75,000 km^2 annual rain forest removal for the whole world during 1981–1985 (cf WRI 1986). The latest data from Amazonia suggest that this worldwide estimate should be substantially increased.
11. Although striking, the hamburger analogy is not especially appropriate in this case, since Brazil does not export beef to the United States (Browder 1988).

REFERENCES

Anderson, A B (1987) Management of native palm forests: A comparison of case studies from Indonesia and Brazil. In H L Gholz, ed, *Agroforestry: Realities, Possibilities and Potentials*, pp 155–167. Dordrecht, The Netherlands: Martinus Nijhoff.

Antônio, I (1989) Sarney defende os dados maquiados sobre Amazônia. *Folha de São Paulo*. p C-3, 8 May 1989.

Barbira-Scazzocchio, F (1980) From native forest to private property: The development of Amazonia for whom? In F Barbira-Scazzocchio, ed, *Land, People, and Planning in Contemporary Amazonia*, pp iii–xvi. Cambridge, England: Centre of Latin American Studies Occasional Publication No 3.

Branford, S and Glock, O (1985) *The Last Frontier: Fighting Over Land in the Amazon*. London: Zed Books.

Browder, J O (1988) The social costs of rain forest destruction: A critique and economic analysis of the 'hamburger debate'. *Interciencia* 13(3):115–120.

Brown, Jr K S (1982) Paleoecology and regional patterns of evolution in Neotropical forest butterflies. In G T Prance, ed, *Biological Diversification in the Tropics*, pp 255–308. New York: Columbia University Press.

Brown, S and Lugo, A E (1982) The storage and production of organic matter in tropical forests and their role in the global carbon cycle. *Biotropica* 14(3):161–187.

Buschbacher, R J (1986) Tropical deforestation and pasture development. *Bioscience* 36(1):22–28.

Caufield, C (1984) *In the Rainforest: Report from a Strange, Beautiful, Imperiled World*. Chicago: University of Chicago Press.

CODEBAR/SUDAM (1986) *Problemática do Carvão Vegetal na Area do Programa Grande Carajds*. Belém, Brazil: Companhia de Desenvolvimento de Barcarena (CODEBAR) and Superintendência do Desenvolvimento da Amazônia (SUDAM).

Davis, S (1977) *Victims of the Miracle*. Cambridge, England: Cambridge University Press.

Denevan, W M (1976) The aboriginal population of Amazonia. In W M Denevan, ed, *The Native Population of the Americas in 1492*, pp 205–233. Madison: University of Wisconsin Press.

Ducke, A and Black, G A (1954) Notas sobre a fitogeografia da Amazônia Brasileira. *Boletim Técnico do Instituto Agronômico do Norte* (IAN, Belém) 4:1–40.

Ewel, J J (1977) Differences between wet and dry successional tropical ecosystems. *Geo-Eco-Trop* 1(2):103–111.

Fearnside, P M (1982) Deforestation in the Brazilian Amazon: How fast is it occurring? *Interciencia* 7(2):82–88.

Fearnside, P M (1985) Environmental change and deforestation in the Brazilian Amazon. In J Hemming, ed, *Change in the Amazon Basin: Man's Impact on the Forest and Rivers*, pp 70–89. Manchester, England: Manchester University Press.

Gomez-Pompa, A, Vazquez-Yanes, C and Guevara, S (1972) The tropical rainforest: A non-renewable resource. *Science* 117:762–765.

Goodland, R J and Irwin, H S (1975) *Amazon Jungle: Green Hell to Red Desert?* New York: Elsevier.

Goulding, M (1980) *The Fishes and the Forest: Explorations in Amazonian Natural History*. Berkeley: University of California Press.

Gradwohl, J and Greenberg, R (1988) *Saving Tropical Forests*. London: Earthscan Publications Ltd.

Haffer, J (1969) Speciation in Amazonian forest birds. *Science* 165:131–137.

Hecht, S B (1982) Agroforestry in the Amazon Basin: Practice, theory and limits of a promising land use. In Hecht, S B, ed, *Amazonia: Agriculture and Land Use Research*; pp 331–371. Cali, Colombia: Centro Internacional de Agricultura Tropical (CIAT).

Hecht, S B (1985) Environment, development and politics: Capital accumulation and the livestock sector in eastern Amazonia. *World Development* 13(6):663–684.

Hecht, S B, Anderson, A B and May, P (1988) The subsidy from nature: Shifting cultivation, successional palm forests, and rural development. *Human Organization* 47(1):25–35.

Hecht, S B, Norgaard, R B and Possio, G (1988) The economics of cattle ranching in eastern Amazonia. *Interciencia* 13(5):233–240.

Hemming, J (1978) *Red Gold*. London: Macmillan.

INPE (1989) Avaliação da Alteração da Cobertura Florestal na Amazônia Legal Utilizando Sensoriamento Remoto Orbital. São José dos Campos, Brazil: Instituto Nacional de Pesquisas Espaciais.

Janzen, D H (1986) *Guanacaste National Park: Tropical Ecological and Cultural Restoration*. San Jose, Costa Rica: Editorial Universidad Estatal a Distancia.

Mahar, D (1979) *Frontier Development Policy in Brazil: A Study of Amazonia*. New York: Praeger Publishers.

Mahar, D (1988) *Government Policies and Deforestation in Brazil's Amazon Region*. Washington DC: World Bank Environment Department Working Paper No 7.

Meggers, B J (1985) Aboriginal adaptation to Amazonia. In Prance, G T and Lovejoy, T E, eds, *Key Environments: Amazonia*, pp 307–327. New York: Pergamon Press.

Moran, E F (1981) *Developing the Amazon*. Bloomington: Indiana University Press.

Myers, N (1979) *The Sinking Ark*. New York: Pergamon Press.

Myers, N (1980) *Conversion of Tropical Moist Forests*. Washington, DC: National Academy of Sciences.

Myers, N (1984) *The Primary Source: Tropical Forests and Our Future*. New York: Norton.

Myers, N (1986) Tropical forests: Patterns of depletion. In Prance, G T, ed, *Tropical Rain Forests and the World's Atmosphere*, pp 9–22. Boulder, CO: Westview Press.

Nair, P K R (1985) Classification of agroforestry systems. *Agroforestry Systems* 3:97–128.

Oren, D C (1988) Uma reserva biologica para o Maranhão. *Ciência Hoje* 8(44):36–45.

Pires, J M and Prance, G T (1985) The vegetation types of the Brazilian Amazon. In Prance, G T and Lovejoy, T E, eds, *Key Environments: Amazonia*, pp 109–145. New York: Pergamon Press.

Posey, D A (1983) Indigenous knowledge and development: An ideological bridge to the future. *Ciencia e Cultura* 35(7):877–894.

Postel, S and Heise, L (1988) Reforesting the earth. In Brown, L R, ed, *State of the World 1988: A Worldwatch Report on Progress Toward a Sustainable Society*. pp 83–100. New York: Norton.

Prance, G T (1985) The changing forests. In Prance, G T and Lovejoy, T E, eds, *Key Environments: Amazonia*. pp 146–165. New York: Pergamon Press.

Rankin, J M (1985) Forestry in the Brazilian Amazon. In Prance, G T and Lovejoy, T E, eds, *Key Environments: Amazonia,* pp 360–392. New York: Pergamon Press.

Salati, E (1985) The climatology and hydrology of Amazonia. In Prance, G T and Lovejoy, T E, eds, *Key Environments: Amazonia*, pp 18–48. New York: Pergamon Press.

Schmidt, R C (1987) Tropical rain forest management. *Unasylvia* 156(39):2–17.

Setzer, A W, Pereira, M C, Pereira Jr, A C and Almeida, S A O (1988) *Relatório de Atividades do Projeto IBDF-INPE Ano 1987*. São José dos Campos, Brazil: Instituto Nacional de Pesquisas Espaciais (INPE) Report Number INPE-4534-RPE/565.

Silveira, E A de and Nestlehner, W (1987) Cerco de labaredas. *Isto E*, 9 September 1987, pp 30–33.

Sioli, H (1973) Recent human activities in the Brazilian Amazon region and their ecological effects. In Meggers, B J, Ayensu, E S and Duckworth, W D, eds, *Tropical Forest Ecosystems in Africa and South America: A Comparative Review*, pp 321–324. Washington, DC: Smithsonian Institution Press.

Sioli, H (1984) The Amazon and its main affluents: Hydrology, morphology of the river courses, and river types. In Sioli, H, ed, *The Amazon: Limnology and Landscape Ecology of a Mighty Tropical River and its Basin*, pp 127–165. Dordrecht, The Netherlands: Dr W Junk.

Smith, N J H (1982) *Rainforest Corridors: The Transamazon Colonization Scheme*. Berkeley: University of California Press.

Spears, J S (1984) Role of forestation as a sustainable land use strategy option for tropical forest management and conservation and as a source of supply for developing country wood needs. In Wiersum, K F, ed, *Strategies and Designs for Afforestation, Reforestation and Tree Planting*, pp 29–47. Wageningen, The Netherlands: Pudoc.

Tardin, A T, Lee, D C L, Santos, R J R, de Assis, O R, dos Santos Barbosa, M P, de Lourdes Moreira, M, Pereira, M T, Silva, D and dos Santos Filho, C P (1980) *Subprojeto Desmatamento*. Sao José dos Campos, Brazil: Instituto Nacional de Pesquisas Espaciais (INPE) Report Number INPE-1649-RPE/103.

Uhl, C (1983) You can keep a good forest down. *Natural History* 4(83):71–79.

Uhl, C and Buschbacher, R (1985) A disturbing synergism between cattle ranching burning practices and selective tree harvesting in the Eastern Amazon. *Biotropica* 17(4):265–268.

Uhl, C, Buschbacher, R and Serrão, A (1988) Abandoned pastures in eastern Amazonia. I: Patterns of plant succession. *Journal of Ecology* 76:663–681.

Uhl, C and Parker, G (1986) Is a quarter-pound hamburger worth a half-ton of rain forest? *Interciencia* 11(5):210.

Weaver, P (1979) Agri-silviculture in tropical America. *Unasylva* 31(26):2–12.

Wood, C and Schmink, M (1978) Blaming the victim: Small farmer production in an Amazon colonization project. *Studies in Third World Societies* 7:77–93.

WRI (1985) *Tropical Forests: A Call for Action*. Washington, DC: World Resources Institute (WRI).

WRI (1986) *World Resources 1986*. New York: Basic Books.

PART III
INTERNATIONAL TRADE

Chapter 7

The International Tropical Timber Organization: Kill or Cure for the Rainforests?

Marcus Colchester[1]

> 'I do not like experts,' he said. 'They are our jailers, I despise experts more than anyone on earth.'
>
> 'You're one yourself, aren't you?' 'Therefore I know. Experts are addicts. They solve nothing! They are servants of whatever system hires them. They perpetuate it. When we are tortured, we shall be tortured by experts. When we are hanged, experts will hang us . . . When the world is destroyed, it will be destroyed not by its madmen but by the sanity of its experts and the superior ignorance of its bureaucrats . . .' (John le Carre, *The Russia House*)[2]

INTRODUCTION

Until recently two international initiatives – the Tropical Forestry Action Plan (TFAP) and the International Tropical Timber Organization (ITTO) – were regarded as the means to deal with vanishing rainforests. Reliance on these institutions to save the rainforests is evident in the latest resolution of the European Communities Council of Ministers for Development Cooperation.[3]

Since then, the Tropical Forestry Action Plan has come under heavy criticism to the extent that even the original proponents of the plan are admitting that it was conceptually flawed to start with and is failing to achieve its stated objectives.[4] Proposals to reform the TFAP process are now the subject of further controversy.[5]

Governments had been hoping that the TFAP would provide a coherent framework for dealing with the global crisis of tropical deforestation in an internationally coordinated manner. The more specific task of curbing the destructive impact of timber extraction in the tropics is seen as the task of the ITTO. Now, 13 years after the organization was first proposed, the prospects of it achieving this task seem increasingly remote.

THE IMPACT OF LOGGING

The world's tropical forests are disappearing faster than ever. Every succeeding study shows a startling acceleration in the process. According to Norman Myers in a study completed for Friends of the Earth last year, some 142,000 km^2 of forests were destroyed in 1989, and a further 200,000 km^2 seriously degraded.[6] The latest results from the United Nations Food and Agriculture Organization (FAO) survey estimates annual forest loss at 170,000 km^2,[7] while a study by the World Resources Institute puts the figure even higher, at some 204,000 km^2.[8]

Logging has played a significant role in this devastation. In 1985 the Tropical Forestry Action Plan task force estimated that some five million hectares of tropical forest are logged over every year. As Judith Gradwohl and Russell Greenberg note:

> The problems with management for timber products are two-fold. First, the logging operations themselves are often highly damaging, and this becomes more of a problem with increasing mechanization. Heavy machinery compacts the soil and, in most operations, the vegetation surrounding the trees being harvested is also damaged. Up to two-thirds of the non-marketable trees in some areas are damaged or destroyed when marketable trees are extracted. Ultimately this can destroy young individuals of economic tree species and preclude the regeneration of the forest. Logging roads and skidder trails further contribute to soil compaction and erosion. The second major problem lies in the decline in economic value of forested land after the most valuable trees have been removed.[9]

Yet the impact of logging has been much wider than this. The reason is that logging opens up previously isolated areas to colonists, land speculators and ranchers who move in and clear previously intact and inaccessible forests.

For example, Norman Myers has estimated that for every cubic metre of harvested timber, approximately a fifth of a hectare of forest is destroyed by farmers who press in close behind the logger. The Australian Wilderness Society cite an FAO estimate that 70 per cent of forest cleared by landless settlers is made possible by logging roads.[10] This is a phenomenon particularly familiar in Africa. In 1973, Cote d'Ivoire, for example, gained 33 per cent of its export returns from a timber industry that made the country Africa's largest and the world's fifth timber exporter, with an annual output of 3.5 million m^3 of timber. Today, however, according to Professor Hans Lamprecht of the University of Gottingen 'the forest industry of Cote d'Ivoire is on the verge of ruin – the forests have been plundered, and 70 per cent of the forest already destroyed.'[11]

Indeed such is the overall impact of logging that Robert Repetto of the World Resources Institute ranks commercial logging as the top agent of deforestation.[12] Globally, states Robert Goodland of the World Bank, 'settlement along logging roads and peasant agriculture may be the main causes of tropical moist deforestation.'[13]

The extent to which tropical forest logging should take the blame for forest loss is, of course, hotly disputed. On the one hand, many foresters see their role as a technical one of managing forests in splendid isolation from the prevailing realities outside the forests. For them, the invasion of lands, which have been set aside for logging as 'Permanent Forest Estate', is not their department. As Lamprecht puts it 'no silvicultural system can be given the blame for inadequate protection.'[14]

Logging as a major cause of forest loss also varies regionally and locally. In South East Asia and Africa logging is pre-eminent as a cause of forest loss, while in Latin America the timber industry has not yet penetrated too far into the Amazonian forests which are suffering first and foremost from settlement along roads opened up for colonization and to promote 'national integration' and 'development'.

The suicidal consequence of this reckless waste is the demise of the timber trade itself. Many countries are exhausting their environmental capital at such a rate that the timber industry faces total closure. According to the World Bank, of 33 countries which were net exporters of tropical timber in 1987, only ten will have any timber left to export by the year 2000.[15] At present rates, Sarawak, the world's number one exporter of tropical sawlogs, will have logged out its primary forests within 7 to 11 years.[16]

But the negative effects of logging go much further than a squandering of biodiversity and the exhaustion of timber supplies. Logging often has direct and shattering effects on local people whose lands are taken over for logging. Yet, through the tropics, foresters and loggers show scant concern for the local people. Indonesian Forest Minister Hasrul Harahap makes this clear in his remark 'In Indonesia the forests belong to the State not to the people'. The fact that these were their ancestral lands did not give them the right of ownership, he said. 'They have no right to compensation' when logging destroys the forests that they depend on.[17]

As Edward Goldsmith has not tired of pointing out, modern 'development' is founded not so much on the creation of wealth as on the transfer of the control of natural resources out of the hands of the poor and powerless.[18] Modern forestry, a classically colonial endeavour, provides one of the most explicit examples of this process.[19]

As Jack Westoby, former Director of Forestry at the United Nations Food and Agriculture Organization, summarized in 1975:

> International aid in forestry has done a useful job in identifying for foreign capital those forest resources suitable for exploitation. In many cases, it has borne a substantial part of the cost of making inventories of those resources. In not a few cases it has compiled the data, and helped provide the justification, for international funding agencies to provide loans to create some of the infrastructures needed to assist the penetration of foreign capital . . . De facto, though this was not its intent . . . it has assisted some irresponsible governments to alienate substantial parts of their forest resource endowment.
>
> Over the last two decades, massive tracts of virgin tropical forest have come under exploitation, in all three underdeveloped regions. That exploitation, with

> a few honourable exceptions, has been reckless, wasteful, even devastating. Nearly all the operations have been enclavistic, that is to say, they have had no profound or durable impact on the economic and social life of the countries where they have taken place . . . Local needs are not being met; the employment opportunities are trifling . . . A significant part of the exports, as logs and as primary processed timber, is exported 'within the firm', and transfer values are fixed to facilitate the accumulation of profits outside the country. . . . the contribution of forestry towards improving the lot of the common people has been negligible so far.[20]

Moreover the growth of these 'enclavistic' timber-based economies has severely damaged the evolution of the democratic institutions of Third World countries. In Sarawak, for example, the corrupting influence of the timber trade has promoted the domination of the economy by nepotistic, patronage politics. This has undermined democratic principles and caused an increasing marginalization of rural people, who find they can no longer rely on their political representatives to defend their interests.

An unseemly public squabble within the ruling political parties in Sarawak, during the 1987 elections, revealed how senior politicians have persistently rewarded their political allies, families and friends with logging concessions. The practice of dealing out logging licenses to members of the State legislature to secure their allegiance is so commonplace in Sarawak that it has created a whole class of instant millionaires.[21] The Commission of Enquiry in Papua New Guinea has revealed a similar decay in standards of public service due to the logging industry and, in fact, the process is very widespread having formed a crucial component in the 'crony capitalism' of the Philippines under Ferdinand Marcos.[22] In Indonesia logging concessions continue to be one of the perks enjoyed by the ruling military clique.

THE ORIGINS OF THE ITTO

When the Japanese originally tabled a resolution at the United Nations Conference on Trade and Development (UNCTAD) for the creation of an International Tropical Timber Organization in 1977, they had in mind a commodity agreement, of the kind adopted for jute and rubber, which would be strictly confined to trade considerations. However, in discussions, it soon became clear that tropical timber could not be treated in such a closely defined manner. Tropical timber, coming as it does from a wide variety of tree species growing over a vast area of the world's forest, cannot be dealt with as a single commodity.[23]

The protracted negotiations at UNCTAD soon took on a more complex character. The usual group of civil servants who deal with trade negotiations were joined by foresters brought in to elucidate the technical complexities of the industry. A crucial shift in the debate came about with the intervention of the UK-based policy research organization, the International Institute for Environment and Development (IIED), which forcefully argued that the agreement could not limit itself to the technical and commercial concerns of timber extraction and trade but must also provide for the other crucial ecological and genetic services provided by forests. This was an argument that the western countries, under increasing pressure from environmental organizations back home, could not afford to ignore.

The International Tropical Timber Agreement (ITTA), signed in November 1983 after six years of wrangling, thus emerged as a unique trade agreement. Its most significant articles not only set the ITTO the task of promoting the trade in tropical timber but also gave the ITTO the apparently contradictory duty of encouraging the sustainable use and conservation of tropical forests and their genetic resources, and maintaining ecological balance.[24]

As far as the national governments playing a part in the ITTO were concerned, establishing these terms in the ITTA was only a small part in laying the framework of the ITTO's work. Equally important was to work out who would control the organization. Much of the internal struggle during the six years between 1977 and 1983 focused on agreeing a voting system, the present structure of which reveals very clearly where the acting parties think the priorities lie.

According to the system established in the ITTA, votes are divided equally between 'producers' in the tropics and 'consumers' in the industrialized world. Consumer countries' votes are apportioned according to the degree they are involved in the tropical timber trade, giving Japan, as the world's largest importer of tropical timber, by far the largest vote of any single nation.[25] Producers similarly get votes according to the amount of timber they export, with only a secondary weighting given to the actual amount of forest that a country has. The net result is that the more a country destroys tropical forests, the more votes it gets. The voting structure ensures that the ITTO's primary role of promoting the timber trade heavily outweighs its secondary conservation role. As soon as the ITTA had been signed, and before it entered into force on 1 April 1985, a further two-and-a-half-year political battle immediately ensued. This time the objectives were to define who should gain the influential role of Executive Director of the ITTO and where the secretariat was to be seated. As the horse-trading and jockeying for power developed it was tacitly agreed that, to ensure balance, the producer and consumer blocs should share the roles: if one provided the seat for the secretariat, the other should provide the first Executive Director.

The front-runners soon emerged, with the United Kingdom, the Netherlands, Brazil, Indonesia and Japan leading in the race to provide the seat, and with the Malaysians and Dutch having the most likely candidates as Executive Director. The Dutch, in particular, took strong exception to the proposal to have Japan host the secretariat, apparently fearing that the ITTO's conservation goals would be subverted by the heavy, vested interests of the Japanese timber trade.

As the negotiations continued regional groupings emerged, the two blocs of North and South themselves divided, to a substantial extent, between East and West. Divided in this way no group was able to secure the required two-thirds majority of votes to allow a definitive solution. As the ITTC limped from one meeting to another, the prospect of the ITTO actually getting down to work became frustratingly remote.

The definitive change came when the Japanese Government offered several million dollars to fund the ITTO, as well as free office premises and secretarial functions, and, in addition, undertook to underwrite the costs of bi-annual Council meetings both in Yokohama and overseas. At the same time, Japan effectively secured the support of a number of producer countries, particularly in Latin America and Africa. The close and existing trade ties between Japan and the South East Asian nations provided them with more natural allies in that region. In exchange the Japanese agreed to promote the appointment of the Malaysian forester, Dr Freezailah bin Che Yeom, to the post of Executive Director.

As Charles Secrett, then Tropical Rainforest Campaigner of Friends of the Earth recalls 'the gossip was that they (the Japanese) were setting up preferential trading deals to secure Third World votes to support Japan's position'[26] The effective, if well-disguised, domination of the ITTO process by the Japan–South East Asia axis needs to be constantly borne in mind when the ITTO is being evaluated.

UNREAL EXPECTATIONS?

In 1986, after nine years of negotiations, the ITTO could finally get down to work. In his opening address to the International Tropical Timber Council (ITTC), Dr Freezailah likened his task to a canoeist swirling down a foaming river with a two-handled paddle, one end of which represented conservation and the other utilization. Only by reconciling the two objectives, he argued, could the long term future of the timber industry be assured, while reliance on any one end would inevitably lead to mishap.[27]

For their part, many non-Governmental organizations (NGOs) welcomed the ITTO, perceiving that it offered a real opportunity to curb the excesses of the logging industry. Some even hoped that a strict enactment of the ITTA would imply that the ITTO would assist in the establishing of national parks as a way of realizing its conservation objectives.

But the main reason that these NGOs have supported the ITTO was their belief in the concept of sustainable logging. Many organizations – such as Friends of the Earth (FOE), the International Union for the Conservation of Nature and Natural Resources (IUCN), and IIED – saw the achievement of this objective as a key means to saving the forests an believed that the ITTO provided a means of enforcing it.[28]

As Jeffery Sayer of the IUCN noted:

> The basic IUCN philosophy on tropical forest conservation, is, to put it in a nutshell, that you might reasonably expect to protect perhaps 10–15 per cent of all the moist forests of the tropics but, in the context of the population growth that one can expect and the development of economic activities in these regions, it is probably not realistic to expect to protect totally a very much larger area. This

> means that many tropical forest species and many of the ecological functions that come from forests will only be maintained in the future uses of land outside totally protected areas, which involve the retention of large areas of natural or near-natural forest. One of the best uses of forest that can in theory retain most species of animals and plants is natural forest management for timber production.[29]

The IIED has gone even further arguing that 'it is a fact that sustainable management is technically feasible'[30] and that 'the sustainable management of natural forest for timber production is onc of the keys to forest conservation and to the timber trade'.[31]

Other organizations have been more cautious and have thus found themselves promoting what are potentially internally inconsistent arguments. Thus the Worldwide Fund for Nature (WWF) on the one hand admits that 'an important question which has yet to be answered is whether tropical forests can be managed sustainably (ie without significant ecological impoverishment) while yielding adequate revenue to producer countries'. Yet, at the same time, the WWF argues that the ITTO 'provides a forum for establishing trade mechanisms which discriminate in favour of sustainably produced timber'.[32]

THE MYTH OF SUSTAINABLE LOGGING

The fact that practically everyone who addresses this thorny topic has their own notion of what 'sustainable' actually means, provides one of the main sources of confusion in this debate.

As made popular by the Brundtland Commission, the phrase 'sustainable development' refers to the means by which development is made to meet the needs of the present without compromising the ability of future generations to meet their own needs.[33] Since the needs of future generations are undefinable and the future potential for wealth generation of species and ecosystems are equally unknowable, the term apparently implies that total biological assets are not reduced, in the long term, through use.

In terms of tropical forests, such sustainable use would include not just maintaining timber resources and conserving biological diversity, but also maintaining the ecological functions of forests such as soil quality, hydrological cycles, climate and weather, and downstream fisheries. It should also imply maintaining supplies of other forest products – game, fruits, nuts, resins, dyes, basts, constructional materials, fuelwood etc – essential to the livelihoods of local people. Logging, which inevitably simplifies forest ecosystems, can never be sustainable in such terms. As Lee Talbot, an ex-director of the IUCN and now environmental officer for the World Bank's Africa Environment Unit (AFTEN), points out 'in practical terms, no commercial logging of tropical forests has proven to be sustainable from the standpoint of the forest ecosystem, and any such logging must be recognized as mining, not sustaining the basic forest resourcc.'[34] It is all the more surprizing therefore that the IUCN continues to refer to such logging of forests as 'sustainable', without defining how much biodiversity can be lost or must be maintained to qualify for this epithet.

SUSTAINED YIELD MANAGEMENT

A much more limited notion popular amongst foresters – that of 'sustained yield management – is often confused with the wider concept of sustainability. 'Sustained yield management' refers only to logging prescriptions which do not remove more volume of timber than a forest is capable of regenerating on a continuing basis.

Yet, even this much more limited concept is doubtfully realisable. As the Rainforest Information Centre has noted 'sustained yield has become the Philosopher's Stone of tropical forestry. No one has yet adequately defined it or attained it, yet everyone talks of it, and policies are prepared based on a concept that is entirely theoretical and never achieved.'[35]

In 1988, an IIED-led survey of the timber industry carried out for the ITTO concluded that 'the extent of tropical moist forest which is being deliverately managed at an operational scale for the sustainable production of timber is, on a world-scale, negligible.' Indeed the study found that less than one-eighth of 1 per cent of tropical forests, where timber extraction is occurring on a commercial basis, were being logged sustainably.[36]

Even this result has now been seriously challenged. Dr Aila Keto of the Rainforest Conservation Society of Australia carried out a detailed field study of the principal example in the IIED report of 'sustained-yield management' – the logging of the rainforests in Queensland. She found that the case should be excluded on two grounds. The first was that far from being commercially viable, forestry in Queensland was heavily subsidized with public money – just like many logging operations in the temperate forests in the United States. Moreover, she found that the actual sample plots on which the estimated rates of timber regrowth were based, were hopelessly unreliable. In sum, the evidence that commercial logging in Queensland was practised on a sustainable basis does not exist.[37] The data, on which the IIED study had based its findings, is, she claims, 'at best bad science or, at worst, scientific fraud.'[38]

Many other professional foresters are also very unsure that sustainable logging in natural forests is achievable. As Hans Lamprecht has noted 'the pressing question of to what extent a natural tropical forest ecosystem may be modified for economic reasons without seriously impairing its ability to function and survive cannot be answered on the basis of the present state of knowledge.'[39]

Dr G Budowski from the Forestry Institute in Costa Rica comments:

> The tropical forester hears and reads about 'selection' or 'shelterwood' systems, yet does not know of a single good practical case where such logging has not resulted in degradation. It may well be worthwhile to destroy the myth that claims that [natural] forests can be successfully managed on a sustainable yield basis, because there is a lack of evidence or good case studies.[40]

Similarly, referring to the tropical timber industry in Africa, forestry researcher Claude Martin writes:

> Unwavering belief in the possibility of sustainable rainforest usage [for timber] has brought absurd results, especially in West Africa. In the area, forest management systems were introduced very early, including Nigeria, whose forest is severely degraded, but the belief has been kept alive till the bitter end because it looks so beautiful on paper, in law statutes that were created decades ago more in Europe than in Africa. The population of these countries, yesterday and today, has been forgotten.[41]

The major funding agencies are also getting cold feet about the notion. Lee Talbot of the World Bank notes 'it appears that true sustainability from natural tropical forests has yet to be proven.' Noting that most logging in Africa has led to serious deforestation and that the funding of logging in intact forests contradicts the World Bank's policy commitment to 'sustainable development', he observes that 'obviously, the Bank . . . should not support further exploitation of [intact forest areas] for timber production.'[42]

The ITTO review set out what it believes are the essential conditions for achieving sustainable logging: a firm political resolve by government to achieve sustainability; secure permanent forest estate; an assured and stable market; adequate resources; and good research and information to allow sound planning, silviculture and management. As one member of the ITTO team noted, 'if any one of these conditions . . . is not fulfilled, then the tropical moist forest will continue to be pillaged for short-term gain.'[43]

However, as the ITTO study also admits, economic, social and political factors pose even greater obstacles to achieving 'sustained yield logging' than the technical or institutional problems. Corruption; politicians' tendency to act to maintain power in the short term rather than ecological balance in the long term; the fact that the political process is increasingly dominated by the market demands of urban elites rather than the rural poor; the lack of continuity and funding in forestry departments; the unpopularity and marginal status of forestry services; the fact that civil services are notably unresponsive; all these and more provide formidable obstacles to sound management.

Another major problem is that, so long as the timber market is partially supplied by the first cut from virgin forests and from lands which are being cleared for agriculture, sustainable production will be financially uncompetitive and therefore unviable. As the ITTO report notes, the low rate of return and the long payback period makes sustainable logging an unattractive private investment. Forest mining on the other hand can be very lucrative. Logging today is dominated by 'the logger who passes through the forest once and buys hotels in Hong Kong on the sale of raw logs.'[44]

The main reason that logging has led to wide-scale forest destruction is, as we have noted, that logging opens up previously isolated and inaccessible areas to colonization. In Africa, as the ITTO report notes, the heterogeneity of the forests coupled with the costs of transport has meant that much logging in the interior has been very selective, taking out only the best quality timber over very wide areas. Yet, paradoxically, this has made it more, not less, destructive: 'permanent forest estate' opened to logging has quickly become permanently deforested as the settlers flood in along the logging roads. So long as settler pressure continues to be the main problem facing tropical forestry today, sustained yield logging will remain an unrealisable dream.[45]

Unfortunately, many foresters continue to ignore these outside factors as issues beyond their area of concern and expertise, but this does nothing to isolate the forests themselves from these forces. On the contrary, so long as foresters continue to take a narrow view of their professional responsibilities, the more vulnerable the forests entrusted to their care will be.

FROM LOGGING TO PLANTATIONS

Robert Goodland and others at the World Bank believe that truly sustainable logging in natural forest can only be achieved at an extremely low level of cutting, to the point where plantation forestry, even of slow growing hardwoods, is likely to become financially competitive.[46] These conclusions are indirectly reinforced by the findings of Roger Sedjo and Kenneth Lyon, of Resources for the Future, who concluded in a recent study that industrial wood prices are unlikely to rise for the foreseeable future.[47] This tendency may be reinforced by the recent cut in tariffs on tropical timber products negotiated in the current Uruguay Round of the General Agreement on Tariffs and Trade (GATT). Goodland and his colleagues argue that the World Bank should halt its financing of logging in natural forests and switch to promoting plantation forestry as the only truly sustainable alternative.[48]

Even this faith in plantation forestry may be misplaced. As Lamprecht warns us 'on no account should it be lost from sight that a high degree of diversity is indispensable if the ecosystems of moist tropical forests are to retain their ability to function.'[49]

Conservationists who have too readily clutched at the notion of sustainable logging as a way of salvaging some biodiversity should proceed with caution. They may find that in the end their support has been used as justification for tree monocultures, a prospect as remote from their hopes of preserving biodiversity as clear cut logging.

The IUCN, it seems, has just woken up to this realization. As Jeffery Sayer noted in testimony to the British House of Lords, 'we now fear that what is going to happen in the tropics is that governments will find that natural forest management is too difficult and will gradually move away from it . . . I think it is significant that, in Indonesia, the Reafforestation Guarantee Fund which contains an enormous amount of money – it runs into millions of dollars – and which was intended to be used to compensate concession-holders who did apply the selective management regulations . . . will be used instead to fund plantations.'[50]

The emphasis on plantations also carries with it many other risks. Plantation programmes, of the kind often funded by the World Bank and other international development agencies, have run into serious criticism for being ecologically and socially inappropriate. Plantation schemes have been established on lands claimed and used by tribal people and planted with species which provide local communities with little employment or benefit. In eastern India, for example, the battle between teak and sal has come to symbolize the conflict between commercial forestry and forests for people.[51]

SUSTAINED DISSIMULATION

Despite the findings of its own studies, which show that sustained yield logging is being achieved practically nowhere, and despite the fact that the ITTO has yet even to agree on a definition of sustainability, sustainable logging remains the ITTO's immediate goal. During the 1990 session in Bali the Council moved further into the realms of fantasy by announcing 'Target 2000' whereby all trade in tropical timber was to be supplied from sustainable logging within ten years.[52]

The resolution which was proposed by a united caucus of producer nations was at first received with incredulity by the consumer country delegates, who, striving to remain true to a reasonable definition, felt that restricting the trade to sustainably produced tropical timber was practically equivalent to the trade's extinction. Despite these hesitations, the resolution was passed.

Official delegates winged their way back home to proudly inform their governments about the important steps being taken at the ITTO to save the world's rainforests. But the real implications of this illusory lurch forward at the ITTO become clearer if we look at the producers' and traders' notions of sustainability. For what we discover, if we look at their rhetoric, is that logging as practised already is sustainable. 'Target 2000' then becomes nothing more than a formula for business as usual.

For example, Arthur Morrell, a prominent member of the UK Timber Trade Federation who has been trade adviser to the Britain's Department of Trade and Industry and is an official delegate with the UK mission to the ITTO, mendaciously asserts that 'nobody disagrees that sustainable yield forest management is technically feasible'. Entirely contrary to the ITTO's own findings, he claims that some 126 million hectares of forests in Indonesia and Malaysia are being sustainably logged, citing the fact that the Malaysian Timber Industries Board issues 'sustainability certiicates' as evidence. Morrell would also add to this figure the 42,000 km^2 in Ghana and, taking the Brazilian President at his word, would like to add in Brazilian Amazonia too.[53]

Faced with a growing call in western nations for a ban on the import of tropical timber, the timber producing countries are engaged in a major publicity drive to revive their image. Leading the way are Indonesia and Malaysia, who in June 1989 entered into an official agreement to jointly counter-campaign against the northern environmentalists.[54] Malaysian Government publications, issued as part of this new campaign, claim that the forests there are managed under 'renewable cycles' of 30–55 years, assuring timber production 'in perpetuity'. Reforestation of fast growing exotics on the forest margins, they claim, 'promotes the multiple uses of forest land on a sustainable basis'.[55]

Indonesian timber tycoon Bob Hasan has provided US$2 million to spend on advertising in the western press and on lobbying in the US Senate and Congress and in the European Parliament.[56] He also hosted a lavish banquet for the ITTO delegates at the Bali Beach Hotel, second only in splendour to the entertainment laid on by President Suharto who, at the Indonesian Government's expense, flew the entire 400-strong conference for a day of pomp and splendour 300 miles west in Jakarta, where the meeting was told by the ITTO's chairman that Indonesia's commitment to sustainable forest management was an example to the world.

One prominent US$45,000 advertisement titled 'Tropical Forests Forever', printed in the *New York Times*, extols the virtues of Indonesian logging.[57] Glossy handouts provided by the Indonesian Government likewise claim that the regeneration of 'saplings under the Indonesian Selective Felling and Planting System is sufficient for the continuation of sustainable forest exploitation.'[58] According to Indonesia's Minister for Forestry, Hasrul Harahap, forest resources in Indonesia are managed under a sustainable system, whereby logging operations are followed immediately by replanting and other reforestation programmes.[59]

These are just some examples which painfully illustrate how far the rhetoric of sustainability has drifted away from reality, which we know is very different. For example, the conclusion of a recent IUCN study of logging in Indonesia was that:

> concessions were generally managed very poorly. The Indonesian Selection System which legally should be applied by all concessionaires, all 500 of them, were not applied . . . they were almost totally either ignored or simply flouted, and in many cases the concessionaires were not really aware of what they should be doing. The knowledge of the [forests'] ecology and species was not there, either in the forest departments or amongst the concessionaires.'[60]

The IUCN's findings only echo what the ITTO's own investigations have themselves revealed.[61] As Simon Counsell of Friends of the Earth notes, 'the ITTO is being used as a stalling mechanism to prevent effective change. Member countries are quick to point to their membership to demonstrate their concern for the environment. Yet are persistently failing to act on the ITTO's findings when these prove inconvenient.'[62]

REGULATING THE TRADE

As might be predicted, given the way the organization is controlled, the ITTO has moved extremely slowly to institute any effective mechanisms for achieving the promised transition to sustainability. Any moves to suggest regulations on the trade in unsustainably produced timber have been hotly resisted and even moves to monitor the trade viewed with hostility.

A crucial pre-condition to achieve the transition, whether by incentives or trade restrictions, is for wood in the market to be labelled as to its origin, thereby allowing buyers and customs officials to be more discriminating. In 1989, when the British Government and Friends of the Earth (UK) proposed an experimental project to test the feasibility of such a labelling scheme, the producer nations led by Indonesia and Malaysia blocked the project from getting approval until it was substantially modified. 'What we were hoping for was a project to specifically test the feasibility of a certification and labelling scheme to trace forest products right from source to the consumer but now, we fear, the project will be so theoretical as to be largely irrelevant to the trade and to consumers' explains Simon Counsell.[63]

Environmental organizations in South East Asia have called for a complete ban on logging in natural forests as the only means of checking a trade that is ruining their countries' ecologies. Such demands have been supported by non-Governmental organizations, like Friends of the Earth in Europe, and the Rainforest Action Groups and Network in Australia and the USA, which have been pushing, outside the ITTO, for a ban on the import of unsustainably produced tropical timber products. On 25 October 1990 the European Parliament adopted a resolution calling for a moratorium on imports of tropical hardwoods from Sarawak, Malaysia. The ITTO has vigorously rejected any such restrictions on free trade. Proposed legislation at the European Parliament and the US Senate, which would impose such trade regulations, has been bitterly condemned by timber producing nations, like Malaysia and Indonesia. The supposedly neutral Executive Director of the ITTO has been quick to take up the cudgels on their behalf.

In his opening statement to the ITTO in Bali in 1990, Dr Freezailah accused those non-Governmental organisations that were pressing for a halt to the trade in unsustainably produced timber, of carrying out a 'simplistic', 'short-sighted' and 'misleading' campaign to restrict the export of tropical timber. He charged the northern NGOs with being oblivious or uncaring of the complexities of tropical deforestation and for promoting a strategy that will seriously harm Third World countries, particularly the poor. Poverty, Dr Freezailah claimed, is the real cause of deforestation. The NGO campaign of 'threat' and 'vilification' 'is built on colossal misconceptions, confusion and obsessions . . .threats of fire and bromstone, alarmism and vindictiveness.'[64]

The official ITTO position is that 'import regulation may result in negative effects in producing countries in terms of their commitment to sustainable management of tropical forests. The regulation will at the same time promote substitution with temperate timbers and synthetic products. Such a development combined with the anti-tropical timber campaign now being mounted by certain conservation groups, will kill the trade in tropical timber. Once the trade is reduced or killed, it will eliminate the leverage of the international community to influence policies and developments regarding tropical forests.' The ITTO argues that once tropical forests lose their value 'as earners of revenue and foreign exchange' governments will be tempted to clear them to earn revenue by other means.[65]

Some organizations, such as the IUCN and the IIED, support this position, the IIED being opposed to any regulation on the grounds that 'any appearance of coercion will be deeply resented.'[66] But regulation of the trade is strongly favoured by the WWF which has supported moves to introduce selective trade bans in the European Parliament. Entirely contrary to the ITTO's view, Francis Sullivan of WWF(UK) believes that 'without regulation the trade has no future, because either there will be no forests left or no one will buy the timber anyway.' According to this view, therefore, it is the lack of immediate and effective regulation which poses the greatest risk to the timber trade. Accordingly WWF(International) is calling for a total switch to sustainably produced timber by 1995, to be imposed by trade restrictions if the ITTO cannot transform the trade by other means. The WWF is already looking to use the Convention on International Trade in Endangered Species (CITES) to halt trade in already vulnerable hardwood species.[67]

Sullivan also rejects the argument that forests will be cleared faster by Governments if they cannot harvest revenue from them. History bears his contention out. For example, in Brazil, the Government has promoted the conversion of forests for agriculture for political reasons, with no attempt to realize the forests' economic value, whereas, in other areas, logging and hence forest destruction has intensified exactly because companies have found means to profit from them. The argument that forests can be saved only by making them lucrative is entirely spurious.[68]

Some professional foresters agree with this assessment. Hans Gregersen and Allen Lundgren of the University of Minnesota and Gary Lindell of the United States Department of Agriculture, note that 'in fact, higher prices (for timber) have not led to sustained yield management, if anything, they led in the recent past to more rapid timber mining. This is partly because, while sustainable yield management may have become more profitable, it is even more profitable for the industry to take all merchantable timber, which often results in damage to residual stands to a point where sustained yield management becomes impossible.'[69]

Moreover, it has become clear that not all producer countries are as opposed to regulation of the trade is Indonesia and Malaysia. In April 1990, Papua New Guinea's Prime Minister, Rabbie Namaliu, noted 'it would be a major benefit if these (consumer) countries could support our own policy requirement for sustainable yield production by imposing their own restrictions upon tropical timber not supplied from genuine sustainable yield projects.'[70]

Consumer countries, too, are likely to move ahead unilaterally in imposing selective bans if the ITTO continues to lag behind. The Dutch Parliament is already in the process of considering a draft ministerial proposal to halt the import of all unsustainably produced timber after 1995, having reassured itself already that such a measure would not contravene the GATT. Many local councils in the Netherlands, Germany and the UK have also taken the decision not to use tropical hardwoods in their construction programmes, while some Australian Trades Unions have refused to handle tropical timber. They are now pressuring the Australian government to place a ban on imported rainforest timber.[71] In the absence of selective bans or at least accurate labels defining a product's provenance, a complete consumer boycott of tropical timber products seems inevitable. As Britain's Prince Charles noted in his

'Rainforest Lecture' in 1990, without a labelling system which assures the buyer that his product has been sustainably produced 'a cautious consumer is almost certainly going to be more inclined to avoid tropical hardwoods altogether rather than risk contributing to their unnecessary demise.'[72]

What the ITTO favours instead of such disincentives is a voluntary code of improved forest management which national governments would be urged to adopt in their own interests. Initially attempts were made at the ITTO to jointly agree binding guidelines for 'best practice', but producer governments have reacted against this too as a violation of national sovereignty. In May 1990, the ITTO agreed on non-binding guidelines, characterised by the NGOs as 'guidelines for guidelines'. Many NGOs are very sceptical of the value of such prescriptions by themselves, noting that in many countries, like Malaysia and Indonesia, sustainable management regimes are already theoretically in place, but the situation on the ground is one of rapid deforestation. What is needed, they argue, is for a combination of incentives and disincentives to force the trade to improve its practice. Increasingly they are looking outside the moribund ITTO to achieve these objectives.

WHAT ABOUT PEOPLE?

Conspicuously absent from this whole debate about the timber trade has been any mention of people. The ITTO is so dominated by the thinking of conventional forestry that it has been unable to find ways of dealing with such a politically sensitive question. Issues such as community-based forest management, community participation in planning, the land rights of forest dwellers, even the human dimensions of sustainability, have barely received consideration. Nor, until recently, have the NGOs done much to redress this critical short-sightedness.

The question was only raised for the first time in the 1988 meeting in Rio de Janeiro when Friends of the Earth adopted language drafted by the human rights group Survival International urging that the rights of forest dwellers to their lands should be respected in the handing out of logging concessions. At the November 1989 meeting of the ITTO, Survival International argued that the ITTO must also include the concept of sustaining forest peoples' livelihoods in its working definition of sustainability.[73] The organization has repeatedly called on the ITTO to respect the rights of forest peoples to the use and ownership of their traditional lands, a call which has now been taken up by most of the other NGOs attending ITTO functions.[74] This concern was echoed by Prince Charles who counts himself 'among those people who find it disturbing that [the ITTO's] Articles of Agreement make no mention of the rights and needs of indigenous forest dwellers.'[75]

The issue of forest peoples' rights finally forced its way into the ITTO's agenda in 1989, as international indignation about the escalating conflict between loggers and native people in Sarawak became too heated to ignore. At the May 1989 meeting, in Cote d'Ivoire, the ITTO adopted a resolution to send an official investigative mission to Sarawak with the aim of assessing 'the sustainable utilization and conservation of tropical forests and their genetic resources, as well as the maintenance of the ecological balance in Sarawak . . . with a view to ensuring their optimum utilization . . .'[76]

As publicly presented, the idea of the mission originally came from the ITTO, promoted by the Malaysian Executive Director, Dr Freezailah. It transpires, however, that in fact it was the Malaysian Government itself that requested the visit and Sarawak's Chief Minister travelled personally to the Cote d'Ivoire meeting to negotiate the terms of reference for the mission. Evidently the Malaysian authorities hoped that such a mission would be an exercise in damage limitation, a way of demonstrating to the outside world their willingness to deal openly with the controversy, while controlling just what the mission might actually look into. Many NGOs feared that the mission was a smokescreen which would obscure the lack of real progress being made within Sarawak to resolve the conflict of interests between loggers and forest dwellers.

Sahabat Alam Malaysia (SAM), the environmental organization which has most closely supported the native resistance to logging, was quick to voice concern about the mission. 'Our fundamental objection to the mission is that it is unable to function objectively because of its vested interest' noted SAM's President, Mohammed Idris. 'A truly independent study of forest management in Sarawak would require terms of reference which do not presume that timber is the primary value of the tropical forest'.[77]

Subsequent events proved how well founded these misgivings were. In the first place the 'terms of reference' of the mission did not explicitly direct it to investigate the social impact of logging, much less the legitimacy of the native peoples' grievances. Moreover the ten-member team, composed as it was of foresters and economists, was ill-prepared to look into such matters. Alerted by Survival International, concerned citizens all around the world deluged the ITTO with letters urging that the mission's mandate was expanded to include an investigation of the tribal peoples' grievance and that lawyers and social scientists were included on the team.[78] The ITTO responded with vague, and as it turned out misleading, assurances that 'obviously the mission will take a holistic view of its mandate and proceed on as wide a spectrum of consultations as possible.'[79]

Yet, even within the constraints imposed by the official resolution, the mission could have done a better job. As Survival International noted, the terms of reference allowed the mission to interpret the concept of 'optimum utilization' as referring to all forms of forest use, not just the extraction of timber. The concept of 'sustainability' should include a concern for sustaining local livelihoods, particularly as the official Forest Policy of Sarawak itself sets the 'prior claims of local demands' above the 'export trade in forest produce.'[80] In the event the mission chose to narrowly interpret its task as 'to assess the sustainability of forestry', thereby marginalizing not only human considerations but also alternative forms of forest use.[81] The mission made no serious attempt to evaluate the importance of non-timber products in the local economy, although many of these products are vital to the lives of local people.

Even as the mission progressed, NGO concerns were sharpened. Mission members made clear that a priority in their minds was that the mission itself should be replicable. In other words the mission could not probe too deeply into issues that might reflect badly on the Sarawak government, because, if it did, other producer nations would be very unlikely to allow similar ITTO missions into their own countries in the future. This 'lowest common denominator' approach clearly implied that social considerations would get short shrift.

Moreover, the first of the mission's three visits to Sarawak was entirely chaperoned by the Sarawak Government. Careful guidance by government officials continued during the course of the investigation. The ITTO's Executive Director also accompanied the mission on each of its visits to the Sarawak. Thus, although the mission did meet with native people, these meetings nearly all took place in formal settings, usually in government buildings with officials looking on. Despite receiving many written invitations, the mission visited none of the settlements which had complained most about the impact of logging. Of the two communities that the mission did visit, one was a model village seen during the first government guided tour, while the second attracted the mission because of its vicinity to a proposed national park. Lamely the mission report concluded that it had been unable to investigate any of the grievances noted by the native people.[82]

Despite these limitations, and before the investigation had even concluded, one mission member felt authorized to write to the US Government alleging that the suggestion that there was rural opposition to logging 'is contrary to the findings of the ITTO mission. Throughout the country, one group after another including rural community leaders told us they favored logging . . .'[83]

Meanwhile, from the Sarawak Government's point of view, the mission was proving a publicity success. Long before the mission actually reported to the ITTO, banner headlines in the local newspapers shrieked 'Top marks for our forest management' citing the mission leader as stating that Sarawak's forest management system is 'one of the best in terms of policy'.[84]

NGOs were quick to criticize the mission report when it appeared. In its opening chapter the report revealingly refers to the 'native peoples question' as an 'awkward difficulty' and it was clearly not one to which the mission members gave much priority. Of some 260 publications to which the mission referred there was not a single anthropological text. The report made no attempt to assess how many people live in the forests or are dependent on them for their welfare.[85] In 1989, there were at least 94 reported deaths in logging accidents, the highest number in 17 years according to the authorities. Yet this issue was not discussed in the body of the report. Apparently not even the sustaining of human lives featured in the mission's concept of sustainability.[86]

Despite the fact that what the native people had been demanding was a mechanism to secure their land rights, the report made no such recommendations. While noting that extensive areas under 'native customary rights' exist in Sarawak and that these rights present a problem to the management of Permanent Forest Estate (PFE), the report simultaneously advocates a considerable expansion of PFE without providing a means of resolving conflicting land claims. 'This is a formula for further increasing land disputes in Sarawak' stated Harrison Ngau, a native Kayan who ran the SAM office in Sarawak until his election as Member of the Malaysia Parliament in October 1990.[87]

Lawyers working with the native people in Sarawak were also very disappointed by the mission's cavalier attitude to legal issues. Deprived of a lawyer on the mission team, even though this had been strongly urged before the final team membership had been announced, the report was obliged to note that it 'was not competent to address land rights matters'. Notwithstanding, the report went on to make categorical statements on native people's land rights. Chee Yoke Ling, a lawyer at the Penang-based Asia-Pacific Peoples Environment Network, is concerned that these statements could prejudice future attempts to resolve the legal problems in Sarawak. 'An examination of the Document reveals fundamental misunderstandings of the law on the part of the mission. In their meetings with native leaders the mission members adopted a lecturing tone and made misinformed statements about the nature and extent of the natives' legal rights,' she said.[88]

Many observers believe that one of the main problems in Sarawak is the way that the handing out of logging concessions is tightly controlled by politicians, who share them out amongst themselves to maintain their wealth and power. With logging so directly benefiting the political elite, the forestry service is ill-placed to insist on careful management or a concern for local communities.[89] Yet this was not an issue that the mission chose to investigate. Indeed, according to one mission member the possibility that timber concessions 'are concentrated in the hands of friends or family members of political elites [does not] serve any purpose relevant to the study . . . if true it might even favour rather than discourage good forest management, since it would then be necessary only to convince the political elite, who then presumably could force their cronies into line.'[90]

NGOs have also criticized the mission report in the area of its professional competence. The mission only recommended a 30 per cent reduction in the annual cut, simultaneous to the introduction of across-the-board improvements in the forestry service. According to WWF(International), instituting such improvements could take up to half a century to have effect by which time there would be no primary forests left. 'Using the mission's own figures for forest timber yields, a 60 per cent reduction seems to be what the mission should have called for' noted Christopher Elliott of the WWF.[91]

Despite all these serious shortcomings, what the mission did reveal about Sarawak's forests has shown that environmentalists' concerns were amply justified. At present rates, the primary forests of Sarawak will be all logged out within this decade. The mission found ample evidence of poor logging management and practice, and consequent forest degradation and ecological problems.[92]

Many NGOs considered the Sarawak study to be a test case of the ITTO's competence. What it reveals very clearly is how the ITTO subordinates human considerations to the interests of the timber industry. The mission adopted a very narrow view of its task, assessing the sustainability of forestry in Sarawak rather than adopting a broad approach which would have more fairly attempted to reconcile the conflicting interests in the forests. The mission is a clear example of what Vandana Shiva calls the 'violence of science', whereby the narrow view of experts rides roughshod over the needs and views of local people.[93]

LOGGING IN SARAWAK: THE FACTS

- Some 70 per cent of Sarawak's forests have been leased out to loggers. The forests are being cut at some 300,000 hectares per year at which rate the primary forest will be practically all logged out in 5 to 10 years. Logging concessions are handed out by the Chief Minister who is also the Minister for Forestry. Nearly all State Assemblymen have logging interests. One of the largest concessions is held by the State Minister for Environment and Tourism.
- Logging roads have been pushed across areas cultivated by native people and their burial sites have been disturbed. The intensive logging of predominantly primary forests has caused a serious decline in game. Mean intakes in protein have declined from 54 kg/person/year to only 12 kg/person/year according to a WWF study, while a Sarawak Government study shows that, in recently logged areas, there is a three-fold increase in serious malnutrition in native communities, affecting some 31 per cent of the population.
- The accelerating forest loss is also leading to increased soil erosion and rapid surface run-off, causing the visible pollution of streams and rivers, muddying drinking and bathing waters, blocking piped water supplies and causing stocks of fish to crash, so further impoverishing the local diet. Government figures reveal that some 60 per cent of Sarawak's rivers suffer such pollution. The destruction of the native peoples' environment is denying them access to forest produce, for making baskets, for constructing canoes and long-houses, for their medicines, arrow poisons and blowpipes, for resins, fruits and dyes. Logging is also directly affecting native people working in the lumber camps, where mortalities are some 21 times the rate found in Canada. About 220,000 native people in Sarawak depend on the forest. Their ways of life are being undermined by the logging, which is destroying the forests that they depend on.
- Since March 1987, native people have repeatedly blockaded logging company roads to prevent the destruction of their lands. The Government has done nothing to address the native peoples' grievances or curb the logging. Instead, the army, the police and forest guards have sought to intimidate the local people. There have been many arrests, most under a new law passed in November 1987, which makes the obstruction of logging roads a criminal offence. The latest blockades went up in August 1990. At the time of writing there have been nearly 350 individual arrests.

THE FUTURE OF THE ITTO

Under the terms of the ITTA, the ITTO is an independent organization controlled by a council made up of member countries, that has a set term of authority of five years. The Council is empowered to extend its term by up to two terms of two years each if it so decides, The ITTC has already invoked this power once and, if it does so again, its authority will lapse in 1994 when the ITTA is either renegotiated and extended or the ITTO is dissolved. Renegotiation of the ITTA would, under normal circumstances, take place under the auspices of UNCTAD, in Geneva. However, timber interests in Japan and South East Asia are keen to control the process as much as possible and are pushing to have the ITTA renegotiated at a special meeting of the ITTC itself in Yokohama. In this case a second UN Conference would be necessary to ratify the second ITTA.

Either way one can be pretty sure that the ITTO meetings in 1993 and 1994 will be completely taken up by the renegotiations, leaving precious little scope for the ITTO to make any substantive progress. Governments would do well to look elsewhere if they are hoping to achieve any rapid transformation in the tropical timber trade.

At the 1989 meeting in a joint statement to the Council, the NGOs expressed their disappointment that 'despite 15 years of negotiations, the ITTO is still not able to decide what to do to manage tropical forests, or even how to define sustainable management. During these years of vacillation 160 million hectares of tropical forest have been destroyed.'[94] This article has attempted to demonstrate how this lack of progress is structurally inherent in the ITTO. Hobbled by political compromise and dominated by the interests of the timber trade, the ITTO's narrow scope is reinforced by scientific expertise which hesitates to address the wider problems plaguing forestry and the tropical forests.

In the words of Jack Westoby:

> Here we come to grips with what I consider is the worst crime that can be laid at the door of foresters; they have conducted themselves as conscientious, loyal and obedient public servants . . . and in so doing they have failed in their civic responsibilities . . . the forester, like any professional scientist or technician, has a responsibility to the hand that feeds him. But this is not the end of his responsibilities. He also has a responsibility to the community-at-large, to society, to the public.
>
> Take for example, the professional foresters of the developing countries of South East Asia. In many of these countries the forest resource is being recklessly pillaged in response to overwhelming local political pressures. The local forester, isolated, lacking political allies, is powerless to check this process – indeed is often an accessory. His situation is not helped when he sees the [expatriate] forester . . . condoning practices that he would never dream of allowing on his home ground. If the expatriate forester were more conscious of his social responsibility, were truly alive to his basic ethic, he would feel an obligation to counter and expose unacceptable practices.[95]

This chapter was originally published in 1990 as a report by the World Rainforest Movement.

ACKNOWLEDGEMENTS

A number of individuals have helped in the preparation of this document and I would like, in particular, to thank them for their assistance and comments, Nicholas Hildyard of *The Ecologist*, Chee Yoke Ling of Sahabat Alam Malaysia, Carolyn Marr of Survival International, Francis Sullivan of the WorldWide Fund for Nature (UK), Herman Verhagen of Milieudefensie, Patrick Anderson of Greenpeace International, Simon Counsell of Friends of the Earth (UK) and Charles Secrett.

NOTES AND REFERENCES

1. Director of the Forest Peoples Programme, World Rainforest Movement, 8 Chapel Row, Chadlington, OX7 3NA, UK. Tel: 060876 691 Fax: 060876 743 Email: GEO2:WRM. International Secretariat: 87, Cantonment Road, 10250 Penang, Malaysia.
2. John le Carre, *The Russia House*, Bantam Books, New York, 1990, p 250.
3. Resolution adopted at the 1407th session of the European Community Council of Ministers for Development Cooperation, Brussels, 29 May 1990.
4. Winterbottom, Robert, *Taking Stock: the Tropical Forestry Action Plan after five years*, World Resources Institute, Washington DC, 1990; Food and Agriculture Organisation, *The Tropical Forestry Action Plan: Report of the Independent Review*, Rome, 1990.
5. Colchester, Marcus and Lohmann, Larry, *The Tropical Forestry Action Plan: what progress?* 2nd edition, World Rainforest Movement, Penang, 1990.
6. Myers, Norman, *Deforestation Rates in Tropical Forests and their Climatic Implications*, Friends of the Earth, London, 1989.
7. FAO op cit.
8. World Resources Institute, *World Resources 1990–1991*, Oxford University Press, 1990.
9. Gradwohl, Judith and Greenberg, Russell, *Saving the Tropical Forests*, Earthscan, London, 1988.
10. Lamprecht, Hans, *Silviculture in the Tropics*, GTZ, Eschborn, 1989, p 102 citing Norman Myers, *Conversion of tropical moist forests*, National Resources Council, Washington DC, 1980; Wilderness Society, *The Tropical Timber Debate*, ms, 1989.
11. Lamprecht op cit.
12. Cited in Goodland, R, Asibey, E, Post, J and Dyson, M, *Tropical Moist Forest Management; the Urgent Transition to Sustainability*, World Bank, 1990, p 4.
13. Ibid.
14. Lamprecht op cit.
15. World Resources Institute, World Bank and United Nations Development Programme, *Tropical Forests: a Call for Action*, Washington DC, 1985.
16. Chin, S C, *Managing Malaysia's Resources for the Sustained Production of Resources* ms; ITTO, 'The Promotion of Sustainable Forest Management: a case study in Sarawak, Malaysia', Report Submitted to the International Tropical Timber Council, ITTC(VIII)/7, 7 May 1990.
17. Japan Times, 5 October 1989.
18. Edward Goldsmith.
19. Shiva, Vandana, 'Forestry Crisis and Forestry Myths: A critical review of the Tropical Forestry Action Plan', World Rainforest Movement, Penang, 1987.
20. Westoby, Jack, *The Purpose of Forests*, Basil Blackwell, Oxford, 1987, p 264–265.
21. Colchester, Marcus, *Pirates, Squatters and Poachers: the political ecology of dispossession of the native peoples of Sarawak*, Survival International and INSAN, Kuala Lumpur, 1989, p 31.
22. Marshall, George, *The Ecologist* 20(4) in press; Anderson, James, 'Lands of Risk, People at Risk: perspectives on Tropical Forest Transformation in the Philippines', in Little, Peter,

Horowitz, Michael and Endre Nyerges, *Lands at Risk in the Third World: local level perspectives,* Westview Press, Boulder, 1987, p 249–267.
23. Hpay, Terence, *The International Tropical Timber Agreement: its prospects for tropical timber trade, development and forest management*, IIED, London, 1986.
24. The International Tropical Timber Agreement, 23 November 1983.
25. Nectoux, Francois and Kuroda, Yoichi, *Timber from the South Seas: an analysis of Japan's Tropical Timber Trade and its Environmental Impact*, WorldWide Fund for Nature, Gland, 1989.
26. Secrett, Charles, personal communication, 3 August 1990.
27. Hpay op cit.
28. Secrett, Charles, 'International timber trade organises . . . and so do NGOs', Friends of the Earth International newsletter, December 1986.
29. Emphasis in original; HMSO, 'Tropical Forests', Report of the House of Lords Select Committee on the European Communities' Policy on Tropical Forests with Evidence, Session 1989–1990 11th Report, HMSO, London, 1990, p 78.
30. Poore, Duncan, *No Timber Without Trees*, Earthscan, London, 1989, p 213.
31. Ibid p 226.
32. Worldwide Fund for Nature, *Tropical Forest Conservation and the ITTA*, Gland, 1987, p 11.
33. UNWCED, *Our Common Future*, Oxford University Press, 1987.
34. Talbot, Lee, *A Proposal for the World Bank's Policy and Strategy for Tropical Moist Forests in Africa*, World Bank, 1990, p 3.
35. Rainforest Information Centre, 'The World Bank Tropical Forestry Action Plan for Papua New Guinea: a critique', Lismore, ms, 1990, p 24.
36. Poore op cit.
37. Keto, Aila, Scott, Keith and Olsen, Michael, 'Sustainable Harvesting of Tropical Rainforests: a reassessment', Paper presented to the Eighth Session of the International Tropical Timber Council, 16–23 May 1990, Bali, Indonesia.
38. ECO LXXV(1) p 2. (ECO is a newsletter produced by environmental NGOs at intergovernmental meetings). Very recently, the conclusions drawn by Keto *et al* have been challenged by members of the Queensland Forest Service, who argue that the data base is more sound than she believes. Vanclay, J K, Rudder, E J, Dale, G and Blake, G A 'Sustainable Harvesting of Tropical Rainforests: a reply to Keto, Scott and Olsen', Queensland Forest Service, ms.
39. Lamprecht op cit p 112 see also World Rainforest Movement, *Tropical Deforestation*, Penang, 1990, p 80f.
40. Cited in Anderson, Patrick, 'The Myth of Sustainable Logging', *The Ecologist* 19(5):166–168, p 166.
41. Martin, C, 'West and Central African Rainforest: hardly used but already destroyed', in Peter Steuben (ed) *Clearcutting Paradise*, Fotus and Carten, cited in Anderson op cit.
42. Talbot op cit p 6.
43. Reitbergen, Simon, in Poore op cit p 72.
44. Palmer, John, in Poore op cit p 161.
45. Colchester, Marcus, 'Guilty until proved innocent', *The Ecologist* 20(3): 114–115.
46. Goodland *et al* op cit.
47. Sedjo, Roger A and Kenneth Lyon, 'The Long-Term Adequacy of World Timber Supply', *Resources for the Future*, Washington DC, 1990, p 180.
48. Goodland *et al* op cit.
49. Lamprecht op cit p 112.
50. HMSO op cit p 77.
51. Shiva op cit 18.
52. WWF News 65 (May/June) 1990 p 1; the WWF had meanwhile announced its own target of achieving this transition by 1995.
53. Morrell, Artur, *Timber Trades Journal*, June 1990:17. It should be noted that the UK Timber Trade Federation has a very impoverished notion of sustainability, their concern being only

that the forest maintains 'its essential nature, which in our terms means the tree cover' HMSO op cit p 44.
54. Karya, Suara, 28 June 1989.
55. Malaysia, 'A Living Heritage', Forestry Department, Kuala Lumpur, 1989.
56. 'Indonesia Declares War on Conservation Lobby', *Down to Earth* No 5, October 1989.
57. *New York Times*, 18 August 1989.
58. Indonesia, 'Indonesian Tropical Rain Forests: forests for sustainable development', Ministry of Forestry, Jakarta, 1989.
59. *Jakarta Post* 17 May 1990.
60. HMSO op cit p 77.
61. Poore op cit.
62. Counsell, Simon, personal communication, 9 August 1990.
63. Counsell, Simon, personal communication, 7 August 1990.
64. ECO LXXV(2) p 1.
65. HMSO op cit p 152.
66. Ibid p 169.
67. Sullivan, Francis, personal communication, 7 August 1990.
68. Davis, Shelton, *Victims of the Miracle*, Cambridge University Press, 1976.
69. Gregersen, Hans, Lundgren, Allen and Lindell, Gary, 'Contributions of Tropical Forests to Sustainable Development: the role of Industry and the Trade', Paper prepared for the ITTO Seminar on Sustainable Development of Tropical Forests, Bali, Indonesia, 19 May 1990, p 3.
70. 'Opening address by Prime Minister, Namaliu, Rabbie L, CMG MP for the Tropical Forest Action Plan Round Table Conference, Islander Hotel, at 2.00 pm Monday 2 April 1990'.
71. *Timber Trades Journal*, 11 August 1990:3.
72. HRH The Prince of Wales, 'The Rainforest Lecture', Friends of the Earth/Royal Botanic Gardens, Kew, 1990, p 18.
73. Survival International, 'The ITTO and Indigenous Peoples', paper presented to the ITTC November 1989 and reproduced in HMSO op cit p 172–175.
74. ECO LXXIV and LXXV.
75. HRH the Prince of Wales op cit p 10.
76. ITTO OP CIT P 2.
77. Statement by Sahabat Alam Malaysia at the November 1989 ITTO meeting.
78. Survival International 'Natives fear fact-finding mission will ignore their rights', *Urgent Action Bulletin*, September, 1989, see also HMSO op cit p 172–175.
79. ECO LXXIV(2) p 3.
80. Letter from Survival International to the ITTO, 14 September 1989.
81. Earl of Cranbrook, 'Sarawak forestry', *New Scientist*, 30 June 1990.
82. ITTO op cit.
83. Wadsworth, Frank H, 'Comment on Congressional Staff Study Mission to Malaysia', nd; the bias behind these comments may be judged by the author's reference to Sahabat Alam Malaysia, in this communique to the US Government, as 'the extreme environmental left'.
84. Borneo Post 1 April 1990.
85. ITTO op cit.
86. Apin, Teresa, 'Environmentalists criticise International Timber Organisation's Report', *Third World Network Features*, July 1990.
87. Ibid.
88. Ibid.
89. INSAN, 'Logging Against the Natives of Sarawak', Kuala Lumpur, 1989; Colchester op cit ref 19.
90. Wadsworth op cit.
91. WWF, 'Massive Cuts in Logging Needed to Save Sarawak's Forests', Press Release, 22 May 1990.
92. ITTO op cit.
93. Shiva, Vandana, *Staying Alive: women, ecology and development*, Zed Press, London, 1989; Shiva, Vandana, *The Violence of the Green Revolution*, Dehra Dun, 1989.
94. ECO LXXIV(5) p 1.
95. Westoby op cit p 302, 238.

Chapter 8

Long Term Trends in Global Demand For and Supply Of Industrial Wood

Mike Arnold

INTRODUCTION

Recent concerns about the state of the world's forests have tended to concentrate on issues of depletion and degradation, and the attendant loss of environmental values. It has been widely argued that harvesting to meet demands for industrial wood constitutes one of the main forces leading to reduction of the forest resource. Underlying these arguments that the world's stock of wood resources is being progressively depleted is an implicit, and often explicit, assumption that harvesting is taking place at a rate in excess of renewal, because the world is entering a period of growing shortage of wood and fibre.

The purpose of this paper is to review recent long-term studies of demand for and supply of wood and wood products in the world as a whole, in order to explore what light these shed on the likely future evolution of the global supply–demand balance.[1] In this first section of the paper the present patterns of consumption and supply are briefly reviewed, and the nature of the information that is available about future trends is identified. The two following sections examine trends in demand and production, and the final section the evolution of the balance between the two, and the implications for future trade and prices.

Present patterns of consumption and production

Consumption of the main industrially processed products of wood and wood fibre is heavily concentrated in the industrially more developed parts of the world. The countries of North America, Europe, the USSR, Japan and Oceania (Australia and New Zealand) together acounted in 1986 for nearly 80 per cent of sawnwood consumption, more than 85 per cent of wood-based panels consumption, and more than 90 per cent of wood pulp consumption.

Global production of wood in the period 1985–87 is estimated to have averaged 3255 million m^3 annually, of which 1350 million m^3 was wood for industrial processing, and is concentrated in the same regions.

Nearly three-quarters of industrial wood output occurs in the countries of the north temperate zone, with roughly 40 per cent in North America and 20 per cent in Europe. The USSR and the tropical region each account for about 15 per cent of total output, and the countries of the south temperate zone for about 8 per cent.

Table 8.1: Global annual production of industrial wood and wood products, 1985–87 (million unit)

	Saw & veneer logs (m^3)	Pulp wood (m^3)	Sawn wood (m^3)	Veneer and ply (m^3)	Particle board (m^3)	Fibre board (m^3)	Wood pulp (mt)
North America	372.0	168.1	154.2	22.3	11.6	6.1	73.2
Europe	155.8	114.6	83.3	4.8	23.7	4.2	34.3
USSR	159.6	39.7	100.4	2.7	7.0	3.6	10.4
Japan	19.0	12.1	29.2	7.4	0.9	0.9	9.4
Other Asia	149.0	10.0	70.9	11.6	1.2	1.4	3.7
Latin America	63.9	31.6	28.5	1.9	1.8	1.0	6.0
Africa	21.7	7.1	7.9	1.2	0.5	0.2	1.7
Oceania	16.0	11.6	5.5	0.2	0.9	0.3	2.0
Total	957.0	394.7	479.8	52.1	47.7	17.5	140.6

Source: FAO, 1989

More than 70 per cent of all wood for industry that is harvested at present is in the form of saw and veneer logs, of which more than 70 per cent is harvested from coniferous species.[2] However, over the past decade output of non-coniferous logs has grown faster than output of coniferous logs, and production of pulpwood has grown twice as fast as production of logs.

Over the ten years preceding the 1985–87 period global production of industrial wood grew by 13 per cent. Growth in production in most regions has been at similar rates, except in the USSR and Japan where it has been slow, and in the south temperate belt where it has been rapid. Broad geographical patterns have therefore not been changing significantly.

The patterns of production of processed products (Table 8.1) largely reflect those of roundwood production, and of wood products consumption. North America has accounted for much of the recent growth in production of sawn softwood and softwood plywood, and tropical Asia for the expansion in output of sawn hardwood and hardwood plywood. Growth in wood pulp production has been concentrated in North America and Europe, and in some of the south temperate countries. Growth in particle board output has been largest in North America and the USSR; and growth in production of fibreboard in the USSR, Asia and Latin America (FAO, 1989).

Present trade patterns and trends

In terms of the raw material needed to produce them, the volume of wood and wood products entering trade is roughly equivalent to a third of world production of industrial wood. Trade is predominantly in processed products, with a steadily increasing proportion of the output of most of the main wood products being traded. Trade in roundwood is mainly between countries around the Pacific Rim and within, and into, Europe, and is increasingly in the

Table 8.2: World trade in forest products, 1980
(million m^3 roundwood equivalent)

To: / From:	Canada	Nordic region	Other Europe	USA	USSR	SE Asia	Other[2]	Total
Other Europe[1]	21	65	67	17	20	6	13	209
USA	79	1	–	–	–	2	1	83
Japan	9	1	–	24	7	28	2	71
Developing countries	7	9	3	14	3	20	4	60
Other	3	14	7	8	6	3	6	46
Total	119	90	77	63	36	59	26	471

Source: ECE/FAO, 1986
[1] Excluding nordic countries
[2] Of which 22 million m^3 from Africa and Latin America

form of pulpwood rather than logs (FAO, 1989).

Table 8.2 summarizes the principal trade flows. At this level of aggregation the pattern of trade has not changed radically over a period of at least 20 years. Trade is heavily concentrated within three intra-regional flows: from Canada to the United States, from the nordic countries to the rest of Europe, and within the rest of Europe. This group of European countries is by far the largest net importer, accounting for nearly a half of all imports over the past 20 years, and the United States and Japan are the second and third largest. Canada is the largest exporter, in terms of equivalent volumes of roundwood, followed by the nordic countries, the United States and the countries of South East Asia. Together, they account for 70 per cent of the total wood volume traded. With the exception of the United States, a very large proportion of their production enters trade. These regions are therefore of particular importance when looking at future supplies.

Predicting future trends and patterns of demand and supply

Attempts to project future events over such lengthy periods as are necessary in forestry are by their very nature extremely precarious. Demand for the products of forestry largely depends on the evolution of economic activity, which itself cannot be predicted with any degree of confidence for other than short periods ahead. Technological change, which constitutes the other main demand shifting factor, is even more difficult to forecast.

Analysis within the forest sector, even over the short term, faces a number of other problems. For much of the global forestry estate even basic information on forest inventory and growth is lacking, so that the scope for global analysis is restricted. Even in the better researched and documented regions the information needed to translate physical into economic relationships is weak or absent.

Most forecasting even in the main producing areas still rests on separate projections of consumption and removals. Some national and regional studies have recently moved to the stage of simulating market equilibrium, but with trade still treated as exogenous to the model (eg USFS, 1989). A start has also been made to trying develop spatial equilibrium models which simulate trade as well as production and demand, notably the Global Trade Model (Kallio *et al*, 1987) developed at the International Institute for Applied Systems Analysis (IIASA) and the Timber Supply Model (Sedjo and Lyon, 1990) from Resources for the Future (RFF).

However, it is important when interpreting what is available from these exercises to recognize that they are designed not to provide forecasts, but for system analysis, that is to explore how a system would deviate from a base scenario under particular circumstances; the base scenario being usually essentially an extension of past trends. It is also necessary to emphasise that they are first generation models. Work is under way to refine them,[3] but should not be expected to provide anything more than some broad indications of the possible direction of future changes.[4]

FUTURE DEMAND

Trends in consumption of wood

For the world as a whole, and for each of the main industrialized regions, the rate of growth in consumption of industrial wood has been slowing down – worldwide from 3.5 per cent annually during 1950–60 to 2.2 per cent from 1960–70, and 1.1 per cent from 1970–80 (Sedjo and Lyon, 1990), though the late 1980s was a period of more rapid growth in use. This long term reduction in growth in consumption reflects, in addition to changes in economic growth rates, a number of long term trends.

One is that wood products are used in applications and markets where their use appears to have been approaching maturity, such as housing in much of Europe, North America and Japan. A second is improvements in the durability, quality and efficiency of wood products which extend their useful life so that replacement is required less often. A third is technological change in processing and manufacture which results in less material being needed for a unit of product, or for a particular application. A fourth is substitution. Wood industries have been as successful as most in enhancing the properties and competitive position of their products, but advances in market share of particular wood products have usually been at the expense of other wood products.

In part such changes reflect, and are dictated by, changes in the user industry. For example, the shift in the main user countries to mechanized and automated manufacturing in order to control labour costs, favours dimensionally stable and easily machined products such as plywood and particle board over products such as sawnwood which have to be hand finished or fitted in many applications. Other changes have their origins within the wood products sector. Thus, the continuing diversification and improvement of fibre and particle-based products in part reflects pressures to take advantage of lower cost raw materials.

The net result of these trends is a progressive shift in the structure of wood products consumption from solid wood to reconstituted wood and wood fibre products. In aggregate, consumption of sawnwood is growing only slowly; with per capita use falling in most of the developed world. Growth in consumption of panels has been rapid, but has slowed down in high per capita use countries. Consumption of paper and paperboard on the other hand continues to grow strongly.

In general these trends are expected to continue. However, the scope for substitution of sawnwood by wood-based panels in North America and western Europe has already been largely exploited, so that use of both of these products may grow in future more in line with change in the user sectors. Within the panel sector it is expected that there could be considerable shifts, with flakeboard and other reconstituted products making inroads into structural plywood use in the United States, and medium-density fibreboard capturing market share from other wood products in Europe. Consumption of plywood is therefore expected to grow more slowly than consumption of other panels.

Because of technical change in the wood products sector, consumption of wood raw materials has grown less rapidly than consumption of the processed products. A particularly important development in this respect has been the rapid progress in most industrialized countries in raising the proportion of the raw materials for reconstituted wood products industries that comes from logging, sawmilling and other residues, and from recycling of paper.

In the United States, the proportion of softwood growing-stock left in the forest as logging residue fell from 37 per cent in 1962 to less than 10 per cent in 1986. Similarly, the proportion of pulpwood consumed which was provided by residues and chips grew from 29 per cent to 38 per cent between 1970 and 1980 (USFS, 1989). In Europe during the same period the share of pulpwood supplies coming from these sources rose from 22 per cent to 29 per cent, and recycling of waste paper raised the proportion of paper-making fibres provided in this way from 26 per cent to 32 per cent (ECE/FAO, 1986).

There are many parts of the world where this shift towards full use of the available wood and fibre base has yet to happen, including major users such as the USSR. However, in both the United States and Europe it appears that most available residues are by now in use. It must therefore be expected that a much greater proportion of future growth in raw material needs must be met from roundwood harvesting. Thus, the CINTRAFOR global projections show residues falling from 29 to 21 per cent of pulpwood supplies between 1985 and 2000 (Cardellichio *et al*, 1989); and the forecasts for the United States (Table 8.3) show the use of residues declining not only as a proportion of pulpwood consumption (from 39 per cent in 1985/86 to 21 per cent in 2040), but also in absolute quantities – due to only limited growth in sawnwood and veneer/plywood production, and increasing competition for residue use as fuel as real energy costs rise (USFS, 1989).[5]

Table 8.3: US forest products consumption in 1986 with projections to 2040 (million m^3)

	1986	2000	2020	2040
Sawnwood				
Softwoods	110.4	106.9	128.1	133.1
Total	131.5	130.7	156.7	164.3
Structural panels				
Softwood plywood	18.2	15.0	17.6	20.1
Total	21.3	22.4	29.1	34.2
Pulpwood				
Softwood roundwood	128.0	176.9	209.2	241.8
Total roundwood	207.0	305.2	391.9	451.7
Total	338.6	406.4	515.5	574.9

Source: USFS, 1989

However, in the mid 1980s only half as much waste paper was being recycled in the United States as in Europe or Japan. As pressures rise to make it mandatory, for environmental reasons, recycling in the United States is expanding rapidly, and it is expected that it will rise towards the levels of nearly 50 per cent in Japan and 51 per cent in the EEC. This could result in growth in demands on pulpwood harvests in North America being significantly less than was projected in the study reported on immediately above (ECE/FAO, 1990). At the global level, though, it has been projected that in the period to 2000 consumption of pulpwood will grow nearly three times as fast as the use of saw logs (Cardellichio *et al*, 1989).

Another important shift in the demand for wood raw materials is that towards greater use of lower grade and size material, and towards hitherto lesser used species. This is largely due to the shift from solid to reconstituted wood products, and within the latter towards fibre- and particle-based products. However, for some products, notably sawnwood, the use of lower quality raw material is causing a reduction in yields.

Technological advances in the pulp as well as the panel industries have broadened the range of use of hardwoods, and advances in sawmilling have expanded their use in that industry. As a consequence, in most industrialized countries consumption of non-coniferous roundwood has recently been growing faster than that of coniferous wood, and this trend is expected to continue in the future. At the global level, the CINTRAFOR study has predicted that in the period to 2000 average annual growth rates for consumption of non-coniferous roundwood will be more than three times as fast for sawlogs, and nearly twice as fast for pulpwood (Cardellichio *et al*, 1989).

Demand forecasts

Demand for forest products is determined principally by change in population, income and price, and by technological change. As each of these parameters can have markedly different impacts in the main use sectors – construction, furniture, packaging – it is preferable to carry out the analysis separately for each. However, past consumption can be disaggregated to end uses only in the more developed countries, and even in these useful forward estimates of likely change in the end use sectors exist, or can be developed, for only short periods into the future. The usual approach to estimating future demand in the long term is consequently to relate aggregate demand of a product to change in just population and income, through application of income elasticities derived from analysis of past change. Price elasticities adjust the estimates for price effects, and further adjustments are made for anticipated future technological change.

The precision of such forecasts is therefore heavily dependent on the availability and accuracy of data for the base period, on the capacity of the analytical model to correctly represent the relationships between consumption or demand and the underlying parameters, and on future population and economic change following the paths anticipated in the projections. Experience even over the short term shows that this cannot be assumed. Thus the ECE/FAO forecasts developed in the early 1970s of European consumption, in 1980 fell short of actual consumption in that year by 25 per cent for fuelwood and 3 per cent for sawnwood, and exceeded consumption by 28 per cent for panels and by 14 per cent for paper and paperboard (ECE/FAO, 1986). A 1973 government forecast of 1983 total wood products consumption in Japan exceeded actual consumption by 44 per cent, and a revised 1980 projection for 1986 by 25 per cent (Nectoux and Kuroda, 1989). When carried through the 50 years of long-term forecasts, such weaknesses in simulation or projection exercises can lead to forecasts which differ from each other, and actual performance, by very large margins – as is demonstrated by some of the results reviewed in this paper.

As has been noted earlier, it is important to keep in mind when interpreting the forecasts and projections reviewed below, that they are based on different assumptions about the underlying parameters and different analytical methods and models. The forward estimates from the different studies therefore cannot be compared directly one with another.

In the United States, the long-run demands for all major wood products are projected to grow over the next five decades (Table 8.3). Projected consumption of sawn softwood will be 1.2 times higher than consumption in 1986 by 2010 and 1.3 times higher by 2040. Consumption of sawn hardwood is projected to go up by margins of 1.7 and 1.8 times over the same periods. Consumption of paper products, fuelwood and pallets is expected to grow particularly rapidly. Within the panels group, fibre-based products are expected to grow most strongly until 2010, when consumption of softwood plywood, which is expected to decline in the near future, begins to rise again (USFS, 1989).

In Europe, consumption of paper and paperboard and of wood-based panels is expected to grow about twice as fast as consumption of sawnwood over the period 1980 to 2000 (Table 8.4).[6] Consumption of sawn hardwood grows considerably faster than sawn softwood; particle board is the fastest growing panel (and fibreboard the slowest); and printing and writing paper grows much faster than newsprint or other paper and paperboard (mainly packaging). Consumption of fuelwood also grows, particularly in northern Europe. Growth in the other products is fastest in southern Europe (ECE/FAO, 1986).

Table 8.4: Consumption of forest products in Europe in 1979–81 with estimates to 2000 and 2020 (million units)

	1979–81	**2000**		**2020**	
		Low	*High*	*Low*	*High*
Sawnwood (m^3)	102.3	119.0	140.8	123.0	148.0
Wood-based panels (m^3)	35.6	49.6	58.5	52.0	60.0
Paper and paperboard (m^t)	49.2	67.2	92.0	68.0	95.0

Source: ECE/FAO, 1986

In the USSR, the period to 2000 is expected to be one of rationalization and intensification in order to increase the productivity of wood use. A much larger proportion of the wood harvest is used as fuel or in solid form (sawnwood) than is the case in most other industrial countries. Consumption of both has been declining for some time, and change in the structure of wood use in the direction of replacement of sawnwood by panels and paperboard, increased use of residues as industrial raw material and further reduction in the use of roundwood as fuel, is expected to continue to the end of the century (ECE/FAO, 1989).

In Japan, consumption of wood peaked in 1970, dropping subsequently by about 10 per cent to the level of about 90 million m^3 at which it has remained throughout the 1980s. In the latest Forestry Agency projections (MAFF, 1987), it is projected to grow (in roundwood equivalent terms) to a level 7 per cent higher than in 1986 by 1994 and 15 per cent higher by 2004 (Table 8.5). In the more recent CINTRAFOR Pacific Rim study, a somewhat faster growth in consumption is forecast – 20 per cent from 1986 to 2000 (Cardellichio *et al*, 1989). All the projected growth is in panels and paper and paperboard, reflecting shifts away from the traditionally heavy use of sawnwood in housing and other construction.

The projections reproduced in Table 8.6 show substantial increases in consumption in the developing regions of Africa, Asia and Latin America in the period 1986 to 2000. Consumption and growth in consumption of all main products is concentrated in South America, the Far East and China, but growth rates in all developing regions exceed those in the industrial regions. In the

Table 8.5: Forecast of Japan's demand for wood products to 2004 (million m^3 roundwood equivalent)

	1984	**1994**	**2004**
Sawnwood	45	43–45	42–45
Plywood and panels	15	17	19
Pulpwood	31	35	40
Other	4	4	5
Total	94	99–101	104–108

Source: MAFF, 1987

period 1986 to 2000, this group of countries as a whole could account for more than a half of the projected increase in world sawnwood consumption. Growth rates in developing country consumption are projected to be fastest for panels and slowest for sawnwood (FAO, 1988).

The results of four recent global projections of demand are summarized in Figure 8.1. As is to be expected, the results show a very considerable spread, with consumption of industrial wood growing from less than 15 per cent to about 40 per cent over the period from about 1985 to 2000, and from one-third to three-quarters over the period 1985 to 2030 or 2040.

Only two studies estimate global demand over 50 years. The IIASA model (Kallio *et al*, 1987) projects quite a high rate of growth in demand throughout the period.[7] Because it incorporates rates of change which are rising exponentially in the latter part of the model, the estimates of demand for 2030 are probably on the high side of possible outcomes. The RFF model (Sedjo and Lyon, 1990) projects past growth trends in aggregate consumption, subsuming

Table 8.6: FAO global forest products consumption outlook projections to 2000 (million units)

	Sawnwood and sleepers (m^3)		**Wood-based panels (m^3)**		**Wood pulp (m^t)**	
1986	*1986*	*2000*	*1986*	*2000*	*2000*	
North America	139.9	142.2	42.4	66.8	65.6	84.7
Europe	95.3	111.3	36.3	58.8	36.9	49.2
USSR	91.2	109.8	12.6	19.8	8.7	12.5
Japan	34.3	36.6	9.7	17.2	11.3	18.0
Other Asia	67.5	114.5	9.8	22.1	5.5	9.1
Latin America	29.7	52.5	5.0	12.4	5.3	9.0
Africa	10.5	14.0	2.0	3.4	0.9	1.4
Oceania	6.8	7.7	1.4	2.3	1.8	2.3
Total	475.2	588.6	119.3	202.7	136.0	186.2

Source: FAO, 1988

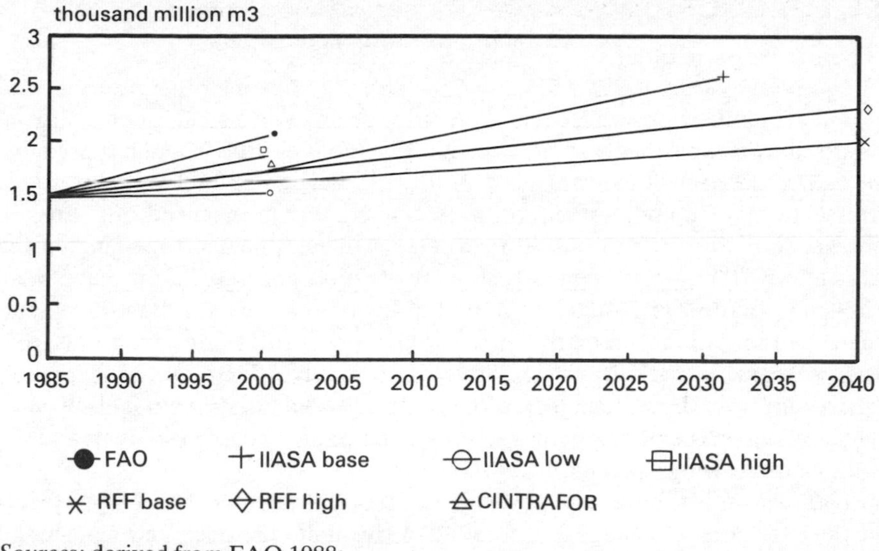

Sources: derived from FAO 1988;
Kallio *et al* 1987; Sedjo and Lyon 1990;
Cardellichio *et al* 1989

Figure 8.1: Trends in global consumption (comparison of long term projections)

the effects of change in income, technology and so on into a single relationship, which is applied to projections of population change. Growth in demand is assumed to decline steadily, to zero by 2040, from an annual rate of 1 per cent in the base scenario and 2 per cent in the 'high' scenario. Even the latter gives slow growth over the latter part of the period, and the long-run estimates could therefore be towards the low end of possible outcomes.

Though there is thus a great deal of uncertainty about rates of growth in consumption or demand beyond the end of the century, a number of features of the future can be discerned with reasonable confidence. The first is that there is no evidence to suggest that there will be fundamental changes in the place of the main wood products in the economy. None is likely to be displaced from its principal end uses, nor are there likely to be major break throughs which would radically alter the patterns of use. Second, with so much of consumption occurring in the mature markets of the industrialized regions, growth and change is likely to be moderate; with much of such growth as does occure likely to happen in the developing regions. Third, the continuing shift from solid to reconstituted products, towards greater input flexibility in processing technology, and towards increasing the efficiency with which raw materials are used, and re-used, will all continue to reduce the need for roundwood for a given end use and to extend the range of available wood biomass which can be used.

FUTURE SUPPLY

Trends in production

Roughly half of the world's forests which lie in the north temperate zone, and which presently supply more than four-fifths of the world's output of industrial wood, have been broadly stable for a considerable period. Though there have been significant fluctuations back and forth in the use of land at the interface between forestry and agriculture, these have roughly balanced out. Loss of forest land to urban and infrastructure developments, and increasing allocations to non-timber uses such as protection and recreation, have been reducing the timber production potential in some areas. On the other hand, increased investment in planted forest and in silvicultural interventions, and the growing availability of relatively productive land withdrawn from agricultural use, have been increasing timber production potential elsewhere. Throughout the north temperate zone net increment exceeds removals and the volume of growing stock has been increasing.

The reverse is the case, however, in the tropical regions. The area of forest in these regions is being rapidly reduced through clearance and transfer to agricultural uses, and there is little investment taking place aimed at increasing productivity of the remaining forest areas or in creating planted forests. In the south temperate zone, in contrast, investment in forests has taken place on a large scale.

During the future period being considered here, there will be a major shift in global production from old growth to planted and managed forests. Old growth resources which have been important sources of production to date, notably those on the west coast of North America and in the tropics, will be worked out, and large planted resources in the south and west of the United States, western Europe, Japan, Oceania and the southern parts of Latin America will be brought into production, together with second growth resources such as the hardwoods of the northern United States.

These shifts will have implications for the quality of wood supplies. Forests under management are seldom grown with premium wood quality as an objective; most are directed towards quantity rather than quality. It is unclear what the impact upon markets and users will be of the resulting decline in supplies of the best qualities of timber, veneer, etc; or what the acceptability of some of the plantation supplies will be.

The shift to greater dependence on planted forests raises other issues. One is the long term availability of the land on which planting is taking place, or is planned to take place. In North America and Europe much of the new planting is taking place on sites previously used for agriculture or pasture. Though the pressures to limit production of agricultural products in some areas and to put land under tree cover for environmental reasons in others presently point firmly towards the trend towards transfer to forestry use continuing and even accelerating (for example see CEE, 1988), in the long period being considered here circumstances could arise which could reverse these trends.

Growing demands for environmental and other non-wood outputs of the forest estate are also likely to constrain wood production in many parts of the world, as logging is prohibited in some areas and others become subject to more complex and costly silvicultural and management requirements. United States Forest Service projections already assume that environmental uses will have priority over wood production on all public forest land (USFS, 1989).

Another factor that could affect future supplies is related to ownership. A large part of the resourcc in such major producing areas as north Europe and the eastern United States is in the hands of non-industrial owners (as is the case elsewhere eg Japan). The predicted increases in output are therefore sensitive to the behaviour of such owners. Until relatively recently these were mainly farmers, who worked their forest holdings as part of the farm household enterprise, but changes away from these ownership patterns have been occurring. In Sweden, for example, where it is estimated that annual production from non-industrial private forests could be increased by 10 million m^3 within present sustainable yield limits, the proportion of holdings apparently operated as part of a farm enterprise dropped from nearly 70 per cent in 1951 to 33 per cent in 1981 (Eriksson, 1989). Non-farmers have been found to have less economic and financial inducements to cut, and they account for most of the untapped potential harvest (Lonnstedt, 1989).

In addition to these economic issues, uncertainties exist about a number of factors bearing on the biological productivity of the forest resource. A shift to large-scale single species planted forests increases the danger of disease or prests on an epidemic scale. Of more concern, because of their potential magnitude, are the possible effects of atmospheric pollution and of global warming.

Production forecasts and projections

Most studies model timber production from growing stock inventory data, adjusting starting values to take account of growth and removals in each future period. They therefore in effect shift a short-run supply curve through time in response to adjustments in the level of timber inventory. Very little of what has been attempted, and is summarized below, addresses longer term timber supply problems, such as those associated with transition from old growth existing natural stocks to managed and planted forests. Also, few studies have been able to develop economic supply curves; most are limited to predicting physical availability. For substantial parts of the world's forests, lack of information has prevented the development even of usable physical production estimates, so that they have to be dealt with outside the models.

Figure 8.2 summarises results from recent national and regional studies which have attempted production projections for the regions in which the main increases in output are expected.[8]

In North America, net growth presently exceeds removals in hardwoods by margins of more than 100 per cent. In both the United States and Canada net loss of forest land is occurring and other areas are being withdrawn from wood

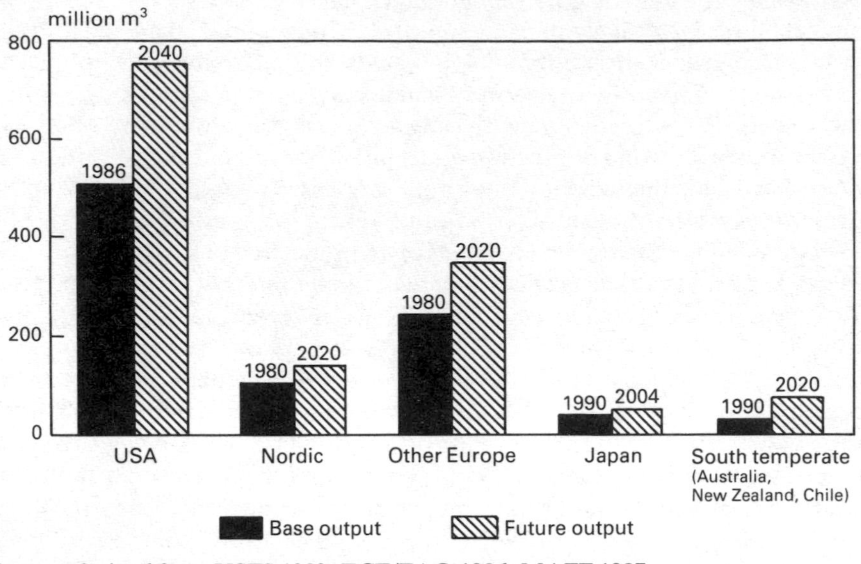

Source: derived from USFS 1989; ECE/FAO 1986; MAFF 1987

Figure 8.2: Future roundwood output (selected regional projections)

production for environmental, recreational and other reasons, but in the United States new planting on a very large scale is adding to net growth.

Because of environmental and other non-marked constraints on use of public forest land, it is assumed in United States projections that all growth in output must come from private forest areas. By about the turn of the century the remaining United States old growth softwood resource in the west will have been harvested, and further expansion of production will have to come from the predominantly planted resource in the south and west, and the very large inventory of predominantly hardwood second growth volume which has built up in the north and east. It is projected that softwood harvests will rise from 331 million m^3 in 1986 to 445 million m^3 in 2040, an increase of 36 per cent, and that hardwood harvests will increase from 178 to 320 million m^3 (79 per cent). The largest increases will be in the south (USFS, 1989).

In Canada, by contrast with the United States, the greater part of the resource continues to be in the form of old growth stands, with the proportion being highest in the far west and lowest in the east. It appears that increases in production of softwoods from public lands could be achieved in much of the east only in conjunction with greater silvicultural intervention, or a revision of allowable cut levels to draw in more of the resource currently classified as economically inaccessible. In the central prairie provinces, in contrast, there are large surpluses of both softwood and hardwood available.

The situation in British Columbia is less clear. Current harvest levels are close to the prevailing level of allowable cut, implying very little surplus.

However, the allowable cut is calculated on a resource base from which more than half of the forest land has been excluded primarily on the basis of economic judgements (eg an assumption that there would be no future demand for hardwoods). Sensitivity analysis indicates potential increases in harvest over 20 years of about 24 per cent through increased utilization of the resource base (ECE/FAO, 1990). Nevertheless, the consensus of opinion appears to be that, in the near future at least, there will be little if any increase in output in the province. At the national level, the most recent projection predicts further growth in Canada's roundwood harvests, mainly in the central region, with the rate of increase dependent on demand and price (ECE/FAO, 1990).

In Europe, the forest resource continues to increase. In the region as a whole, forest area grew by about 2.5 per cent in the 1970s, net annual increment by 8.9 per cent, and growing stock by 12.3 per cent. The increases were concentrated in the Nordic countries and in the countries along the western edge of the region where investment in plantation forests has been heavy (France, Ireland, Spain and the United Kingdom), but all countries recorded some growth. Continued planting in forest areas, natural extension of forest areas and the growth in planting outside forest as land is taken out of agricultural use, are expected to result in continued increases in the region's production potential.Europe's net annual increment on exploitable closed forest has been predicted to grow from 504 million m^3 in 1980 to between 540 and 566 million m^3 by 2000, and by a further 24 to 48 million m^3 by 2020 (ECE/FAO, 1986).

Removals have been forecast to rise from 350 million m^3 (under bark) in 1985 to between 391 and 438 million m^3 by 2000, and between 431 and 490 million m^3 by 2020. The growth in harvests is concentrated in those parts of the region, identified immediately above, where the resource has been expanding most strongly. As this resource expansion is predominantly in coniferous species, the latter account for much of the increase. Nevertheless, in Europe as a whole, hardwood removals are predicted to grow as rapidly, maintaining its one-third share. Much of the growth in hardwood production is in southern Europe (ECE/FAO, 1986).

In a subsequent study by IIASA it has been estimated that the sustainable potential biological harvest in Europe would be 534 million m^3 (over bark) from the present forest area, and 564 million m^3 if the forecast expansion of forest area to 2020 is taken into account (Nilsson *et al*, 1990). The latter level would allow harvests to be raised by up to about 140 million m^3 above the level of actual output in 1987, an increase similar to the higher level ECE/FAO forecast.

Growing stock is also reported to be increasing in the USSR. Annual roundwood production peaked in 1975, and in 1985 at 368 million m^3 was no higher than it was in 1960, seemingly largely because of problems in the logging and transport industries. However, the proportion of hardwoods in the harvest has grown, and the proportion of fuelwood has fallen sharply. A recent sector study by the Soviet authorities (ECE/FAO, 1989) predicts that the volume of wood harvested annually will not change greatly in the period to 2000. However, the total output of industrial wood is predicted to rise by more than a quarter between 1985 and 2000. Considerable increases are also planned in output of processed forest products, based on increased use of residues, better allocation of the roundwood available and other measures to increase productivity and efficiency. Production from Siberia and the Soviet Far East has been growing in recent years, but it is predicted that in the period to 2000 output will continue to be concentrated in the European and Uralian regions. Whether there will be a substantial rise in the longer term in output from the large resources in the far north and northeast, where transport, environmental and productivity problems have been encountered in the past, is not clear.

In Japan, a heavily forested country, investment in plantations in the 1950s and 1960s on a large scale has created a very substantial resource from which domestic production could be expanded in due course. The Forestry Agency (undated) has estimated that in 20 years time the potential output from these long rotation planted forests will have reached 89 million m^3. Much of the resource is in private ownership, and high costs and low prices in markets which continue to be dominated by imported supplies have not yet made production of other than high quality timber attractive. However, though domestic output is bound to remain responsive to such factors as the impact of exchange rates on the cost of imports, the presence in the sawlog resource of a growing share of the large sized high quality timbers required in traditional construction, and pressures on the pulp industry to limit its dependence on external raw material supplies, are likely to result in increased domestic production in the future. In the government's 1987 Basic Plan for Japan's forestry resources, the country's production is predicted to increase from recent levels of between 30–35 million m^3 to between 40–43 million m^3 by 1994, and up to 52 million m^3 by 2004 (MAFF, 1987).

The situation in the tropical regions is very unclear. The data base is very poor or non-existent, and the heterogeneous structure and selective felling form of harvest practised in most tropical closed forests do not permit the use of models developed elsewhere. It has been estimated by FAO/UNEP that during 1981–85 about 11.3 million hectares were being deforested annually in the tropics, of which some 7.5 million hectares were closed forest (Lanly, 1982).[9] The same source estimates that about 20 per cent of the remaining closed forest has been subjected to clearing under shifting cultivation in the past, and a further 20 per cent has been logged over. In Africa and Asia the two are linked; the road infrastructure created for harvesting creating access for slash and burn farmers. In those two regions most deforestation is thus occurring in forest areas of commercial value. In Latin America clearing of forest is primarily to create pasture for livestock production.

The lack of reliable data severely hampers the task of analysis and projection for the tropical resource. An early attempt to try and model and project long-term tropical wood production resulted in projections of decline in output in Asia from the 1990s, but a rapid increase in production in Africa and Latin

America to several times 1980s levels by 2000, followed by a steady fall in output in both regions (Grainger, 1986). More recent studies have produced much lower projections. In the CINTRAFOR study, it is estimated that in the period to 2000 production from the existing forest resources in most of the main producing areas in South East Asia will be no more than maintained at present levels, that output in West Africa will also remain at about present levels, but that production will grow by about 20 per cent in Latin America (Cardellichio, *et al*, 1989). Production from existing tropical resources is expected to decline in the early decades of the next century, but the main Asian producers expect recent and planned investments in planted forests to support a renewed growth in production, as does Brazil, but this will be a quite different resource to the old growth forests on which the trade in tropical hardwoods has been based.

Particularly fast rates of growth in output are expected in the south temperate zone, based on the large areas of high yielding plantations which have been established and are continuing to be established there – in particular in Chile, New Zealand and Australia. Over the past decade, output of industrial wood from the south temperate zone grew by nearly 50 per cent, by far the fastest rate of growth of any region. Recent forecasts predict a growth in output of industrial wood from plantations between 1990 and 2020 by two-and-a-half times in Chile to 29 million m^3, by a similar margin in New Zealand to an output of 25 million m^3, and by more than double in Australia to about 20 million m^3. Considerable uncertainty still attaches to both quantities and quality, and the acceptability of what will be predominantly utility grades of radiata pine lumber in the market, and recent forecasts have tended to revise projected outputs downward (Hunter, personal communication). Nevertheless, it is quite clear that even the already existing planted resource will become a major source of supply in the period under review.

Possible impacts of atmospheric pollution and global warming

Much uncertainty surrounds the possible impacts of atmospheric pollution on forest productivity. It is unclear to what extent various pollutants which could be harmful to trees are actually affecting forests, and if they are, what orders of magnitude of decline in productivity result, or could result.

The impact of damage on supplies is likely to be two-fold: short term disruptions to wood markets, as production from salvage and sanitation fellings is added to supplies, and a longer term reduction in sustainable output. In a recently completed study by IIASA the impact of sulphur and nitrogen compounds on sustainable output from forests in Europe was explored. This was based on the extent of forest areas with deposits in excess of 'target loads', and on what is known about damage cycles at higher levels of deposit for different species, age classes etc. For Europe as a whole, this resulted in an estimate of long term loss of potential wood supply of about 85 million m^3 (over bark), equivalent to a reduction of 15 per cent in the base case rate of increment. The extent of the decline varied considerably within Europe, from 8 per cent of the rate of increment in the Nordic countries to 25 per cent in

Eastern Europe (Nilsson *et al*, 1990).

However, an analysis of the possible consequences of a range of forest damage scenarios for western Europe, using the Finnish global simulation model, found that even major changes in increment or sanitation/salvage fellings are likely to have only modest impacts on production, trade and use of forest products. The results indicated that a 50 per cent reduction in increment by 2010 would reduce harvests in the region in that year by only 10 per cent. A 100 per cent shift in the supply curve of coniferous timber in western Europe, due to mortality and damage, increased coniferous roundwood harvests by only 14 per cent by 2000, and output of processed coniferous products by considerably less (Seppala *et al*, 1990).

A number of factors contribute to this. The effects of pollution take effect only progressively. Sanitation fellings are substituted for scheduled removals, roundwood imports into the region are reduced, and increased consumption and export of forest products in response to the lower prices resulting from the supply increase help absorb increased industrial production. Trade therefore diffuses and dampens the effects of the change in supply in one region, by absorbing some of the impact in other regions. This is illustrated in Table 8.7, which summarizes the estimated changes in production which result under the scenario with the greatest increase in supply in western Europe.

Table 8.7: Projections of changes in coniferous roundwood harvests and processed product output by 2000 as a result of severe acid rain damage in western Europe[1] (million unit)

	Coniferous harvest (m^3)	**Coniferous sawnwood production** (m^3)	**Paper and board production** (tons)
Western Europe	+12.7	+1.2	+3.6
Finland	−1.0	+0.2	−0.3
Sweden	−0.9	−0.2	−0.7
North America	−4.9	−2.2	−2.3
Latin America	−1.5	+0.1	−0.2
Rest of World	−0.5	+2.6	−0.1

Source: Seppala *et al*, 1990

[1] The scenario assumes that damage results in a 100 per cent shift in the supply of coniferous timber in the region

Global warming is likely to be generally beneficial in terms of improved forest growth. However, drier summers and other effects of climate change could produce undesirable effects on growth of some species, and the shifts in the climatic boundaries within which particular species will grow could adversely effect production in a given country or region.

Again the high degree of uncertainty as to the magnitude or timing of global warming, and about the climatic changes that could result in different parts of the world, have prevented any meaningful work on the possible impact on wood supplies. An analysis as part of the IIASA Global Trade Model study (Kallio *et al*, 1987), which assumed that the effect of warming would be confined to the boreal forests, that it would have a pronounced effect on growth in these forests, and that this effect would have fully worked its way through to increased production within the next 50 years, produced a predictable projection of lower prices, increased global consumption and lower production outside the boreal zone. However, as the authors point out, this is an extreme scenario, and it is likely that in reality any impact on existing forest yields over the time horizons being considered in this study would be small. However, if planting of trees was adopted on a large scale, in order to take up and store CO_2 – as has been under consideration in the United States – it might significantly affect wood availability before the end of the period (Kaiser, personal communication).

FUTURE BALANCE BETWEEN SUPPLY AND DEMAND

Supply and demand are brought into balance through the effects of price changes; rising prices dampening demand and stimulating output, while falling prices have the reverse effect. Trade is of course a key component of this balancing process. In this section these dimensions are examines first through the global wood supply–demand simulation models, and then by examining trade and price trends separately in some more detail.

Simulation of global wood supply–demand relationships

The projections under the base case scenario in the IIASA model (Kallio *et al*, 1987), which is associated with a relatively high growth in demand, show global output rising to about 2.6 thousand million m^3 annually by 2030. A very large part of the growth in removals in this estimate is accounted for by the United States and Canada, which by 2030 are predicted to account for 45 per cent of industrial wood removals. Non-coniferous species account for a large part of United States growth. Growth is also faster than average in the south temperate regions (Brazil, Chile and Australia–New Zealand). In all other regions removals increase only moderately. Under a lower growth scenario (with a growth rate of income per capita 50 per cent lower than in the base case scenario) growth in removals is about one-third less than in the base scenario. This is reflected in roughly similar reductions in the main industrial categories, with the exception of non-coniferous sawlogs which show much less decrease.

Under the RFF model[10] (Sedjo and Lyon, 1990), which assumes a relatively slow growth in demand, global supply rises to about two thousand million m^3 by 2040. Under this model the largest increases in industrial wood removals come from the United States (predominantly from the south) and the south temperate zone, which by the end of the 50 year period is second only to the southern United States in magnitude of production. Canada shows an overall decline, with western Canada declining substantially over the first 30 years but increasing again later. In the Nordic region of Europe removals expand initially but fall back later. South East Asia maintains its level of removals.

Under a higher demand scenario (demand growing initially at 2 per cent rather than 1 per cent annually, but still declining gradually to zero by the end of the period), which involves twice as large an increase in global removals over the period, to 2.3 thousand million m^3, production in eastern and western Canada and in north Europe is considerably higher, as higher prices extend the margins of what is economically accessible. Production in the United States also grows more rapidly and to higher levels.

Comparison of these projections with each other, and with the regional and national forecasts examined earlier, predictably shows considerable differences in the magnitudes involved. For example, in the base case projections from the IIASA and RFF global models roundwood supply in the southern region of the United States grows to levels far above that arrived at in the much more detailed study by the United States Forest Service (USFS, 1989). Nevertheless, the broad patterns of change are essentially similar – the main increases in supplies are accounted for by areas with maturing planted or second growth forests, but most of the regions with large resources of natural forests continue to be major producers and account for the greater part of the additional increases in output that would be called forth by a more rapid growth in demand and prices.

Future trade trends

Trade is the most difficult component of the supply–demand balance to analyze and predict. It has not proved possible to model and project trade separately; in most exercises it is treated as a residual, to be adjusted subjectively in the light of the results of separate production and consumption projections. The global simulation models developed to study trade in conjunction with production and demand have not yet reached the stage where they produce useful estimates of particular trade flows; but are of considerable value in exploring how total exports, patterns of supply and prices might change under different conditions.

The results from the IIASA model show the shares of Canada and the Nordic producers in sawn softwood trade increasing, while the United States becomes a major exporter of sawn hardwood and of plywood/veneer. The United States and south temperate countries gain share in pulp exports, Canada in newsprint, the Nordic countries in printing and writing papers, and the United States in packaging paper and paperboard. Western Europe's imports of most product groups decline, the exception being printing and writing paper (other than newsprint).

The recent joint study for North America by the United States and Canadian forest services (ECE/FAO, 1990) forecasts net outflows from that region in the period to 2005 (Table 8.8). Substantial increases in exports of sawn softwood, paper and paperboard and building board are predicted, and the United States also expects exports of softwood plywood to develop as supplies from tropical producers diminish, and exports of sawn hardwood as production of this

Table 8.8: Forecasts of North American exports to 2005[1] (million units)

	1985	1995	2000	2005
Coniferous logs (m^3)	16.9	14.6	13.5	13.5
Coniferous sawnwood (m^3)	10.6	12.6	13.7	15.1
Non-coniferous sawnwood (m^3)	0.7	1.2	1.2	1.2
Coniferous plywood (m^3)	0.7	1.0	1.1	1.2
Paper (tonnes)	1.6	2.9	3.3	3.4
Paperboard (tonnes)	2.8	3.8	3.9	4.1

Source: ECE/FAO, 1990
[1] Exports of the USA and Canada to the rest of the world (ie excludes trade between the two countries)

product expands in the east of the country. The large export of roundwood at present is expected to decrease, as the west coast old growth resource is worked out. However, the CINTRAFOR projections indicate increased pulpwood flows within North America, from Canada to the United States (Cardellichio *et al*, 1989).

The main recent outlook study for Europe, the European Timber Trends Study, estimated that the region's net imports would grow by a margin equivalent to 25–45 million m^3 of wood between 1980 and 2000, but that it would be unlikely to grow further by 2020 unless demand expanded considerably faster than postulated in the base scenario (ECE/FAO, 1986).

Subsequently, a number of studies in the Nordic countries have been exploring how the region's trade, within Europe as well as into it, is likely to develop. A Finnish study (Seppala *et al*, 1990) predicts declining trade in sawn softwood after 2000 as self-sufficiency within western Europe increases, a rapid rise in Nordic exports of printing and writing papers to the rest of the region, but a decline in the intra-regional flows in newsprint, as Canadian imports gain share, and in other paper and paperboard in which the United States is expected to increase its share of the European market. Intra-regional trade in pulp also declines as Nordic producers add value by concentrating on paper and board production (Bystrom and Lonnstedt, 1989).

A continued rise in Europe's trade in roundwood in the near future is also predicted (Seppala *et al*, 1990), based on outflows from eastern Europe and the USSR. However, the economic changes now taking place in these countries clearly could alter their supply availability for export. The CINTRAFOR study forecasts declining roundwood outflows from the USSR and eastern Europe, as growing demand for pulp and declining residue availability increase pressures on pulpwood supplies everywhere, and predicts that the diversion of sawlogs to pulp use which is already taking place in some countries could continue, possibly curtailing Nordic exports of sawn softwood (Cardellichio *et al*, 1989).

Japan plans to reduce its imports, both as a share of aggregate consumption and in absolute terms, in favour of domestic output. As was noted earlier, this shift has not yet started to take place. However, its import pattern has recently been radically restructured, following the reduction in availability of tropical roundwood supplies, primarily towards roundwood from North America. As supplies of roundwood from old growth resources on the west coast of North America diminish a shift towards a greater share of processed products within the import total is expected. Already Japan has become a large importer of tropical hardwood plywood, and North American analysts expect it to become an importer of pulp and paper on a much larger scale than at present (ECE/FAO, 1990). However, the phasing of such a shift is uncertain, because of the potential availability of coniferous pulpwood from Chile, New Zealand and the far eastern region of the USSR; and of hardwood chips from a number of sources. As in the recent past, the balance between raw materials and processed imports is likely to be materially affected by exchange rate fluctuations.

The future evolution of tropical forest products exports is particularly difficult to predict. Aggregate trade is presently about one-quarter lower than it was at its historical peak in 1979. The sharp fall thereafter was partly due to depressed demand at the beginning of the 1980s in the main importing countries, and partly to the restrictions in log exports by some of the main producers, which sharply cut back imports into Japan (by far the largest importer). Exports of processed products have been growing, but face limited market prospects in the main user regions. Tropical hardwood sawnwood is mainly used in Europe and North America in applications such as joinery in which it faces keen competition from both temperate sawn hardwoods and sawn softwoods. The largest user of tropical hardwood plywood has been the United States where the market for the product peaked in the 1970s and has been declining since. Japan, another major user of non-coniferous plywood (manufactured in Japan from imported logs) has expanded its imports as tropical hardwood roundwood supplies have declined, but is also shifting away from its use.

It therefore seems likely that present outflows to industrialized countries will be reduced due largely to higher quality woods in the course of the future period being considered in this chapter. If environmentally inspired consumer boycotts of produce from unsustainable sources were to take effect even this trade could decline. Recent studies suggest that, even in the period to 2000, the bulk of tropical hardwood production is likely to be increasingly directed to domestic consumption and to exports to other developing countries. The CINTRAFOR study predicts that in this period only Brazil among the major suppliers will significantly expand its exports, with outflows from the main producers in South East Asia being maintained at about present levels under the stimulus of rising prices, and exports from West Africa declining as domestic consumption grows (Cardellichio *et al*, 1989). In the longer term, Southeast Asia producers expect to be able to expand exports of plantation grown wood products.

Much of the greatly increased output from the planted forests in the south temperate zone should be exported. As was indicated earlier, they are expected to become a major source of pulp, or pulpwood, and probably of coniferous sawnwood as well.

To sum up, it appears likely that in future trade will play a declining role in the supply of sawnwood in the industrialized regions, given mature markets and growing self-sufficiency, but will become more important for some fast growing developing country users, such as China. Trade in the more highly valued pulp and paper products, on the other hand, should continue to increase its share of the principal supply patterns, with the trade structure evolving to reflect both changing raw material availability and shifts in the product strategies of the major producers. However, as was pointed out in the discussion of the possible impacts of heavy damage from atmospheric pollution, changes in trade are likely to be the main way in which unexpected disruptions of supplies, or markets, are absorbed. Therefore trade could show greater deviations from what has been predicted than production or consumption.

Price trends

At the global level, the base scenario in the RFF projections shows a very gradual rise in real prices of roundwood (aggregate of sawlogs and pulpwood) over 50 years to 2040, at an average rate of just under 0.2 per cent annually. Under the high demand scenario this rises to 1.2 per cent (Sedjo and Lyon, 1990). The base scenario in the IIASA projections to 2030 shows real prices rising globally more rapidly, as is consistent with the faster growth in demand assumed. Prices of sawlogs rise faster than prices of pulpwood, but prices of processed products rise very little (with the lowest rises being in pulp and paper products). In the low demand scenario even the prices of roundwood are almost constant over time (Kallio *et al*, 1987). The CINTRAFOR global projections to 2000 highlight relatively faster rises in prices for coniferous pulpwood and non-coniferous sawlogs. Again product prices are broadly stable over time (Cardellichio *et al*, 1989).

The evolution of real prices has been most extensively documented and analyzed for the United States softwood resource. A number of recent studies for the pine resource in the southern US, the largest supply region, indicate that prices there will increase during the next 30 years, but at a decreasing rate and at rates that will be lower than those that occurred in the past. It has been argued that the historical and forecasted changes in these prices are consistent with a shift from mining old growth timber to investment in forest management as prices rise to a level at which forestry is profitable because of wood production alone (Binkley and Vincent, 1988).

In the most recent United States Forest Service study (USFS, 1989), the real price of coniferous sawlogs in the United States south grows at an average of 1.03 per cent annually to 2030, and at 1.14 per cent in the west, with much of the rise expected to happen in the period immediately after the turn of the century as supply shifts from old growth to plantations. Non-coniferous sawlog prices grow more slowly initially but are expected to rise after 2000, as hardwood stocks are brought into use (with quality hardwoods attracting higher prices throughout). Plywood prices rise more slowly than sawlog prices, because of the expected substantial increase in the recycling of paper.

There is less information about expected future prices in Europe. The evidence in the ECE/FAO study seems to indicate that, in general, in contrast to the United States, real prices for forest products did not rise in Europe between the mid 1960s and the early 1980s (ECE/FAO, 1986). A recent study using the revised Finnish global model (Seppala *et al*, 1989) predicts for the period to 2010 increasing real prices in western Europe for coniferous pulpwood – because of the slow growth in supplies of residues – but declining prices of coniferous logs. Real prices of processed products increase only very slowly (with the fastest rises being for construction panels and sawn hardwood). Nordic country analysts point to the growing supplies of roundwood in parts of Europe without corresponding processing capacity, the growing use of roundwood imports to achieve flexibility of supply at a given price level, and continuing technical development towards wood and fibre saving in the processing industries, as pressures likely to keep roundwood prices from rising (Lonnstedt, personal communication).

There is thus predictable variation among the price projections and forecasts reviewed. These vary with the underlying assumptions about the rate of growth in demand, and with differences in the forecasts about such factors as the extent to which increased recycling of paper will limit increases in demand for pulpwood, and the speed with which depletion of old growth resources will limit the supplies of woods of particular qualities and characteristics. However, none of the studies suggests that on average real prices are likely to grow at rates in excess of those in the past; indeed most suggest lower rates of growth. Moreover, with further technological advances and improved efficiency at the harvesting, processing and distribution stages of production, delivered prices for wood are likely to continue to grow more slowly than stumpage prices, and product prices more slowly than wood prices – which should exhibit little if any growth.

CONCLUSIONS

As has been pointed out the basis for drawing any conclusions about the long term future of wood supplies is extremely weak. Nevertheless, a number of broad trends and conclusions do emerge from the main studies that have been reviewed here. These are summarized on the following page.

In brief, the evidence, such as it is, suggests that if global demand continues to grow at rates consistent with historical growth in consumption, supplies are likely to expand without appreciable real increases in prices. Though some regions would experience a worsening domestic supply situation, the world as a whole would not. Moreover, it suggests that, even if demand were to grow relatively rapidly, requiring more substantial increases in prices of roundwood in order to stimulate sufficient supplies of wood, prices of processed products

SUMMARY OF PROJECTED LONG TERM TRENDS

Demand

- Global demand for industrial wood has been projected to grow by margins of 15 to 40 per cent over 15 years, and by 35 to 75 per cent over 50 years.
- Consumption will continue to be concentrated in the developed world, but a substantial part of the increase in consumption will occur in the developing countries (in particular in sawnwood).
- Continued shifts from solid to reconstituted products will contribute to sawnwood exhibiting the slowest, and pulp and paper products the fastest, growth in consumption.
- Demand for roundwood will grow more slowly than demand for products as technology and increased processing efficiency continue to extend its use.
- Use of non-coniferous species will grow faster than coniferous species.

Production

- Though the tropical resource will continue to decline, the much larger and expanding temperate resources can support expansion of supplies in line with projected global demand without appreciable real increases in prices.
- There will be a major shift from old growth to planted and second growth resources: the tropics and the west coast of North America will become less important in the global total, and planted resources in the United States south and in the south temperate zone (and in the EC and Japan), and the second growth hardwood resource of the north-eastern USA, more important.
- Large resources in central Canada, northern Europe and the far eastern regions of the USSR can also support expanded production if demand dictates.
- Even if the rate of growth in demand results in increasing real prices of roundwood, there is likely to be little if any rise in product prices.

Trade

- The two largest importing regions of western Europe and Japan could become more self sufficient, and parts of the developing world less sufficient.
- Import of roundwood is likely to decline as a share of supply, but pulpwood trade could increase where wood residue supplies tighten.
- Import of sawnwood is also likely to become less important in developed world supplies but more important in the developing world, absorbing much of declining tropical wood exports.
- Trade in pulp and paper is likely to become more important in global supplies, with North America, northern Europe and the south temperate zone accounting for much of the expansion in exports.
- Adjustments through trade are likely to diffuse and dampen the impacts of even large adverse disruptions in supply or demand, such as might occur due to damage to forests from atmospheric pollution.

could be expected to rise very little, if at all.

This evidently is at variance with the widely held view which also permeates many official reports on the subject (for example CEE, 1988), that the world is facing a growing shortage of industrial wood. It is possible that this view has been too much influenced by the decline in the tropical resource, and underestimates the contribution of the stable and growing temperate resources. The importance of technical change in extending the resource and containing increases in costs is also widely underestimated.

As has been noted, there are a number of possible outcomes which could bear adversely on supply, such as damage from atmospheric pollution. Nevertheless, exploration of the sensitivity of the supply situation to changes of different kinds which has been carried out in several of the studies reviewed suggests that even large adverse disruptions would not result in appreciable long term increases in costs and prices. This is because adjustments through trade tend to dampen and smooth the impacts of change, with roundwood production and the processing industries able to adjust to much of the rest of the consequences of even major changes.

A final point that should be made is to note the extent to which the strong supply situation is the result of past investment in wood production. A continued growth in the sector at the same pace beyond the 50 year period cannot be assumed unless there is a continued inflow of capital into managed resources able to provide wood and fibre without substantial rises in real costs. Also, much of the additional roundwood supplies which will become available over the next decades, even in western Europe, are in regions where there is not at present a corresponding industrial processing capacity. The pattern of supply and use, and to some extent of price, is therefore likely to be affected by the extent to which investment is made in processing and manufacturing to complement the changing structure of roundwood supplies.

This chapter originally appeared in 1991 as 'The Long-term Global Demand for and Supply of Wood' in '*Forestry Expansion: A Study of Technical, Economic and Ecological Factors*', a report by the UK Forestry Commission.

NOTES

1. As fuelwood, and other industrial wood (principally poles), are traded in only very small quantities, they are considered in the rest of this paper only to the extent that they affect supplies of categories of wood and wood products which are traded.
2. In the report the term softwoods is used interchangeably with coniferous products and hardwoods with non-coniferous.
3. Much of the second generation work starts from the IIASA model and is developing analytical models tailored to the situation of particular countries and regions. Two which are of particular interest in the present context, because they are developed within a global framework, are the CINTRAFOR Global Trade Model (Cardellichio *et al*, 1989) being developed at the Center for International Trade in Forest Products, University of Washington, primarily to analyze forest products markets in the Pacific Rim region, and the MESTA model developed at the Finnish Forest Research Institute to study the implications for the Finnish forest sector of market and production developments elsewhere in the world (eg Seppala *et al*, 1990).

4. The limitations that will remain even after further work have been well summarized as follows: there will always be unrealistic features of the model structure and assumptions; the underlying data are of poor quality or non-existent in many regions of the world; the model cannot be statistically validated; many economic relationships defy quantification; and structural change remains difficult to model and even more difficult to predict (Cardellichio and Adams, 1988).
5. Real energy costs are projected to grow five-fold by 2040 in the United States ins the US Forest Service outlook study (USFS, 1990).
6. The projections to 2025 shown in the table simply extend the per capita consumption levels estimated for 2000, and apply them to population forecasts for 2025 (ECE/FAO, 1986).
7. The 'low' growth projection shown in Figure 8.1 assumes per capita incomes grow at only one half of the rate assumed in the base projection; the 'high' growth projection assumes per capita incomes in developing, but not developed, countries grow at rates 50 per cent higher than those in the base scenario.
8. As was noted earlier, the different studies employ different methodologies, and are based on different assumptions about the evolution of demand, costs, and other supply-shifting factors. The projections for the different regions are therefore not necessarily comparable with each other.
9. Preliminary results from a more recent FAO study have produced much higher estimates; 52 countries in which the annual rate of deforestation had been estimated to be 9.2 million hectares in the earlier assessment show a rate of 16.8 million hectares in the more recent study (Dembner, 1991).
10. This model incorporates only some of the producing regions, which together presently account for somewhat more than half of global output of industrial wood. Output in the remaining 'non-responsive region' – which includes western Europe and most of the tropical producers as well as the centrally planned countries – is assumed to grow initially at a rate of 0.5 per cent annually, gradually falling to zero after 50 years (Sedjo and Lyon, 1990).

REFERENCES

Binkley, C S and Vincent, J R. (1988) Timber Prices in the US South: Past Trends and Outlook for the Future. *Southern Journal of Applied Forestry*, vol 12, no 1, pp 12–18.

Bystrom, S and Lonnstedt, L. (1989) The Swedish Forest Industry in an International Perspective 1970–2000. SIMS Report no 7, Swedish University of Agricultural Sciences, Uppsala.

Cardellichio, P A and Adams, D M. (1988) Evaluation of the IIASA Model of the Global Forest Sector. Working Paper 13, CINTRAFOR, University of Washington, Seattle.

Cardellichio, P A, Youn, Y C, Binkley, C S, Vincent, J R and Adams, D M. (1988) An Economic Analaysis of Short-Run Timber Supply Around the Globe. Working Paper 18, CINTRAFOR, University of Washington, Seattle.

Cardellichio, P A, Youn, Yeo Chang, Adams, D, Joo, Rin Won and Chmelik, J T. (1989) A Preliminary Analysis of Timber and Timber Products Production, Consumption, Trade, and Prices in the Pacific Rim Until 2000. Working Paper 22, CINTRAFOR, University of Washington, Seattle.

CEE. (1988) Community Strategy and Action Programme for the Forestry Sector. Commission Communication COM(88) 255 final, Commission of the European Communities, Brussels.

Dembner, S. (1991) Provisional data from the Forest Resources Assessment 1990 Project. *Unasylva*, vol 42, no 164, pp 40–44.

ECE/FAO. (1986) European Timber Trends and Prospects to the Year 2000 and Beyond. ECE/TIM/30, United Nations, New York.

ECE/FAO. (1989) Outlook for the Forest and Forest Products Sector of the USSR. ECE/TIM/48, United Nations, New York.

ECE/FAO. (1990) Timber Trends and Prospects for North America. ECE/TIM/53, United Nations, New York.

Eriksson, M. (1989) The Change of Ownership Structure in Private Forestry in a Historical Perspective. Department of Forest-Industry-Market Studies Report No 5, Swedish University of Agricultural Sciences, Uppsala.

FAO. (1988) Forest Products: World Outlook Projections. Forestry Paper 84, FAO, Rome.

FAO. (1989) FAO Yearbook: Forest Products 1976–1987. FAO, Rome.

Forestry Agency. (Undated) Forestry and Forest Industries in Japan. Wood-Products Stockpile Corp, Japan.

Grainger, A. (1986) The Future Role of the Tropical Rain Forests in the World Economy. D.Phil thesis, Department of Plant Sciences, University of Oxford.

Kallio, M, Dykstra, D P and Binkley, C S (eds) (1987) The Global Forest Sector: An Analytical Perspective. John Wiley and Sons, New York.

Lanly, J P. (1982) Tropical Forestry Resources. FAO Forestry Paper 30, FAO, Rome.

Lonnstedt, L. (1989) Goals and cutting decisions of private small forest owners. *Scand J For Res* 4:259–265.

MAFF. (1987) Basic Plan for Japan's Forestry Resources and Long-term Demand and Supply Projection for Important Forestry Products. Ministry of Agriculture, Forestry and Fisheries, Tokyo.

Nectoux, F and Kuroda, Y. (1989) Timber from the South Seas: An Analysis of Japan's Tropical Timber Trade and its Environmental Impact. WWF International, Gland.

Nilsson, S, Sallnas, O and Duinker, P. (1990) Forest Decline in Europe – Forest Potentials and Policy Implications: Executive Summary. Draft, April 1990. Biosphere Dynamics Project, IIASA.

Sedjo, R A and Lyon, K S. (1990) The Long-Term Adequacy of World Timber Supply. Resources for the Future, Washington.

Seppala, H, Seppala, R and Kallio, M. (1990) Economic Impacts of Western European Air Pollution on the Finnish Forest Sector. In: Kauppi, P, Anttila, P and K Kenttamies (eds). *Acidification in Finland*. Springer-Verlag. In press.

USFS. (1988) The South's Fourth Forest: Alternatives for the Future. Forest Resource Report No 24, Forest Service, United States Department of Agriculture, Washington.

USFS. (1989) RPA Assessment of the Forest and Rangeland Situation in the United States, 1989. General Technical Report WO-56, Forest Service, United States Department of Agriculture, Washington.

PART IV
FORESTRY DEVELOPMENT AID

Chapter 9

Forestry Development Approaches

James J Douglas

Forestry can be practised in a variety of ways. At one extreme, there is the large-scale plantation and natural forest management by a traditional government department, for purposes (principally) of supplying material to a highly capitalized, technologically-oriented processing sector, or to export. At the other, there is small-scale rural forestry, practised at the household community level by rural dwellers for purposes of fuelwood, basic structural materials, shelter, food and fodder, and the range of minor forest products. Superimposed over the former type, in Lesser Developed Countries (LDC's), is very often the slash and burn practices of shifting cultivators who may or may not have been the historical residents of the forest areas. This phenomenon, and the associated one of unofficial encroachment on forested lands for permanent agriculture, are the most visible manifestations of the land use (and, ultimately, the economic) conflicts and paradoxes that so often characterize the forestry sector in LDCs.

There is as yet no clear consensus as to how the newer ideas on development should be put into forestry practice in LDCs. There is, however, more agreement that what was written about the role of forestry in LDCs in the 1950s and 1960s – and even well into the 1970s in some quarters – was (and is) very largely inappropriate. Nevertheless, large elements of the older view still persist in the forest sector administrations of LDCs. It is necessary, therefore, to examine some of this older material – first to understand why it does not stand the test of critical scrutiny on the basis of more recent general development precepts; second as a basis from which to examine why much of it survives in practice; and finally in order that we can consider some measures for rationalization and adjustment that can be taken in practice.

In addition to the literature on the role of forestry in development *per se* (which in fact is not extensive) there is a great deal of material on the specific techniques of economic evaluation of forestry projects in LDC's. The general literature on project evaluation in LDCs, upon which the specialized forestry material is based, is very large. Although in theory this material is not of direct relevance to the subject under consideration, it is necessary to deal briefly with one or two aspects of it, for reasons which will become apparent.

PROJECT EVALUATION IN FORESTRY

It is, or should be, a commonplace amongst practitioners and theoreticians alike in the project evaluation field that the results and conclusions from its application are highly sensitive to assumptions made about valuation of the various costs and benefits and externalities involved. These, as will probably be already apparent, are in turn dependent on the basic concept or model of the developing economy used. While project evaluation is not intended as a means of deciding upon the broad issues in development, it is often used for this purpose, either because of lack of definite guidelines on the general development strategy to be employed in the economy in question, or because of a lack of awareness or understanding of such guidelines on the part of the analysts and administrators involved. Arnold (undated), in identifying this tendency in the analysis of forestry projects, observes that in such cases, 'the project's relationships to such other objectives as employment, balance of payments, distribution of income, etc has been confined to listing or stating the number of jobs created, foreign exchange earned or saved etc by the project or project design chosen on the basis of its expected economic performance. Thus the project which emerges is not necessarily the one which represents the best balance in meeting the various objectives.'

Arnold notes that the sequence of project preparation from its origin of the technical levels through the administrative and financial stages to its economic appraisal, is very prone to the problem of the analaysis being narrowed or constrained to the extent where the best overall alternative may be prematurely discarded, or never considered. The author would extend this argument to the overall sectoral level: if sectoral activity is composed of a series of projects or programmes which are evaluated through the same sequential system, then the sector strategy as a whole will be unnecessarily constrained, and prone to errors of omission, unless specific steps are taken – usually from above the sector administration levels – to alter this approach. The most common omission, it is suggested, is adequate consideration of income distribution: sectoral project evaluations often either ignore it, or attach neutral weights to consumption gains, no matter where these accrue.

Distribution is paramount. Even when analyzing the difficulties in the theory of welfare economics, Little (1973) is quite clear on the point that a judgement about whether a particular distribution of income is good, or bad, must be made by someone. Little and Mirlees (1974) are highly critical of practitioners of cost–benefit methodology who overlook this requirement. Of the UNIDO (1972) guidelines for project analysis in developing countries (which advocate the use of aggregate consumption as the numeraire in calculation of accounting prices) Little and Mirlees say: '. . . it is not the case that aggregate consumption is the ultimate objective, which might make it a good numeraire. On the contrary, the authors emphasize that the consumption of different economic groups should be given different economic weights.'

The reasons why this is so go well beyond the moral imperative. Where impoverishment and destitution are sector and group specific to an appreciable extent, and where the worst affected groups are potentially important in terms of productivity, then re-distribution is a basic ingredient of, rather than an adjunct or side benefit to, aggregate income growth on a sustained basis. Yet, unweighted income growth criteria have been frequently used in the project evaluation literature. In Gane's (1969) detailed exposition on forestry and development in Trinidad, it is argued that forestry projects can be analyzed on the basis of aggregate gains to consumption. Gane opens his arguments for this viewpoint by reference to classic analyses by Pigou and Little (op cit) which argue that comprehensive direct measurement of welfare cannot be made, and that welfare effects cannot be estimated without value judgements about income distribution. Gane seems to deduce from this that it would be preferable therefore to avoid assigning weights to consumption by the various economic groups. However, it is impossible to avoid doing so, and the weightings implied in the aggregate criterion are no less definite, (but are probably decidedly less desirable), than those that would result from a more subjective weighting. Nowhere, it seems, does Little argue that deliberate allocation of different weights for income changes for different groups in society is undesirable; indeed, he suggests it is unavoidable.

The interesting theoretical development which seems to have arisen from the work of Lipton (1977), Streeten (1975) and others is that positive discrimination in investment capital allocation towards the industry containing the bulk of low income people might, ironically, have led many LDCs to higher aggregate income growth in any event. Whether such a finding would be repeated if applied to the lowest income groups within that sector is as yet imperfectly researched, but the nexus between sustained development and maintenance of demand may eventually lead LDCs in that direction anyway. What Lipton and others seem, indirectly, to have demonstrated, is that the use of the aggregate consumption criterion would have been acceptable, had all the important opportunity costs involved been reasonably calculated or estimated. Gane returns to the matter of opportunity costs, in a more limited way, later in his analysis, but it is surely apparent that his exposition would have benefitted by their inclusion at this point.

Gane develops his defence of the aggregate income criterion by claiming that forestry is unlikely to exercise negative effects on income distribution. He suggests that the sector by its nature favours lower paid workers, and argues that the flow of forest products will not cause income reductions to low income groups, and through import replacement or export generation will create employment for them. These arguments do not in fact stand when the bases of comparison are widened beyond other LDC industrial sectors; failure to remove this sectoral constraint is the basic conceptual flaw in material of this type. If the forestry sector, including the forest processing industries in an LDC, is consuming large amounts of development capital, then it seems, on the basis of Hagen's and Lipton's observations, that negative effects on income, and income distribution, will in fact be occurring – but this only becomes obvious when a wide enough range of possible alternative uses for such capital is considered. It is in fact extremely likely that the flow of forest products, or at least of those manufactured by and for a very narrow socio-economic stratum of society, will reduce the income of poorer groups in society.

The application of project evaluation methodology can only be as good as the broad economic assumptions upon which it is based; it cannot substitute for them. Even in cases where general guidelines on the need to create employment or re-distribute income are passed down from above the sector administration, the effect of these may be marginal, relative to the scale of change needed. It may be relatively easy for a sector administrator to adjust the existing programme of his sector, in accordance with the letter of such edicts. It will usually be a good deal less simple for him to perceive that the problem implied in such edicts might be far more profound – that a strong linkage between the existing economic structure of his sector, and the continued impoverishment of certain groups in society, might exist. In this, he will ultimately be guided by his received opinion of the appropriate role for his sector in the economy. This, if we accept the thrust of Lord Keynes' famous dictum,[1] will be a version of what academics and theoreticians have said about it. Therefore, this chapter will now consider what has been, and is being argued, as the most appropriate role for forestry in the developing economy.

FORESTRY IN DEVELOPMENT: THE ORIGINAL WESTOBY VIEW

The view of forestry in LDCs taken by forest economists and planners has, not surprisingly, been influenced by general concepts and thought on the economics of developing countries. Basically therefore, there have been two distinct phases since the 1950s – the earlier view, based more or less on the industrialization approach to development of the forestry sector, and the later approach, based on concepts of rural community forestry, village level industry and so on. This paper will examine critically how the general economic arguments have been translated into forestry terms, with a view to arriving at a generally acceptable approach to forestry in LDCs.

There has been relatively little written on the specific subject of the role of forestry in poor countries. It is fair to say that the case for the industrialization approach to forestry in LDCs was made in its most complete form by Westoby (1962), in a long article. Until the general concept of development implicit in that work became subject to strong doubt and criticism (and indeed, for some considerable time after this), little development of the arguments and contentions in it occurred.

Westoby delineates his area of interest by suggesting that forests provide a raw material for industries which have 'acquired great importance in developing countries'. He reinforces this constraint on his analysis by excluding fuelwood from consideration altogether, suggesting that it is of secondary importance from the point of view of economic growth. He does point out that the effect of fuelwood may not always be negligible, and suggests that South Asia is one area where it might assist agricultural productivity – a welcome exception indeed (although not developed in his analysis), considering the fact that in most of the developing world fuelwood is the most important forest output for most people, and is destined to remain so for some time to come.

A central theme of Westoby's argument is revealed in the identification of the high degree of structural interdependence between forestry and the industrial sector of the economy. These strong forward and backward linkages with the remainder of the economy demonstrate, according to this argument, the seminal importance of forestry in the development process. Nautiyal (1967) has shown that in fact the forestry- and timber-based industries have quite weak backward linkages. In the author's view, from a reading of the original data of Chenery and Clark which Westoby used, it is not apparent that the forestry sector does have an impressively high degree of inter-industry linkage with the rest of the economy at all. Even if it were so, the high degree of variability of multiplier results (which often seem as sensitive to the definition of the inter-industry model used as to genuine sector differences), from one country to another (and even from one region to another within countries) make the applicability of it to the general development case highly suspect. Westoby uses a 'special case' argument – that of exclusion of the construction sector from intermediate demand (and therefore from 'multiplier effect') – to defend his argument that forestry multipliers would be high if this effect were included. Perhaps so, but similar arguments undoubtedly apply to other sectors, and until someone examines them all on an exhaustive and systematic basis, results for individual sectors are of very limited value.

In any event, a high degree of structural interdependence with industrial sectors of the economy might be the very last thing desired of a primary sector, in an LDC where constraints of technology, capitalization and infrastructure are critical. The creation of demand for intermediate inputs from a range of industrial sectors is no guarantee whatsoever that such demand will be met – not, at least, in the real world of the developing economies. In other words, the existence of strong forward linkages for a given industry, measured on an historical basis (usually from a developed country economy) tells us nothing about whether similar effects can and will arise in the particular LDC we are studying.

Westoby is on firmer ground when he suggests that income elasticities of demand for forest products are high in low income countries, and that therefore any country with fairly rapidly growing income is going to demonstrate rapid demand growth for these products. However, apart from the possibly undesirable distributional aspects which apply to this (see the discussion of Gane's work above) it seems to be, once again, an incomplete argument: high income elasticities of demand undoubtedly characterize the incipient markets for a great many consumer items in LDCs, but it would be most surprising to find this were not so. There is no systematic evidence that forestry products are preferable in this respect. Nor is there any compelling reason for us to accept Westoby's associated argument that this market can be catered for relatively easily in LDCs because the forest industries are either easy to set up, or generate such high rates of return that investment will be quickly recovered from them. Once again Hagen's argument that the early phase of industrialization must only attempt to deal with already existing industries in LDCs is relevant here. Finally, in considering this argument, it is important to bear in mind that the principal task in the early stages of development is not to arrange the structure of growing demand, but to initiate income growth and its appropriate distribution.

The essential difference the authors of *A reappraisal of forestry development in developing countries* (from which this paper is taken) have with Westoby is in the view taken of the dynamics of the development process. Westoby wants the forestry sector to lead the LDC economy where he believes it should go, and he is in some haste to get there. But the mainstream of present thought on development would suggest that evolution towards development will be slow, and must be phased from a basis of effective growth in rural incomes and output, rather than from a superimposition of a modern, industrial sector. In the early stages of development, it is not a gross over-simplification to say that economic activity which is not obviously part of the solution (which involves rural productivity and the reduction of absolute poverty) will be part of the problem. Despite the time lags inherent in forest production it is no more logical to pursue accelerated establishment of the modern sector, or aggregate production as an end in itself in this sector, than it is to do so in agriculture, in the midst of the realities of intensifying rural poverty, a backward social structure and an ineffective capital market. The essential question is not when the forest resource should be mobilized, but how.

As noted earlier, little development of the ideas in Westoby's article occurred until the large shift in the general approach to development penetrated forestry circles. Thus, in FAO (1974) for example, we see reference to the favourable status of forestry as an industrialization base: the high growth of demand in relation to income growth; the strong forward and backward linkages of the sector with the economy; the ease of establishment of forest industries; the role of the sector as nucleus for industrial development; positive foreign exchange and employment effects. Much of the general approach in Westoby's original article also survives in MacGregor (1976), Von Maydell (1976), Gregory (1972).

THE FOREST INDUSTRIES OF BANGLADESH

The forestry sector of Bangladesh is a reflection of the dualistic nature of the economy in general which arises from the 'industrialization first' approach. On the one hand there is the Government forest resource, for the most part occurring in large concentrations and managed principally to supply output to the large scale forest products industries; on the other, there is the village forest resource, which supplies the basic fuel and structural needs of the rural population, and operates more or less independently of Government activities. This chapter is concerned with the evaluation of the forest industries of Bangladesh, in particular the larger scale (principally Government-owned) industries. This review draws heavily upon a report on the economic performance of this sector (Douglas *et al*, 1981).

Bangladesh is currently involved in pulp and paper production (from wood and non-wood furnishes) including rayon and cellophane; sawmilling; ply, veneer and reconstituted boards; and a range of secondary processes – furniture manufacturers, packaging firms and so on.

Before attempting to draw some general conclusions about the forest industries sector, some (very broad) assessment of the major products performance is warranted.

Pulp and paper

Paper making in Bangladesh is completely dominated by the Bangladesh Chemical Industries Corporation, a large Government conglomerate which controls some 30 enterprises, including a newsprint mill, two other paper mills, a pulp mill, and a rayon and cellophane plant. The basic economic performance data for these operations is given in Table 9.1 below.

Table 9.1: Economic and output data: Bangladesh pulp and paper sector

	1975/6	1977/8	1979/80
Production ('000 mt)	39.8	72.5	79.1
Domestic sales ('000 mt)	30.2	50.8	52.3
Operating surplus (Tk 10^6)	−157.2	−111.8	−121.8

Mt = metric tonnes
Tk = taken

In each case, very rapid rises in imported input costs have occurred in recent years. In the case of the newsprint mill, foreign inputs now account for some 65 per cent of total costs. Most of the plant established in Bangladesh is old, and is sub-economic in terms of scale and technology. Yet production from the mills far exceeds domestic demand, and much of the product from the mills is sold for export at prices well below domestic levels (and below production costs, in most cases). If foreign exchange earning (or saving) is seen as the principal benefit of the paper industries (as is frequently implied), then they are a remarkably expensive means of achieving this: in Douglas *et al* (1981), shadow price calculations are done which show that the cost of earning or saving one Taka of foreign exchange in the paper industries costs Bangladesh almost three Taka – this amount is well out of line even if significant over-valuation of the local currency is assumed.

Perhaps more than any other group of industries, the pulp-based sector of Bangladesh illustrates the difficulties of attempting to establish capital- and technology-based industries in an economy of that nature, regardless of what benefits for domestic production may have been adduced at the time. As it stands, the paper-making industries are now high on the list of the largest money losers in the economy. The infrastructural and capital requirements to sustain profitability in industries of this nature are not available in Bangladesh, and will not be for the forseeable future. Raw material resource constraints are also apparent, particularly in view of the priorities for use of the country's dwindling resources.

Manufactured board

The fortunes of the manufactured board sector in Bangladesh are less uniformly bad than in the pulp and paper sector. As can be seen from Table 9.2, it is a much smaller sector overall.

Table 9.2: Economic and output data: Bangladesh manufactured board industries

	1975/6	**1977/8**	**1979/80**
baseboard ('000 m^2)	1264.0	1858.0	1598.0
Production particle board (mt)	1684.0	3098.0	2476.0
ply ('000 m^2)	697.0	1500.0	1200.0
Domestic sales hardboard ('000 m^2)	1189.0	1756.0	1449.0
particle board (mt)	893.0	471.0	
ply ('000 m^2)	697.0	1500.0	1200.0
hardboard	−0.7	−0.1	+0.5
Operating surplus (Tk 10^6) partboard	−5.0	−6.8	
ply	−1.3	+0.1	+1.5

Source: Douglas *et al*, 1981

Both the hardboard and ply industries sell principally on the domestic market (the latter almost entirely to production of tea-chests for the export tea industry). Although both industries have a range of technical and other problems, they are at least managing to make ends meet, and to sell their products at a reasonable price.

The particle board industry is another matter. Presently it is based on jute fibre rather than wood. The product is sold on the domestic market for a loss, and on the export market for an even larger loss. In view of this, it is difficult to understand why the Government has established another (wood-based) particle board plant in the country. When this comes into production, the marketing problem will become even more severe. Moreover, the new plant will have other problems: it is to be based on the Chittagong Hill Tracts resource where, even for existing industries there, severe supply difficulties are already occurring.

Sawmilling

Sawmilling is a highly heterogeneous industry in Bangladesh, ranging from a fairly large Government sawmill, integrated with planing, preservation and other treatment units, through to itinerant pitsaws based in the rural area.

The large Government mill is presently highly unprofitable, partly because of poor design and maintenance, and partly because no market appears to exist for its intended output of large-sized material (for re-processing elsewhere), forcing it to base production on irregular, small-sized orders.

The small mechanical mills of the urban areas of the country, although besieged by the usual problems of poor power supply, irregular raw material availability and so on seem, by and large, to be able to provide the basic products required.

Little is known about the state of the rural sawmilling sector. In Douglas *et al* (1981), a residual estimate shows that it is a large sector. Since it has operated and continues to operate at a significant level without assistance or intervention from the Government, we must assume that it is reasonably profitable. However, from observation, it is an inefficient processor, in terms of sawn recovery, although the lack of preservation facilities must shorten the life of sawn products in the Bangladesh environment considerably.

Interpretations

There are a number of other industries in the Bangladesh forest products sector – match factories, furniture factories, packagers and so on – but the above group predominates, and serves to illustrate some basic factors about the industrialization route to forestry sector development in LDCs:

- There is a broad, inverse relationship between the size and complexity of established industries, and their relative profitability. Production, resource and infrastructure problems affect all industries in the forest sector, but the larger, more capital intensive ones among them are least able to minimize their problems themselves.
- Basic demand problems also seem to affect the larger industries more than the smaller ones. Although the paper-making industries are all now far too small to achieve economies of scale, by present standards they all produce far more than the domestic market can absorb, but do not produce it at competitive international prices. Consumption levels for the range of forest products is extremely low. This seems to lead analysts and planners alike into prediction of substantial potential for market growth. In the view of the author, however, the time to invest in substantial capacity to serve the domestic market is when other conditions in the economy are such that real demand is at a sufficiently high level.

IS THERE A FOREST INDUSTRY OPTION?

The experience of Bangladesh with larger scale forest industries – particularly the paper industry – has been essentially negative, and it is an experience many other LDCs have shared as a result of the industry optimism of the 1950s and

1960s. When these industries were established, Bangladesh lacked (and still lacks) not only the infrastructural and technological base to support them, but also the effective demand for the products. This might perhaps have been less damaging had the products created found a ready export market but, almost inevitably when industries are grafted onto undeveloped economies, the price and quality competitiveness of the resulting output will normally preclude the export option.

However, it would be most unwise to extrapolate a general, absolutist view on the matter from this example. Obviously enough, there are countries in the LDC group where sufficient real demand for these sorts of products exists to justify (technically, at any rate) some consideration of the option to manufacture them – particularly given the fundamental role of paper in education, literacy programmes and so on. In Bangladesh itself, there are some forest products which are highly demanded, which have an important export role (the plywood case), and which are manufactured there in a reasonably efficient manner. Such industries may not be optimal in terms of overall sectoral balances in the economy, but they at least do not make a negative contribution. Their operation gels reasonably well with appropriate technology and basic needs yardsticks, and much of their output is indispensable to the functioning of the economy. The problem is really one of deciding which products to manufacture – in other words, the appropriateness of the product is as important as the appropriateness of its means of production. So far as the paper industry is concerned, an LDC government faced with a decision on this matter might do well to consider (in addition to the important distributional and socioeconomic aspects involved) the following – that

> . . . the soundest and most reliable market for the justification of a capital investment is a growing domestic economy in which paper products in the form of packaging or printing and writing grades have already begun to be established as an integral part of the economy of the nation. Gomez (1978).

Much attention has been given in recent years to the design of more appropriate processing technology for LDCs, particularly in the pulp and paper areas since these are generally the most problematical because of the scale and complexity of paper making as it has evolved in the industrialized world.

In 1976, the FAO established the Pulp and Paper Industries Development Programme, which was aimed at producing workable designs for smaller scale and simpler paper-making technology which would allow LDCs to produce reasonably cheaply. Leslie and Kyrklund (1980) in an assessment of this work, have drawn some relevant conclusions:

- the development of pulp and paper technology is now virtually confined to the industrialized world; economies of scale of production in the West have become so large in paper making that direct transfer of the technology to LDCs is impossible;
- reasonably small scale units, capable of producing around 36,000 tons of

paper per annum, have been tested and shown to be technically efficient;
- mechanical pulping is also possible at an efficient level on a fairly small scale; however, little progress has been made in the area of reducing the viable size of chemical pulping plants.

One interesting aspect of forest industry development which can best be examined by reference to examples is the matter of how the industry in question is designed and run. The first case concerns the establishment of a packaging and subsequently package manufacturing enterprise in Colombia in the early 1950s. As explained by Gomez (1978), serious errors were made with this enterprise:

- The enterprise was, in a mood of general optimism, brought about by high world demand for paper during the Korean war, extended into the paper manufacturing area from the original concept of bag making from purchased papers. This development went not only beyond local expertise, but also beyond that of the United States-based partner firm.
- The investment made was too large, in view of available funds, and installed capacity too great to serve the local market.

After the Korean war, this industry suffered very severe import competition from local packaging manufacturers who used cheaper, better quality imported paper. Had it not been for a general financial crisis forcing the Government of Colombia to implement broad import controls from 1956, it is unlikely the enterprise would have survived at all. In Gomez's view, although the enterprise prospers now, the decisions and risks taken in the 1950s were not justified, and are not vindicated by the fortunate long-term outcome.

An observation on the method of operating the enterprise is relevant: considerable success, apparently, has been obtained through a division of responsibilities between the partners. The US partner handles management training, finance and technology, while the Colombian counterpart oversees the legal, labour and social requirements.

A second paper-making enterprise of interest is one described by Sila-On (1978), the Siam Kraft Paper company of Thailand. This mill was designed as a quite sizeable enterprise by LDC standards: kraft paper and board capacity of 200 tonnes per day; a pulp mill capable of 60 tonnes per day. In all, the mill can meet 25 per cent of Thailand's total paper requirements.

In common with many projects of this nature in the developing world, Siam Kraft has had a less than impressive history. In its first eight years of operation (through the 1970s) the company was on the verge of bankruptcy on three occasions. The Government eventually provided it with a monopoly on kraft paper manufacture in Thailand, and then instituted total import bans on competing imports.

Unlike many of the earlier exercises in LDC paper manufacturer, Siam Kraft was, according to experts cited by Sila-On, equipped with well-built, good quality plant. The technology transferred was not the basic problem: financial structure and commercial management were. These were approached in an unusual way: a large company, Siam Cement, took the operation over in 1974, and has been able to bring financial resources and management expertise to bear on some of the problems. Siam Cement hired a Japanese firm, Honshu Paper, to improve technological performance and know-how. This method of operating allows retention of local control over the enterprise (without stifling its operation with lack of funding), but allows immediate access to technological input and expertise from the Japanese firm. In this sense, it is an example of the approach to technology transfer characterized by direct application of already available research and development expertise, rather than attempting to create technology within the LDC from a low base.

FORESTRY IN DEVELOPMENT: NEWER APPROACHES

The lessons of experience of the industrialization approach to development have, in forestry, shifted the emphasis to other types of sectoral development.

Westoby himself (1962), has fully acknowledged that the approach to development of the forestry sector in poor countries advocated by him and by others, has proved largely to be a failure. However, the practices and policies of this era are still persistent today in LDCs, even if the justification is not.

There is little point in attempting to re-construct the historical process by which the theoretical view of the role of forestry has undergone its radical change. According to the records of a working party meeting of IUFRO in 1975, (in which the papers by MacGregor and Von Maydell cited above appear) on the contribution of forestry to development, strong elements of the industrialization view were still apparent then, and the meeting seemed divided over its view of a paper by Haley and Smith (1976), which specifically criticized the use of indirect benefits, import replacement and disguised unemployment arguments in favour of forestry proposals. Haley and Smith in fact presented what might be termed the economic rationalist viewpoint: the primary purpose of investment in forestry, they remind, is not the growing of trees, but the formation of capital. It is very easy, they suggest, for planners and administrators to lose sight of this; to become overly attentive to the indirect, intangible benefits of what otherwise seems an inferior investment.

Westoby (1978), in his recantation of the views advocated in his earlier work, begins by referring to the inadequacy of the GNP criterion. Westoby's new interpretation of the reasons for underdevelopment is interesting:

> The underdeveloped countries are not underdeveloped because they started late in the development race. They are not underdeveloped because they lack resources. They are not underdeveloped because they lack know-how. They are not underdeveloped because they are overpopulated. They are underdeveloped as a consequence of the development of the rich nations. The development of the latter is founded on the underdevelopment of the former, and is sustained by it. The ties between the affluent, industrialized countries and the backward, low income countries are intimate and compelling. Their nature is such that the

objective impact of most of the so-called development effort to date has been to promote underdevelopment.

This explanation obviously owes a great deal to the tenets of Dependency Theory.[2] Westoby argues that the growing interest in forestry projects by the development establishment in the 1960s had little to do with the idea that forestry had a significant contribution to make to economic and social development, but was rather a result of the rich countries needs for raw material, and the opportunities offered to obtain it in underdeveloped areas. In support of this contention he cites the rise of export of tropical harwoods from three million to over 40 million m^3 between 1950 and 1976 – nearly all of which went to affluent, industrialized nations.

Westoby is pessimistic that mere recognition of the priority that should be given to forestry's role in the rural economies of LDCs by the large multilateral organizations and development banks will achieve real change. In his view, the principal problem is that governments and officialdom in many LDCs are simply not interested in mitigating the circumstances of the poor – that they are, in Dependency terms, part of the urban elite and determined to stay that way. In pursuit of this theme, Westoby claims that the demand for a new economic order in the world is a red-herring, used by LDC governments to excuse their own shortcomings in the area of relief of the poor by reference to machinations of the rich countries against them.

Westoby's final conclusion refers to the need for poor countries to concentrate on producing an agricultural surplus. The role of forestry in this must be to support the traditional rural sector, and in effect must be carried out by rural people themselves. Industrialization must be directed towards satisfaction of basic domestic needs. Measures to reduce income inequalities will in themselves bring about change patterns of demand for forest products, but some intervention and manipulation of the market by government may be necessary (to direct and assist this process). Exporting has a role to play in the forestry sector, but must be subordinate to its domestic role.

Westoby suggests that forestry development along these lines might slow economic growth, as measured by the conventional criteria, but development, inclusive of the social aspect, will be accelerated. Discussing the forest industries that have been set up in LDCs in the past, Westoby observes that: '. . . very, very few of the forest industries which have been established in the underdeveloped countries have made any contribution whatever to raising the welfare of the urban and rural masses, have in any way promoted socio-economic development. The fundamental reason is that those industries have been set up to earn a certain rate of profit, not to satisfy a range of basic, popular needs.'

The Food and Agriculture Organization has also recently entered the field with a detailed policy statement on forestry in LDCs (FAO, 1980). This document also identifies the problem raised by Westoby: that few deliberate efforts have so far been made directly to improve the lot of the poor in LDCs through forestry activities. Forest services, operating on outmoded technical and societal perceptions, have been unable effectively to manage and control the resources under their jurisdiction. The general pressure of population today on forest lands, and the specific need of the poor to occupy and utilize such lands, and whatever grows on them, to sustain life, means that a new approach to management of forests (in densely populated LDCs in particular) is needed.

FAO identifies fuelwood, material for rural building and some of the minor forest food products as being high priority output. Integration of forestry with agriculture, voluntary participation of local communities in forest management and the use of appropriate technology are seen as the important elements in overall approach.

For governments of LDCs, FAO pre-supposes the political willingness to implement the necessary changes. Given this, the principal need is for creation and strengthening of national forest planning capabilities, forestry training and education, and community extension, to involve rural communities in forestry programmes.

For the FAO itself, the required role is perceived to be assistance to governments in planning a rural development strategy for forestry. FAO sees a need for its own operation to become more integrated rather than separated into specialized areas if it is to be effective in this.

The World Bank has issued an important statement of its policy on forest sector development (World Bank, 1978). This document brings the Bank's forestry sector lending criteria into line with the major reorientation of Bank philosophy towards the concept of rural development which came about in the 1970s. The Bank comments on the pronounced deforestation occurring in developing countries at the present time. Apart from the ecological consequences of over-exploitation and deforestation, the Bank identifies the serious consequences for rural welfare of increasing constraints on fuelwood availability, in particular.

The document provides a useful typology for forestry development projects, and it is relevant to briefly re-state these here to emphasize the point that there are many forestry problems, not one, in the various LDCs.

- Wood deficit marginal lands: This type of forestry problem is exemplified in the Sahel region where overgrazing and poor growing conditions have reduced existing vegetation and increased desertification. Without effective settlement, afforestation programmes would be extremely difficult to manage.
- Potential afforestation areas: Similar to the wood deficient marginal lands, but with more favourable forest growing conditions and less population pressure.
- Overpopulated wood deficient areas: Much of the Indian sub-continent comes into this category, which is typified by overcutting of upland forests with consequent erosion flooding and so on. Severe fuel shortages are

already apparent, and institutional problems, with undermanned and poorly trained forest services, exacerbate conflicts with the local populace over forest use.

- Wood abundant poor areas: Countries which are underdeveloped, but which have a large, untapped resource of natural forests; the Congo, parts of Indonesia and New Guinea typify these areas. The major problem is to prevent conversion of forests to agriculture in such a way as to bring about rapid fertility losses in the fragile rainforest soils – a phenomenon which has been cited in many parts of the world.
- Wood abundant areas with severe population pressures: Much of the world's remaining tropical forest is in areas under this category, in Africa and South America. Shifting cultivation poses the major threat to resources.
- Wood abundant rich areas: Typified by Canada, the USSR and Scandinavia.

The World Bank, in analyzing its own role in forestry, emphasizes that its projected lending will be directed at rural forestry to a much greater extent than has been the case in the past: over half of its intended schemes will address this problem, compared to only a quarter or so of projects undertaken in the 1953–76 period. Within this category, a significant part of Bank lending will be for projects of environmental protection – shelter belts, erosion control afforestation and so on, to preserve or accrete land for useful production by rural populations.

It is necessary to take stock of the very different analyses of the forestry problem in LDCs that are implicit in these statements. It is a measure of the degree to which the general approach to development in very poor countries has penetrated forestry circles to note that Westoby, FAO and the World Bank do not differ in their basic assessment of what the major problem with forestry in LDCs is, that is lack of involvement of, and benefit to, the rural poor in the mainstream of forestry sector activity, and a consequent aggravation of the problems of management of forested land and of the income gap constraint on development in these countries. However, the explanations offered as to why this situation has arisen, and consequently the suggestions advanced to correct it, are very different. Westoby's version of the problem is grimly political; the FAO view determinedly apolitical. The Bank seems more aligned to the FAO view, albeit that it does identify a tendency for LDC governments not to support forestry institutions to the extent necessary. This paper attempts some general assessment of which view – if either – is close to reality.

The Westoby view

Westoby begins by suggesting in effect that poor countries are poor because of the manner in which rich countries have made themselves rich. However, the conclusion of the authors' discussion of the New International Economic Order (NIEO) and Dependency Theory (see Chapter 2 of Douglas 1983 from which this paper is taken) is that the trading disadvantages applying to LDCs through dealing with Centre countries do not appear to be as severe as some protagonists of NIEO claim. Nor, more prosaically, are they likely to alter significantly in the near future.

An important interpretation that Westoby makes for forestry, on the basis of his general explanation of underdevelopment, is his claim that the rich countries, running short of wood, have exploited the forest reserves of the poorer nations. This is probably so, but to what extent are continuing overall wood shortages a reality, and what advantages and disadvantages does the answer imply for LDCs?

Impending famines of wood in the world have in fact characterized forestry literature for many years and, to a large extent, still do so. Our view, which has been argued elsewhere (Douglas *et al*, 1977) is that on current evidence, the hypothesis of an imminent, worldwide and sustained shortage of wood is not well supported by the evidence available. It may be true, as is suggested by the World Bank (op cit) that something like 15–20 million hectares per annum of the estimated 1200 million hectares of mature forest in developing countries is being lost to over-exploitation, agricultural incursion and so on each year. This has serious environmental and economic implications for the LDCs involved, but what does it mean in terms of international supply? Many developed countries or regions of the world are now following policies of raw material self-sufficiency. Even Japan, the quintessential raw material importer, now has a very considerable 9.4 million hectares (see Byron, 1979) of man-made forests, (much of it currently in an immature state). Matsui (1980) gives some growing stock data for Japan which imply a standing volume in that country of 20 m^3 per capita – quite an appreciable amount.

It is interesting to note that the Food and Agriculture Organization – previously among the most pessimistic of forecasters in terms of supply shortfalls – now acknowledges that the world supply of industrial wood will meet additional demand to 2000 AD (the forecast period) (FAO, 1981).

Wood is a renewable resource. In fast-growing plantations, rather small establishments can replace quite large areas of natural forest, (so far as wood volume production is concerned, if not for ecological, recreational and other purposes). Some developing countries may well run short of wood themselves, but it would be unrealistic of those of them with surpluses to expect massive trade advantages to come their way for their remaining resource, because of overall shortages in the world. It is undoubtedly true that relative to their own histories of absolute abundance of wood, some rich nations have recently experienced greater need for forest raw material from elsewhere, and have engaged in highly exploitative operations to get it. But, because of the renewability of the resource, and existing reafforestation policies in the West, this sort of shortage is quantitatively and qualitatively different from that for, say, oil. Wood-exporting LDCs would, in our view, be most unwise to expect dramatic results even if they do form consortia or cartels to control supply. The fact that oil-rich countries were able to do so, and simultaneously to overcome a number of trading disadvantages (the presence of multinational, Western based extractive industries; a history of colonial exploitation; concerted pressure from the western trading block against the cartel) should not be taken as a precedent for the wood exporters, albeit that Westoby and the World Bank occasionally seem to come fairly close to suggesting this. Clark (1974) provides an interesting analysis of the characteristic price response of wood in its markets. He shows that a very definitely convex demand function for wood applies. This indicates that the commodity is one which finds a wide range of uses when cheap, but a wide range of substitutes when expensive. While we read much about the high income elasticity of demand for wood products, we seem to see rather less on this demand/price response in the literature on overall supply and demand in the international wood markets. It indicates that demand will adjust to supply well in advance of serious sustained shortages in this market, and the possibility of offsetting supply responses in deficit areas is also, as already discussed, an important factor. Prices for the raw material may rise on the international market but, despite what is indicated by rather simplistic static supply–demand analyses, the authors do not believe that wood-surplus LDCs should expect real price rises to be large, nor should less well endowed LDCs enter the plantation supply market on the assumption of radical price rises. Better control of natural resources in wood-rich LDCs would undoubtedly be to their economic advantage, and there are good theoretical and practical reasons for LDCs to trade in forest products with Centre countries to mutual advantage. This paper merely argues that this should be kept in some perspective. Wood-short LDCs in particular, should pay most attention to their own specific needs when determining resources policy.

In fact, Westoby himself seems ambivalent on the question of exploitation of LDC forest resources. Later in his paper, he suggests that the issue of the New International Economic Order is irrelevant. Now, either the Centre countries have been exploiting LDCs as Westoby first suggests, in which case the latters' demands for a new international trade and aid arrangement are justified (no matter who makes them); or they have not, in which case Westoby's accusation of LDC governments of using such claims to mask exploitation tendencies of an internal origin in such countries becomes relevant. But both cannot apply.

Westoby's interpretation of what has occurred in the past so far as the establishment of forest industries in LDCs is concerned also seems faulty, albeit that his overall conclusion – that these industries have not made any significant contribution to development – although drawn for the wrong reasons, is more tenable. Westoby says that the reason for this is that these industries have been set up to earn a profit. He seems to have this same point in mind when he suggests that forestry development along the lines he suggests might slow the rate of economic growth. The authors would dispute both observations. From analyses of overall results on income growth and development for very poor countries, Douglas *et al* (1983) would immediately reject any proposal which threatens to slow growth but, as Lipton has shown, this should certainly not exclude programmes based on rural development (nor, by extension, forestry schemes within such programmes). And the reason the previously established forest industries in LDCs have proved so detrimental to socio-economic development is not so much that they have been set up to make profits, but rather that they have usually failed to make them. The basic reason for this, as already suggested, is because they were not appropriate industries in the first place. Had they made substantial profits, then their reputation and standing today, and the current view of development overall, would look rather different. Redistribution towards impoverished groups is possible (if not always practised) when industries are making large profits, and are unassisted by government subsidies of one sort or another. It is impossible when they are operating at a loss.

The FAO view

It is rather more difficult to critically analyze the FAO statement on forestry in LDCs, largely because it is a somewhat bland and amorphous document. It offers little direction on implementation of its recommended approach, and even less on how the trade-offs it implies should be managed. That there will be trade-offs in many LDCs should be obvious: they have in the past established (very often on the advice of FAO and other large international assistance agencies) forest growing and processing sectors which, they are now being advised, are inappropriate for their development aspirations. The document speaks of strengthening planning capabilities, broadening the administrative bases of forestry, and so on. While these appear to be positive exhortations, their consequences might be anything but positive. They certainly offer no direction as to how dis-investment of existing industries should proceed, nor do they address the fundamental problem of incorrect or inappropriate perceptions of the role of forestry.

In any event, the document appears somewhat in two minds about the question of industry policy. On the one hand, the FAO is critical of the establishment of forest industries in the past, for reasons of the lack of involvement of the poor in the sector. The, later, it is asserted that there is no need to constrain the poor to simple and small technology, to do so, it is claimed, raises the possibility of unacceptable dualism in the development pattern. The FAO suggests that appropriately designed technology, capital intensive and sophisticated, can be used to directly benefit the poor. Given the absolute shortage of investment capital, the manifest difficulties in establishing capital intensive industries (and demand for products) in LDCs, and the overwhelming need to employ humans, rather than machines in these countries, we would have to disagree. Of all organizations, the FAO should be well aware that the justification for the intensive, technological approach to development in the past has always been that, eventually, the poor would benefit. And all the resulting industry that was set up as a result of former theories on development was, at the time, believed to be appropriate and efficient. FAO's argument here would seem to be a reversion, mid-stream, to the older concepts of development. It is, furthermore, nonsensical to suggest that small and simple technology is socially unacceptable in the sense of promoting dualism. The alternative for the poor, as ought by now to be abundantly clear, is not meaningless involvement in the so-called modern sector of the economy, but rather little or no involvement in anything at all – surely a much more extreme strain of 'dualism'.

The World Bank view

The World Bank document also seems, on occasion, to drift into this type of reasoning. The Bank suggests, at one point, that LDCs have a comparative advantage in the growing of wood, insofar as their better ecological conditions and cheaper inputs allow this. Wood, however, is expensive to transport in the raw form (not sufficiently so, it would seem from the Bank's own arguments, to have prevented large scale external commercial exploitation of forests in the past). The Bank's solution is primary, not secondary processing within LDCs. We shall shortly come to a reasonably successful application of that principle, but the review of the often disappointing performance of large scale forest processing industries later in this chapter should serve to warn of the dangers of too much enthusiasm allowing highly technical, inappropriate processing tails to set about wagging the resources dog all over again.

THE MALAYSIAN EXAMPLE[3]

Malaysia is a case of a developing country where the forestry sector has made significant contributions in both developmental and distributional terms, and for this reason it warrants some attention here.

Malaysia has some distinct advantages: a relatively small population (12 million or so) for its land area, and abundant resources. On Peninsular Malaysia, only about 20 per cent of land area is currently farmed; in Sabah and Sarawak, this figure is only 5 per cent. The remaining area of the country is forested with a resource of variable quality and quantity.

On Peninsular Malaysia, forest utilization and management has been ultimately connected with both the development of a fairly sophisticated and competitive export processing sector, and rural development. In 1981, about 4.4 million m^3 of logs – about 55 per cent of total harvests on Peninsular Malaysia – came from clearing operations for agriculture, and the Government of Malaysia quite obviously sees considerable socio-economic benefit in this sort of operation.

However, of the total log harvest on Peninsular Malaysia, less than 3 per cent is exported in log form. The remainder goes into locally established sawmilling and ply operations, predominantly aimed at the export market. These exports earned Malaysia $US684 million in 1980. At present, integration of secondary and tertiary processing is being developed.

A problem is beginning to arise, in that developments in processing in Peninsular Malaysia have outpaced log availability. The Government is reducing the rate of harvesting, and is attempting to introduce improved management and utilization. Also, what seems to be a highly optimistic plantation programme to establish 188,000 hectares per annum for the 15 years from 1981 of exotic conifer plantation, has been announced.

In Sabah and Sarawak, the emphasis is still on export of log material: 94 per cent of forest exports from Sabah and 82 per cent from Sarawak were in log form in 1981 (total forest exports from these states totalled $US1324 million in 1981). Presently, Sabah is trying to lower the log component of exports somewhat by creating processing facilities there.

Of the 416,000 tonnes of paper consumed in Malaysia in 1980, only 63,000 tonnes were produced locally. Recently, plans to use the softwood plantations on Peninsular Malaysia for paper making were shelved, so that the resource can go to existing wood processors. Plans exist for construction of a 200,000 tonne per annum pulp operation on Sabah: even with construction of this facility, Malaysia would remain dependent on imports for the bulk of its paper requirements.

The Malaysian forestry sector, on the basis of the above brief sketch, demonstrates a number of interesting features:

- A successful export industry has been established on Peninsular Malaysia on the basis of the relatively low level technology required in sawmilling and veneer production.
- The Government of Malaysia has control of forest clearing and allocates land to rural dwellers for agricultural development from the process. Even given the impending resource problems on Peninsular Malaysia, this integration of forestry and agricultural responsibilities has arguably produced a more orderly and equitable distribution of wealth than might otherwise have occurred.
- Although a relatively efficient processor, Malaysia as a whole still generates more export income from log exports than from processed forest output. The question of how quickly the transformation to domestic processing in Sabah and Sarawak should be made remains contentious.
- Malaysia has so far not made significant attempts to replace imports in the paper sector. While pressures in some quarters to do so are mounting, the Government needs to bear in mind that the nation is located close to large paper producers in the South East Asian region.

Without attempting to minimize some of the basic forestry problems faced by Malaysia, it seems reasonable to interpret the overall performance of the sector in a favourable light: control over forest exploitation has been maintained; a rural development/redistribution element has been included in forest operations; specialization of industry activities into areas of comparative advantage (even to the extent of retaining a fairly large raw material component in exports) has established Malaysia in some important markets; and justifiable caution has (so far) been applied to the question of investment in high technology, capital-intensive production processes such as paper-making.

This chapter is taken from Douglas, J (1983) *A Reappraisal of Forestry Development in Developing Countries,* Nijhoff/Junk Publishers.

NOTES

1. 'The ideas of economists and political philosophers, both when they are right and when they are wrong, are more powerful than is commonly understood. Indeed, the world is ruled by little else. Practical men who believe themselves to be quite exempt from any intellectual influences, are usually the slave of some defunct economist. Madmen in authority who hear voices in the air, are distilling their frenzy from some academic scribbler of a few years back. I am sure that the power of vested interests is vastly exaggerated when compared with the gradual encroachment of ideas.'
2. The Dependency Theory attempts to explain the apparent failure of the modernization approach to development adopted in the 1950s and 1960s to bring about progress in Lesser Developed Countries (LDCs), in terms of their trade with Developed Countries (DCs) or in their internal development. The essence of this neo-marxist theory is that the penetration of capitalism into under-developed countries is not a solution to, but rather the cause of, their problems. This is due to the fact that profits generated in the LDCs (the 'Periphery') do not benefit these countries, but are repatriated to the DCs (the 'Core'). Some of the main exponents of the Dependency Theory are Frank (1978), Cardoso (1979) and Amin (1976).
3. The bulk of data referred to in this section comes from a forthcoming paper by S Parsons of the Australian Bureau of Agricultural Economics.

REFERENCES

Amin, S (1976) *Unequal Development*, Harvester Press, Brighton.

Arnold, J E M (undated) *Lessons of Experience in Planning Forestry Development*, Mimeo, Harvard University.

Byron, R N (1979) 'An economic assessment of the export potential of Australian forest products', *Industry Economics Monograph* no 20, Australian Bureau of Agricultural Economics, Canberra.

Cardoso, F M (with Faletto, E) (1979) *Dependency and Development in Latin America*, University of California Press, Berkeley.

Clark, C (1974) 'Optimising the supply and use of market timber through market prices', *Australian National Bank Monthly Summary*, October 1974.

Douglas, J J, Bond, G, Connell, P, Ramasamy, V, Buckley, C and Treadwell, R (1976) *The Australian Softwood Products Industry*, Bureau of Agricultural Economics, Canberra.

Food and Agriculture Organisation (1974) 'An introduction to planning forestry development', FAO/SWE/TF 118, 1974.

— (1981) 'FAO's Medium Term Objectives of Forestry and Programme of Work for 1982–83', *Forestry Planning Newsletter*, FAO, Rome.

— (1980) 'Towards a forest strategy for development', COFO – 80./3. 1980.

Frank, A G (1978) *Dependent Accumulation and Underdevelopment*, MacMillan, London.

Gane, M (1969) 'Priorities in planning', *Commonwealth Forestry Institute*, paper No 43, University of Oxford.

Gomez, G (1978) 'Case History of a South American Paper Mill', *Unasylva*, vol 30, no 122.

Haley, D and Smith, J J G (1976) 'Justification and sources of funding of forestry operations in developing countries', in 'Evaluation of the Contribution of Forestry to Economic Development', *UK Forestry Commission Bulletin*, no 56.

Leslie, A J and Kyrklund, B (1980) 'Small-scale Mills for Developing Countries', *Unasylva*, vol 32, no 128.

Lipton, M (1980) 'Migration from rural areas of poor countries: the impact on productivity and income distribution', *World Development*, vol 8, no 1, 1980.

— (1977) *Why Poor People Stay Poor: Urban Bias in Developing Countries*, Temple Smith, London.

Little, I M D (1973) *A Critique for Welfare Economics*, (second edition), Oxford University Press, Oxford.

McGregor, J J (1976) 'The existing and potential roles of forestry in the economics of developing countries', in 'Evaluation of the Contribution of Forestry to Economic Development', *UK Forestry Commission Bulletin*, no 56.

Matsui, M (1980) 'Japan's forest resources', *Unasylva*, vol 32, no 128.

Nautiyal, J C (1967) 'Possible contributions of timber production forestry to economic development', PhD thesis, Faculty of Forestry, University of British Columbia.

Sila-On, A (1978) 'The Transfer of Technology', *Unasylva*, vol 30, no 122.

Streeten, P (1975) 'Industrialisation in a unified development strategy', World Development, January.

United Nations Industrial Development Organisation (1972) *Guidelines for Project Evaluation*, United Nations, New York.

Von Maydell, H J (1976) 'Effective policies for stimulating investment in forestry industries in countries with tropical forests', summarised in 'Evaluation of the Contribution of Forestry to Economic Development', *UK Forestry Commission Bulletin*, no 56.

Westoby, J C (1978) 'Forest industries for socio-economic development', Guest speaker's address *8th World Forestry Congress*, Jakarta, October 1978.

— (1962) 'Forest industries in the attack on underdevelopment' in *The State of Food and Agriculture*, FAO, Rome.

World Bank (1978) *Forestry Sector Policy Paper*.

Chapter 10

Beyond The Woodfuel Crisis

Gerald Leach and Robin Mearns

Much has been written about the plight of developing Africa and the environmental crisis which underlies it. In many parts of the continent food production lags behind population growth, hunger and famine strike with dreadful persistence, soils are eroding, forests and trees are disappearing at unprecedented rates, and poverty deepens in the countryside and cities.

It is not like this everywhere. Nor are prospects for the future as hopeless as such headlines imply. On the contrary, in many places hard work and creative innovations by people, governments and aid agencies are doing remarkable things to put the land into good shape, increase food production, restore soils and a healthy cover of vegetation, and generally enhance livelihoods on a sustainable basis. How best to support and amplify these efforts has become one of the most urgent and chellanging tasks of our times.

The book from which this chapter is taken, *Beyond the Woodfuel Crisis: people land and trees in Africa*, joins the 'literature of hope' rather than of despair by presenting these challenges and the opportunities they offer, while recognizing the problems to be overcome. However, it approaches them from the narrower perspective of energy and the so-called woodfuel 'crisis' of Africa and other parts of the Third World. It does so for two basic reasons.

First, woodfuel and related energy problems are important and pressing topics in their own right. Since most Africans are poor and can afford or have access to little other than firewood, charcoal, or crop and animal residues to meet their basic energy needs, woodfuels dominate the energy economies of virtually all African countries. In sub-Saharan African they account for 60–95 per cent of total national energy use, with the highest proportions in the poorest countries and in the household sector, even though consumption is small by world standards and amounts roughly to only one cubic metre of wood per person annually, or a mere quarter of a ton of oil equivalent. It will take many years of rising incomes and infrastructure development before such countries can afford alternatives to this massive woodfuel dependence.

The more obvious symptoms of this dependence are well known. In many places woodfuel resources are dwindling because of deforestation which is caused to varying degrees by the need for farming land, and by over-grazing, commercial logging, uncontrolled fires and tree cutting for fuel. As wood resources diminish and recede the costs of obtaining woodfuels for millions of people, whether in cash or time for gathering them, are imposing severe and increasing strains on already marginal household survival and production strategies. These impacts are greatest for the poor and for women, who normally bear the responsibility for fuel provision and use. They are not yet felt everywhere, but they are spreading to more places and more people.

Great efforts will be needed to reduce these impacts, prevent them from spreading, and provide sustainable and adequate energy supplies at affordable costs for fast-growing populations.

The second reason for adopting a woodfuel perspective is that ideas about the nature of the woodfuel crisis and what to do about it are beginning to change quite fundamentally, with far-reaching implications not only for energy specialists and decision-makers but for all the disciplines and institutions which are concerned in some way with the land, environment and sustainable livelihoods – from agriculture, forestry and rural development to urban planning and systems of law governing land use rights.

Over the past 15 years, large policy, planning and donor aid structures have been created and hundreds of millions of dollars committed to addressing the woodfuel crisis directly as a problem of energy supply and demand. The issues appeared quite simple and the remedies self-evident. Where trees and woodfuels were scarce or getting scarcer many direct solutions seemed to offer a good chance of quickly bringing forests and woodlands, woodfuel supplies and woodfuel demand into balance. This would both improve welfare by reducing the costs of obtaining fuels and save the trees.

Governments and foresters set about trying to protect public forests and woodlands from encroachment by woodfuel cutters. Foresters tried to increase woodfuel supplies with peri-urban plantations and village energy woodlots. Energy ministries tried to curb rising consumption by promoting more energy-efficient cooking stoves, or reduce pressures on the forests with more efficient charcoal kilns. Attempts to promote the use of oil and electricity instead of woodfuels became key elements of African and other Third World energy strategies.

While there have been a few successes with these energy-focused efforts, most have failed to turn the tide of wood depletion or prevent growing pockets of fuel scarcity. But as one expensive disappointment has led to another and simple certainties have begun to evaporate, important lessons are now being learned.

Better information is showing that many of the most basic assumptions on which these efforts were based are false or highly misleading: for instance, that the use of woodfuels is normally the principal cause of deforestation, or that the expanding circles of deforestation around cities will inevitably force up woodfuel prices and hence provide a powerful economic rationale for all kinds of afforestation and conservation measures.

At the same time, it is now increasingly recognized that by focusing so closely on woodfuels and the symptoms of their scarcity, these direct approaches looked only at the tip of the proverbial iceberg and ignored the much broader and deeper strains in the environmental, social, economic and political fabric of which woodfuel scarcity is only one manifestation. They obscured the fact that woodfuels are only one of many basic needs and that their provision – for example, by tree growing – is only one aspect of household coping strategies and land use systems on the farm or in the village. Indeed, these top-down and over-specialized approaches often failed to notice that in many places rural (and urban) people were already responding to woodfuel and other land use stresses in ways that are imaginative, innovative and with far lower cost than most project interventions.

The more comprehensive and objective view of woodfuels which is now emerging recognizes that there are no single, simple answers and that the problems surrounding them are inseparably linked to the complex, diverse, extremely dynamic and multi-sectorial issues underlying Africa's broader crisis of population, food, poverty, land and natural resource management. Equally, successful remedies for woodfuel problems must be firmly rooted in these broader contexts. In particular, if planning, projects or other types of intervention are to create lasting successes they must recognize at least three basic factors:

- the need for local assessments and actions and the unhelpful nature of large-scale averages. The 'landscapes' and 'peoplescapes' of Africa, especially, are extremely diverse. Problems and opportunities to solve them are therefore specific to place and to social groups in each place. The aim should be to reach underlying causes rather than heal the symptoms;
- the need for indirect approaches to woodfuel issues and greater participation by local people at every stage to help them to prioritize and solve their own problems. This follows from the first point, and also from the fact that success normally depends on starting and strengthening processes rather than delivering technical packages – on how rather than what things are done;
- the need for decentralized and multi-disciplinary approaches, including the use of competent and trusted grassroots agencies, to facilitate the two first points. However, this does not exclude the need for economic, legal and political initiatives at the macro-level to improve the broad contexts for local, positive change.

One might well say: so what's new? Isn't all this now broadly accepted? Well, accepted maybe – but not yet acted upon. Although these perspectives are in tune with the broad paradigmatic shift which is now sweeping through governments, aid agencies and other parts of the development community, one has to bear in mind the enormous inertia and vested interests which can resist such basic changes to conventional structures of authority, responsibility and knowledge.

If we restrict ourselves to woodfuel issues, it is clear that energy policy makers, planners, analysts and project staff need to redefine their roles, learn new concepts and styles of working, and even surrender their bureaucratic empires and specialist corners to others. New institutional linkages, joint policies and data gathering, and other kinds of alliances between government agencies, extension services and the like – as well as new forms of interventions – are all needed if woodfuel issues are to be addressed in a more holistic and relevant way.

Much the same is true of foresters, who often justify their efforts as direct remedies to the woodfuel crisis while failing to recognize the broad contexts which underlie them. Agricultural and livestock schemes, in turn, have often ignored the needs of foresters (and local people) and have often greatly worsened local woodfuel problems by ignoring them as a planning issue. Narrow specialism, false diagnoses of problems and top–down attitudes are found in all of the many disciplines and institutions which work, directly or indirectly, towards the better management of land and natural resources. Getting off the beaten track and heading for new territory with unfamiliar allies will not be easy either for them or for woodfuel specialists.

The main purpose of *Beyond the Woodfuel Crisis* is to support these changes by showing why they are necessary from a woodfuel perspective and hence what they imply for resolving woodfuel problems. However, it should be obvious from the remarks already made that this aim will take us far deeper into many of the fundamental issues to do with the broader environmental and production crises of Africa and what might best be done about them.

Of course, the book cannot cover every aspect of such a vast subject. Nor can it always point with certainty to effective solutions. Everyone concerned with these large issues is on a steep learning curve: as yet there are few cut and dried answers. What the book does attempt is to review as objectively as possible the main issues, positive options and constraints (as well as areas of profound ignorance) and successful achievements to do with the woodfuel problem in its much broader setting of sustainable land use and natural resource management. It uses actual case studies wherever practicable to back general discussions with on the ground realities.

However, one thing that the book does not try to do is cover all issues of energy for development. Its focus is on woodfuels as basic needs: 'energy for survival'. We feel that dealing with this must take a higher priority than looking beyond survival needs to conventional energy developments or the plethora of renewable energy devices such as biogas plants and windmills that have been so widely proposed as aids to development, especially in the rural areas of the Third World.

This first paper lays the groundwork of the book by making a sometimes harsh critique of conventional views of the woodfuel crisis. It does so in order to ask the basic question: what is the woodfuel problem? Without a clear understanding of what the problem is, or what the underlying causes are, there is little chance of developing appropriate and effective policies and other interventions to meet it.

WOODFUEL GAPS AND THE DEATH OF THE FORESTS

The woodfuel crisis of developing countries was discovered in the mid-1970s when much of the world was gripped by the energy crisis of modern fuels which followed the first oil price shocks of 1973–4. The scale of deforestation across the Third World was already recognized. As energy analysts and anthropologists began to pile up the evidence across the developing world about the huge scale of woodfuel use and the difficulties that millions seemed to be facing in getting enough wood as tree stocks declined, it seemed natural to regard both types of crisis as essentially similar.

The woodfuel problem seemed to be a classic case of rising energy demand outstripping supply. Although the resources in this case were renewable – unlike oil, gas and coal – they were apparently being overused at unsustainable rates. So a numbers game known as woodfuel 'gap theory' was conceived which quickly came to dominate almost every attempt to measure the scale of both the woodfuel crisis and the remedies which would be needed to alleviate it.

For instance, all of the sixty-odd UNDP/World Bank energy sector assessments for African and other developing countries which considered woodfuels in the first half of the 1980s adopted gap theory methods. They were used for the extremely influential UN Food and Agriculture Organization (FAO) study in the early 1980s which estimated that in 1980 just over 1000 million people were living in areas of woodfuel 'deficit' because they were cutting tree stocks to meet their energy needs faster than the trees could regrow, and that this number would almost double to 2000 million by the year 2000.[1] The same idea underpins many more recent and widely quoted reviews of Africa's forestry and woodfuel problems[2–5] and was being used even in 1988 by major donor agencies to justify large-scale forestry projects in Africa.

The basic premise of gap theory, as normally practised, is that woodfuel consumption is the principal cause of deforestation and therefore of mounting woodfuel scarcities. To measure the scale of this impact, one estimates the consumption of woodfuels (and sometimes of timber, construction poles and other tree products) in a given region and compares it with the standing stocks and annual growth of tree resources. The latter may be scaled down to allow for controlled forest reserves, game reserves, and trees in remote places where access is difficult.

Typically one finds that consumption greatly exceeds the annual growth of trees. For instance, studies of the Sahelian countries found that woodfuel use exceeded the growth rate of tree stocks by 70 per cent in Sudan, 75 per cent in northern Nigeria, 150 per cent in Ethiopia and 200 per cent in Niger, with a small surplus of 35 per cent in Senegal.[4]

The next step is to project these present-day gaps. Since consumption has to be met from somewhere, one assumes that the difference – the 'gap' – is made up by cutting into tree stocks. Woodfuel consumption is projected, usually in direct proportion to population growth, and calculations are made of the resulting tree stock each year. As consumption rises and trees are felled, the annual growth falls, the gap grows biggers, and the tree stock is still further depleted. Inevitably, the stock of trees declines at an accelerating rate towards a final woodfuel and forestry catastrophe when the last tree is cut for fuel.

Table 10.1 shows the results of just one of many applications of this method,

Table 10.1: Woodfuel gap forecasts for Sudan

	1980	1985	1990	1995	2000	2004*
		(million cubic metres of tree stock)				
Forest stock	1994	1810	1539	1145	607	57
Forest growth	44	40	34	25	14	1
Woodfuel consumption	76	88	102	121	141	159
Woodfuel gap (3–2)	32	48	68	96	127	158

Note:
* Extrapolated from published data to year 2000
Source: Anderson and Fishwick (1984)[2]

in this case for Sudan.[2] Forest stocks will fall to zero by 2005. A similar exercise for Tanzania, published in 1984, showed that the last tree would disappear under the cooking pot by 1990.[6] There are still many trees in Tanzania.

The final step is to ask what must be done to close the gaps and bring consumption and tree resources into balance. With few exceptions, the answer is afforestation (or demand management by the dissemination of more efficient cooking stoves etc) on a staggering scale. For instance, the World Bank study which did much to legitimize woodfuel gap theory estimated that tree planting in sub-Saharan Africa would have to increase fifteen-fold in order to close the projected gaps in the year 2000.[2] The vast scale of these remedies, and the calamitous consequences if they are not applied, naturally tend to combine to provide strong justifications for large, centrally directed, plantation forestry projects focused on woodfuel provision.

In criticizing these methods we do not intend to imply that there are not serious and growing woodfuel shortages in many places; that woodfuel consumption does not often exceed renewable supplies; that afforestation is not, for many reasons, an admirable objective; nor even that supply–demand analysis, of which traditional gap theory is one model, is not a legitimate tool for resource assessments at the national or regional level.

Our criticisms concern the serious practical and theoretical flaws in gap theory as it has so often been (and still is) applied. By ignoring these flaws, gap methods have done much to exaggerate the scale of the woodfuel problem and foster inappropriate, large-scale, energy-focused remedies at the expense of other actions which could have done much more to improve welfare, reduce deforestation, and generally support sustainable development.

One serious flaw is that the large-scale aggregate perspectives of gap theory help to obscure the fact that woodfuel problems are location-specific and require precisely tuned and targeted interventions, usually on an individually small scale.

The second flaw is that this numbers game is played with weak numbers. While this fault is widely acknowledged, the game continues and its conclusions continue to be taken with great seriousness. Three points deserve emphasizing:

- estimates of woodfuel consumption and of tree resources are rough and in many cases little more than guesses, yet it is the relatively small initial difference between them which drives the gap forecast. This difference is extremely uncertain, thus putting the whole projection in serious doubt;
- estimates of tree stocks are particularly uncertain and are usually gross underestimates of the resources which are actually used for fuels. Most such statistics are held by forest departments, which typically know little or nothing about the volumes of trees outside the forest – for example, on farm lands, fallow lands and village commons – or about scrub, bushes and other forms of non-tree woody biomass resources. These additional woodfuel resources may be very large;
- most gap predictions assume that once a hectare of forest has been cut it becomes dead land, without any natural regeneration of trees or shrubs. Adding in this factor of tree regrowth can soften dramatically the dire predictions of gap forecasts.

A third and more fundamental flaw concerns the forecasting methodology. Consumption is usually assumed to rise in line with population, even while supplies dwindle to vanishing point (see Table 10.1). Everyone acknowledges that this is unrealistic and that as scarcity worsens and wood prices or the labour costs of gathering fuels increase, many coping strategies would come into play. Tree planting would increase, consumers would use fuels more economically, or they would switch to more abundant fuels such as crop residues and so on.

Nevertheless, these inevitable responses are usually disregarded because no one knows how large they will be. In economic jargon, there is virtually no information for woodfuels (or tree growing) on the price elasticities of supply or demand. Any downward adjustments that the gap forecaster made to the rising consumption curve would therefore be entirely arbitrary. Consequently, it is better to leave it alone and present the gap predictions as 'trends continued' scenarios which are designed only to show what large adjustments must be made to demand or supply to bring them into balance.

The crucial flaw to this apparently reasonable attitude is that it greatly exaggerates the need for planned interventions. It implies that all the supply–demand adjustments must be implemented when in fact many of them will be made (and are already being made) naturally by ordinary people without any external assistance. This fault could be corrected by better information, thus putting gap theory on a respectable footing; but until this is done the method must be regarded as a dangerously misleading assessment and planning tool.

Another most fundamental flaw of gap theory, as it is so often used, is its basic assumption that deforestation is driven mostly (or in many models, entirely) by woodfuel consumption. This question is addressed next.

WHERE DO WOODFUELS COME FROM?

How would all gap theory predictions change if woodfuels were only a minor cause of deforestation? In particular, what if most woodfuel supplies arise as a by-product of agricultural land clearances?

If this is the case, woodfuel gap forecasts are facing the wrong way. Rather than being driven by woodfuel demand, they should be driven by data and trends about land use, from which woodfuel supplies arise merely as a by-product.

Many other aspects of the woodfuel crisis are also turned on their head. To arrest deforestation one needs to halt the depredations caused by agriculture rather than by woodfuel consumption. Measures to reduce woodfuel use become much more a matter of improving welfare by cutting consumer costs than attempts to save the trees. Indeed, if all woodfuel use stopped tomorrow, deforestation rates would hardly be altered. Most importantly, the main strategy for halting deforestation would be to intensify cropping and grazing systems so that less new land is needed as populations grow. This puts the onus of maintaining forest cover and woodfuel supplies firmly on the agricultural system – or agriculture plus forestry – rather than energy.

There would also be major implications for the economics of forest use. It is often said that urban woodfuel consumers pay too little because their supplies are 'mined' from the forest with only trivial payments (if any) for the use of these public resources. In turn, this leads to the argument that substantial royalties should be paid by woodfuel cutters for the use of state-owned trees, to reduce pressures on them and to pay for their intrinsic value as well as their less obvious environmental and other social benefits.

Now there may be several good reasons for (as well as severe difficulties in) introducing such measures. At this point it should simply be noted that if land clearance is the main cause of deforestation it is the farmer, not the woodfuel trader, who should pay the resource costs. Since much of this clearing is by subsistence farmers, or smallholders living on the edge of survival, collecting substantial resource royalties from them would be, to say the least, difficult.

Equally important is that if land clearing is the major cause of deforestation the strategic woodfuel questions change. The key question for the sustainability of woodfuel supplies is now: how long can land clearing for agriculture – with its surplus woodfuel bounty – continue before new land runs out? And if the rate of clearing slows as agriculture is intensified, what will this do to woodfuel supplies? How might the supply–demand system adapt? Some interesting answers to these questions can be found in South Asia, where the land clearing frontier was reached some time ago and where most woodfuels now come from managed trees on farm and village land rather than the forest.[7]

Is this also the future for Africa? Chapters 1 and 2 in *Beyond the Woodfuel Crisis* look at the many positive gains to be made by rural tree growing and management, in Chapter 3 at the constraints which can prevent them, and in Chapter 4 at the institutional innovations and linkages which can help to overcome these constraints.

Unfortunately, there is little robust information to answer the critical question of where woodfuels do in fact come from. The best that can be said is that there are several major sources and that the relative importance of each appears to vary greatly from place to place. Each place – particularly each city – must discover the facts for itself in order to plan realistically and effectively. However, of the five main sources of woodfuels described below, there is strong evidence that the last two are in general, on the large scale, by far the most important.

Tree cutting directly for fuel

This is practised especially to make charcoal. It occurs around some cities and in more distant patches of forest close to main roads or rail. Cutting may be intensive, amounting almost to clear felling, but is usually more selective so that only the larger trees of suitable species are felled to leave a fairly complete cover of smaller trees. This may be left to regenerate or, since the charcoalers have made the work easier, cleared for farmland. One can see large areas of degraded woodland like this around Blantyre, Malawi, for example. The larger trees have gone, leaving a dense patchwork of smaller trees interspersed with patches of maize and vegetable crops. The sustainability of this source depends on whether trees are replanted and on cutting rates compared with the rates of natural regeneration in the affected areas.

Although these operations can be a well-organized, round-the-year business, it is important to appreciate that in many places they reflect failures in the agricultural system. Much commercial firewood and charcoal destined for the cities is produced from trees felled by rural people to supplement their incomes, especially in the slack agricultural season, or in years when returns from farming are poor due to drought or low farm prices. To give just one example, tree felling for sale as woodfuel by Communal Land farmers in Zimbabwe greatly increased during the years of drought and crop failure from 1978–81 but then returned to normal levels.[8]

Dedicated woodfuel plantations

Although these are quite common in Asia and there is interest in them as sources of urban woodfuels in Africa, supplies in Africa are at present small except for a few cities such as Addis Ababa, Ethiopia, where woodfuel prices are exceptionally high. Their economic viability depends mostly on whether single-purpose 'industrial' or mixed smallholder methods are used and on the price structures of urban woodfuel transport and markets.

By-product wood

From various tree growing activities: for example, 'lops and tops' from multi-purpose farm trees, or commercial forestry for timber, or specialized farm tree crops such as gum arabic in Sudan and tannin from small woodlots of wattle in Kenya and several southern African countries. Multi-purpose trees and other forms of woody biomass are crucially important sources of woodfuel for rural people; whether or not they can provide urban supplies depends on the complexities of urban market structures and prices.

Dead branches and twigs

These are picked off the ground or cut from the tree. Many surveys confirm that these non-destructive sources account for the great bulk of rural firewood which is taken from state-owned forests and woodlands as well as a good deal of the wood from managed tree and other woody resources on farm lands and village commons.

Surpluses arising from agricultural land clearances

These include rotational fallow or 'slash and burn' farming systems. As discussed below, these sources usually greatly exceed local woodfuel needs, even though many trees may be left standing as part of the farming system while others are burned to provide soil nutrients or simply to clear the land. The extent to which they provide woodfuels for the cities again depends on prices and market structures. Much of the forest devastation which surrounds urban centres must have been due to agricultural land clearance rather than direct cutting for fuel, since it is normally much more profitable to use land within reach of urban markets to grow food for the city than to leave it under trees and sell the wood. The pressures to clear such land of trees, sell any salvage wood to the city, and then farm it, are almost irresistible.

How important is land clearance as a cause of deforestation? When he was chief forestry adviser to the World Bank, John Spears estimated that from 1950 to 1983 it was responsible for about 70 per cent of the permanent forest destruction in Africa. A rough calculation can take this estimate further by showing why land clearance must be a major source of woodfuel. In sub-Saharan Africa, average cropland is close to 0.4 hectares per person.[9] Lands which are typically cleared for farming, such as savannah woodlands and reasonably productive bush, have standing tree stocks of about 20–30 cubic metres per hectare.[10] So with population growth, and without any increase in per hectare farm yields, each extra person's food needs have to be met by clearing some 8–12 cubic metres of wood, or enough to meet a typical annual per capita woodfuel consumption of 1–1.5 cubic metres for 5 to 12 years. Even if some trees are left standing, or are burned, or one assumes standing stocks of only 7–10 cubic metres per hectare typical of degraded savannah or dry bush, there will be substantial surpluses to provide for woodfuels.

More direct evidence of the crucial role of land clearing comes from survey data. In Botswana a rural energy study carried out for the UK Overseas Development Administration[11] made extensive surveys in the eastern region and concluded that the rate of tree felling was many times greater than consumption of the principal wood products: firewood and building poles. Most trees were felled to clear farmland, when they were commonly burned to waste. The second most common reason for felling was for building poles. Fuelwood in rural areas came mostly from fallen branches and the smaller pieces salvaged from land clearances.

In Zimbabwe, a large and detailed survey of wood use, tree planting and forest cover based on aerial photographs in the Communal Lands gave unequivocal confirmation that land clearance has been the main cause of tree loss over the past two decades.[12] In every one of the six classes of land distinguished by their average forest cover, field areas have increased, mostly at the expense of degrading medium-forested land (25–50 per cent tree cover) to woodlands with less than 25 per cent cover.

In Zambia, a recent study concluded that deforestation is primarily caused by agricultural practices.[10] Although trees are cut for firewood and charcoal around the major 'line of rail' urban centres, the main deforestation problems arise from a combination of population pressure and the failure to intensify farming systems. In Luapula, Eastern, Southern and Northern provinces over 20 per cent of forest land has been cleared for farming. This process is probably accelerating, since the more successful farmers are increasing production and incomes by acquiring more land, not by intensifying their production methods.

In the Sudan, a progressive improvement in data about the wood energy system has led to an almost complete reversal in perceptions about the causes of deforestation.[13] The large National Energy Assessment (NEA) completed in 1983 concluded that the national loss of trees amounted to 75 million (solid) cubic metres a year and that 95 per cent of this was due to woodfuel consumption (62 per cent charcoal and 33 per cent firewood). Over the next few years serious faults in this analysis were revealed by the Sudan Renewable Energy Project which was co-funded by the Sudanese government and the US Agency for International Development. Careful surveys showed that charcoal production was nearly twice as efficient and consumption much less than previously supposed. Consequently a large project to improve the efficiency of charcoal making was hastily abandoned, since the traditional producers were already getting higher yields than expected from much more expensive and sophisticated technologies. The NEA had also made a large error when converting woodfuel consumption measured in stacked cubic metres to solid cubic metres 'on the tree'.

When these faults were corrected, estimates of the national tree loss were reduced more than three-fold to around 22 million cubic metres a year, the proportion due to woodfuels fell to 72 per cent, and the contribution from farm clearances and shifting cultivation rose tenfold from 2.3 to 23 per cent. However, if firewood is excluded from these estimates, since little of it comes from tree felling, the changes are even more dramatic. The proportion of the tree loss due to charcoal is now only a third, while mechanized land clearing and shifting cultivation account for 55 per cent. Although these new data are not the last word in accuracy, they undoubtedly come closer to reality than the earlier estimates. Table 10.2 summarizes these changes and gives a very different picture from the woodfuel consumption and deforestation rates shown for the gap prediction for Sudan in Table 10.1.

Striking evidence for urban woodfuels comes from Kenya, where in 1986 a survey was made of the sources of charcoal for four urban centres. It was found that for Nairobi and Nakuru nearly all the charcoal came from clearance of forest or rangeland for agriculture, and the remainder from sustainable wattle and other plantations. For Kisumu 80 per cent came from the latter sources. Mombasa's supplies came from a mixture of land clearing and low-intensity

Table 10.2: Changing views on the causes of deforestation in Sudan

	original estimates (NEA 1983)		revised estimates (Gamser 1988)	
	(million cubic metres of tree loss per year)			
Firewood	24.94	(33%)	13.13	(59%)
Charcoal	46.53	(62%)	2.93	(13%)
Land clearance, etc.	1.75	(2%)	5.04	(23%)
Over-grazing, fire, poles etc	2.13	(3%)	1.12	(5%)
Total tree loss	75.35		22.22	
Excluding firewood				
Charcoal	46.53	(92%)	2.93	(32%)
Land clearance etc	1.75	(4%)	5.04	(55%)
Over-grazing, fire, poles etc	2.13	(4%)	1.12	(13%)
Total tree loss	50.41		9.08	

Notes:
NEA = National Energy Administration
Land clearance = mechanized agriculture plus shifting cultivation
Source: Gamser (1988)[14]

felling on forest or rangelands.[14]

Such findings cannot be generalized: diversity is the rule and the only valid information comes from specific data on the conditions in each country and, more importantly, each district and city. Yet it is clear that for these places at least, deforestation and attendant woodfuel problems depend mostly upon the growing demands for farm and grazing land rather than for woodfuels. Woodfuel supplies are mostly a residue of these pressures or come from more or less sustainable systems or multipurpose tree management elsewhere.

Where this is the case, the obligation for slowing deforestation and maintaining woodfuel supplies at affordable costs falls on a wide range of sectors. Apart from such basic issues as reducing population growth, the main sectorial strategies are:

- for agriculture, to intensify production per unit of land. Much broader issues than deforestation and woodfuels are obviously involved here, including the improvement of food security and rural incomes. Equally, there is a wide range of possible strategies, from conventional technical approaches such as mechanization, irrigation and fertilizers, to the many 'softer' approaches which are now being developed;
- for forestry, to support many of the softer approaches to agricultural intensification and, more generally, to enhance rural livelihoods, by various agroforestry approaches. Trees in farming systems provide vital environmental services, especially the protection and improvement of soil and water resources, as well as supplying essential products such as animal

fodder, construction materials, food, medicines and fuel;

- for forestry, to counter deforestation by growing trees or managing natural forests wherever this is economically justified. The economic sums need to include a wide range of environmental concerns, from the fact that trees fix carbon and help to limit global warming due to the release of carbon dioxide, to watershed and soil protection, the maintenance of genetic diversity and support of wildlife, as well as the benefits and costs of supplying valued wood and other tree products. Providing jobs can also be an important economic rationale for afforestation schemes;
- for energy, to support these strategies indirectly since they help to maintain woodfuel supplies. Otherwise, most energy options centre on demand management in order to increase the welfare of consumers by reducing their energy costs.

Where woodfuel needs are a direct cause of deforestation, all three sectors obviously need to add the provision of woodfuels at affordable costs to these agenda.

GIVING SCARCITY A HUMAN FACE

One of the most basic assumptions of conventional woodfuel thinking is that physical scarcity of wood is the key issue to address. As typified by gap theory, analysts and planners have measured the scale of woodfuel problems in terms of volumes of wood resources and consumption and distances from resources to consumers. In rural areas, the distance and time taken to collect woodfuels are commonly used as the main yardstick of scarcity and the need for remedies. For urban areas, it is commonly assumed that woodfuel prices will rise as forest stocks are depleted and the transport distances from the city to its main woodfuel resources lengthen. Since increasing physical scarcity or distance can impose considerable costs to consumers, the basic aims of woodfuel interventions are to reduce these costs by reducing physical scarcity.

There is, of course, a good deal of truth in these attitudes. However, interventions are most unlikely to succeed if they do not recognize that physical scarcity means nothing unless it is related to the human dimension. As Peter Dewees[15] has cogently pointed out, one has to ask whether these costs are the outcome of physical scarcity itself or of much more fundamental issues such as labour shortages, land endowments, social constraints on access to wood resources, or cultural practices. These human issues are both complex and dynamic and frequently undergo rapid and adaptive change which the outsider may easily miss.

Consider wood gathering, for instance. There is now compelling evidence from time-budget studies for rural women that the time spent collecting firewood can vary greatly from one week or season to the next depending on agricultural and other labour demands; it is often minor even in wood scarce areas compared with time for collecting water, food preparation, cooking and other survival tasks; and it is both adjusted and perceived as a more or less severe problem, in relation to the totality of labour needs and time available.[16, 17] Similarly, it is the total time required for cooking and fuel collection which is one of the critical factors deciding whether women burn scarce but otherwise preferred and easily managed species of firewood or turn to more abundant and easily collected fuels such as crop residues or animal wastes which may take more time and trouble to use in the cooking fire.

The basic issue is therefore one of labour availability, not fuel availability. If spare labour is abundant it may not matter if woodfuel-collecting trips are long or getting longer. If labour is very scarce, even the collection of abundant woodfuel supplies may be perceived as a serious problem. What matters is local perceptions of these questions and the coping strategies that people have evolved or are evolving to deal with them, not the outsider's simple physical measurements.

Labour scarcity on the farm may also be a deciding factor in rural tree growing versus crop production, as exemplified in Chapter 3 of *Beyond the Woodfuel Crisis*. It is even an important factor in setting urban woodfuel prices. It is commonly assumed that these will rise mainly because scarcity rents will be imposed as trees are depleted, and transport costs will rise as haulage distances increase. But as shown in Chapter 6, urban woodfuel prices depend to a considerable extent on the abundance or otherwise of agricultural labour through its effects on the costs of harvesting trees or making charcoal. Many factors other than the physical scarcity of trees are also critical, including the availability of trucks and the demands made on them for all purposes; shortages (and prices) of alternative 'modern' fuels; crop prices and their effects on rates of land clearance; urban wage rates and employment prospects; and the degree of competition in the woodfuel market. Many of these factors have changed and could be changed substantially in order to increase the incentives for urban woodfuel production on a sustainable basis without planting a single tree to reduce wood scarcity.

This brief discussion has hardly scratched the surface of what physical scarcity actually implies in human terms. All that we have tried to do is suggest that many issues are involved and that the diagnoses of woodfuel problems and the design of remedies for them have to reach these underlying dimensions.

In conclusion Figure 10.1 presents a sketch which underlines the basic point that woodfuel issues must be seen and acted upon in their broadest contexts. It makes two main points.

First, conventional woodfuel approaches have usually looked only at the top of the diagram, using simple energy-centred measures to decide whether there were significant problems demanding interventions. It is now increasingly recognized that a more holistic and participatory approach to problem identification, diagnosis and intervention must be used. Whether or not people feel that woodfuels are a significant problem must depend on their views of the costs involved compared with the many other concerns and costs in their lives. If they do not recognize these costs as significant, then a woodfuel intervention that itself has significant costs compared with benefits, or does not

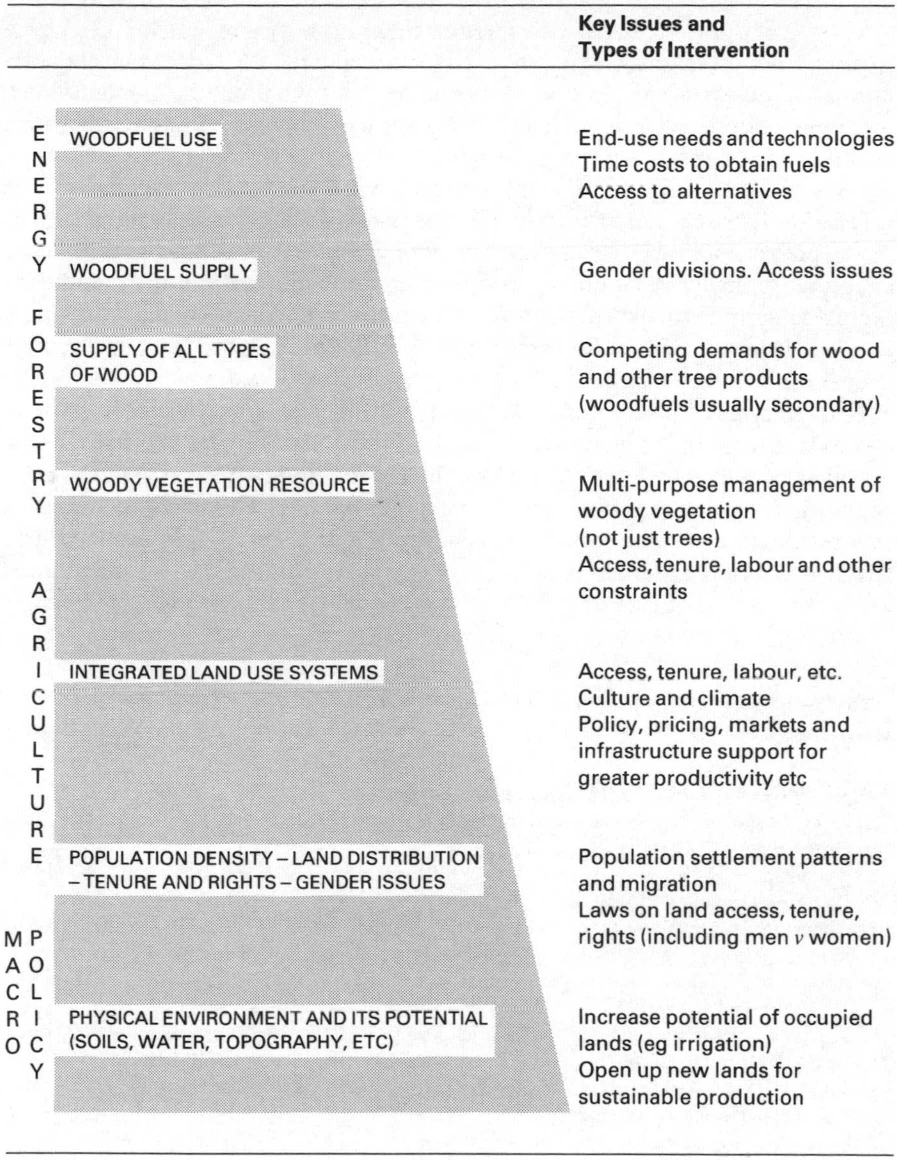

		Key Issues and Types of Intervention
ENERGY	WOODFUEL USE	End-use needs and technologies Time costs to obtain fuels Access to alternatives
	WOODFUEL SUPPLY	Gender divisions. Access issues
FORESTRY	SUPPLY OF ALL TYPES OF WOOD	Competing demands for wood and other tree products (woodfuels usually secondary)
	WOODY VEGETATION RESOURCE	Multi-purpose management of woody vegetation (not just trees) Access, tenure, labour and other constraints
AGRICULTURE	INTEGRATED LAND USE SYSTEMS	Access, tenure, labour, etc. Culture and climate Policy, pricing, markets and infrastructure support for greater productivity etc
	POPULATION DENSITY – LAND DISTRIBUTION – TENURE AND RIGHTS – GENDER ISSUES	Population settlement patterns and migration Laws on land access, tenure, rights (including men *v* women)
MACRO POLICY	PHYSICAL ENVIRONMENT AND ITS POTENTIAL (SOILS, WATER, TOPOGRAPHY, ETC)	Increase potential of occupied lands (eg irrigation) Open up new lands for sustainable production

Figure 10.1: The contexts of rural woodfuels

simultaneously reach other and more urgent concerns, will in all probability be rejected or allowed to collapse into failure.

Second, on this more holistic view, woodfuel issues are merely the tip of a seamless pyramid which reaches down to progressively more fundamental aspects of survival, production and land management. Various disciplines – energy, forestry, agriculture, land use planning and the like – have each staked out layers of this pyramid as their territory. Again, it is increasingly recognized that each discipline must do much more to look at and link with the layers both above and below its own special concerns. At the same time, one needs to look sideways, across the pyramid, at the various issues and conflicts which are found at each level.

Only a highly integrated and multi-disciplinary perspective can hope to succeed in this formidable task. That is why, at the local level, a highly participatory approach is essential. The people who live closest to each of these differing, local pyramids are the people who know most about them and who are most likely to recognize – albeit with external assistance of various kinds – the key leverage points for beneficial change and the brakes which could prevent it.

However, this participatory bottom-up approach is not sufficient because it has little to do with the many constraints and possible levers for positive change which are found at the base of the pyramid: the macro-level issues which are the concern of top–down planning and government policy. The strengthening and co-ordination of policies and actions across these higher levels to match the realities and opportunities at the micro-level is an extraordinarily difficult task but is also one that must be faced if sustainable development is to be achieved.

This chapter is taken from Leach, G and Mearns, R (1988) *Beyond the Woodfuel Crisis,* Earthscan, London.

NOTES AND REFERENCES

1. FAO, *Fuelwood Supplies in the Developing Countries* (Rome: UN Food and Agriculture Organization, 1983).
2. Anderson, D and Fishwick, R *Fuelwood Consumption and Deforestation in African Countries* (Washington, DC: World Bank, 1984).
3. Anderson, D, 'Declining Tree Stocks in African Countries', *World Development*, 14/7 (1986): 853–63.
4. Anderson, D, *The Economics of Afforestation: a Case Study in Africa* (Baltimore and London: Johns Hopkins University Press, for the World Bank, 1987).
5. Spears, J, *Deforestation, Fuelwood Consumption, and Forest Conservation in Africa: an Action Program for FY86–88* (Washington, DC: World Bank, 1986). (Restricted)
6. Nkonoki, S and Sorensen, B, 'A Rural Energy Study in Tanzania: the Case of Bundilya Village', *Natural Resources Forum*, 8 (1984): 51–62.
7. Leach, G, *Household Energy in South Asia* (London and New York: Elsevier Applied Science Publishers, 1987).
8. Mazambani, D, 'Peri-urban Deforestation in Harare', *Proceedings of the Geographical Association of Zimbabwe*. 14 (1983): 66–81.
9. WRI and IIED, *World Resources 1987* (Washington, DC: World Resources Institute; London: International Institute for Environment and Development; 1987).
10. ETC 1987. A composite reference to the collection of reports of the study *SADCC Energy Development: Fuelwood*. The separate titles, all preceded by *Wood Energy Development:*, are: *Policy Issues; A Planning Approach; Biomass Assessment; The LEAP Model; Bibliography of the SADCC Region; Country Reports for the Nine SADCC Member States* (Leusden, Holland: ETC Foundation; Luanda, Angola: SADCC Energy Technical and Administrative Unit).

11. ERL, *A Study of Energy Utilisation and Requirements in the Rural Sector of Botswana: Final Report* (London: Environmental Resources Limited, for Overseas Development Administration, 1985).
12. Du Toit, R F, Campbell, B M, Haney, R A and Dore, D, *Wood Usage and Tree Planting in Zimbabwe's Communal Lands: a Base line Survey of Knowledge, Attitudes and Practices* (Harare: Resource Studies, for the Forest Commission of Zimbabwe and the World Bank, 1984).
13. Gamser, M, *Power from the People: Innovation, User Participation, and Forest Energy Development* (London: Intermediate Technology Publications, 1988).
14. Dewees, P A, 'Commercial Fuelwood and Charcoal Production, Marketing, and Pricing in Kenya'. Working Paper II, Phase I Report, *Kenya: Peri-Urban Charcoal/Fuelwood Study* (Washington, DC: World Bank, 1987).
15. Dewees, P, 'The Woodfuel Crisis Reconsidered' (Oxford: Oxford Forestry Institute, 1988) (Draft paper).
16. Tinker, I, 'The Real Rural Energy Crisis: Women's Time', *Energy Journal*, 8 (1987):125–46.
17. Cecelski, E, *The Rural Energy Crisis, Women's Work and Family Welfare: Perspectives and Approaches to Action* (Geneva: International Labour Office, 1984).

PART V
INDIGENOUS PEOPLES

Chapter 11
Indigenous Peoples and Development

John Bodley

The contemporary issue of tribal peoples and economic development can best be viewed in a broad historical context. The Neolithic world of 10,000 years ago was occupied by some 75 million people, organized into perhaps 150,000 tribal nations. These nations were politically autonomous, decentralized, and economically self-sufficient; they were small-scale communal societies with distinctive ways of life dependent on local ecosystems. The basic features of tribal nations were completely different from those of modern nation states. These differences go far beyond the mere facts of ethnicity, and they reflect a high degree of local autonomy. Particularly significant is the social equality found in tribal societies and the absence of leadership hierarchies and favoured access by special classes to natural resources. Tribal nations were self-regulating systems that provided maximum independence in decision making at the lowest levels of society.

The archaeological record suggests that tribal nations were relatively secure adaptations that effectively satisfied basic human needs, traditional cultures which worked well, even among ethnographically known modern groups. This is a particularly important issue because development changes are supposed to improve people's lives. In many parts of the world archaeological sequences reveal millenia of continuous occupation by tribal societies with only minor shifts in subsistence, which reflect fluctuation in the environment. There was, however, a steady, often barely perceptible trend toward intensification of subsistence effort and increased population density. This long-term trend culminated in dramatic changes in the Neolithic world some 6000 years ago when the first tribal nations became nation states.

The new nation states represented evolutionary 'progress'. They were larger, politically centralized, class-based societies with expansionist tendencies. It is not clear that states were an improvement in terms of satisfaction of human needs, because social hierarchy often meant that social benefits were inequitably distributed. States were also inherently unstable. Their appearance suddenly made the world unsafe for tribal nations. As states expanded in search of new resources, threatened tribes either became states in order to preserve their nationality or were absorbed or extinguished and their surviving populations converted into peasants. The early states spread slowly until the beginning of the industrial era, at which time they controlled nearly half the globe. Development progress over the past 200 years has seen hundreds of tribes extinguished along with millions of tribal peoples.

Today only a handful of independent tribal nations exist, and several million recent descendants of former tribal nations remain as culturally distinct colonized peoples who still aspire to autonomy. They are commonly called 'indigenous peoples' because they are the original inhabitants of their territories. However, precise terminology that takes into account the differing histories, cultures, political statuses, and aspirations of these diverse peoples is lacking. The state often designates conquered tribal nations as 'tribes' for administrative purposes, but they become 'peasants' when dependency and integration are complete and they remain on the land. When they move to urban areas, they are often called 'ethnic minorities'.

Historically, when states began to assume effective control over the territory of tribal nations and to extract resources from it for state interests, tribal nations found it difficult to sustain their original adaptive systems. They suffered demographic upheavals, resource depletion, internal inequality and conflict, and increased pathologies. They were often locked into grossly inequitable and discriminatory economic relationships with the dominant state society, and they were reduced to insecurity and poverty. This pattern of disruption and impoverishment continues to be the dominant outcome of modern development programmes directed at tribal peoples. Indeed, this is one of the major dilemmas of development.

Today there are some 200 million people who are members of recently conquered or still autonomous tribal nations. Of course, states permit few of these indigenous peoples to maintain their decentralized, communal adaptations, but much of the original self-sufficient pattern remains. Today's indigenous people continue to assert their autonomy, and their existence represents a major challenge to state policy makers both because independent tribal nations were so successful, yet so different from states, and because their autonomy claims are basically just.

THE INTEGRATION POLICY

Throughout this century people of good will have advocated a development policy of integrating tribal people into the dominant state society and economy as ethnic minorities while attempting to minimize the human costs incurred in the process. Ideally, this would mean preserving the 'best' features of the traditional culture and eliminating those features that might be considered obstacles to economic progress. Integration has been the predominant solution to the challenge posed by the existence of indigenous peoples when direct extermination or blatant exploitation were acknowledged to be inhumane and inefficient. Various forms of development aid, education, protective legislation, and health and welfare programmes have been implemented to achieve successful minority status for tribal peoples. Clearly, such efforts have had many positive results and have eased much pain; it is not surprising that integration policies continue to dominate the development field. However, when integration is pursued without a clear understanding of the original features of the indigenous community that contributed to its well-being, serious harm can easily be done. A major problem with development policies promoting integration is that their aim is usually to benefit individuals, often at the expense of the community. When development undermines a community's ability to defend and manage its own resources, or when it is imposed by outsiders, genuine benefits can hardly be expected.

For their part indigenous peoples have consistently resisted integration programmes, realizing that they are a further threat to their independent existence. It can be assumed that people only surrender political autonomy under duress. Furthermore, the material benefits actually derived from this kind of development have seldom outweighed the costs. Given the apparent shortcomings of integration approaches to development, it is appropriate to examine the basic assumptions underlying them. These assumptions appear to include at least the following:

- the way of life of indigenous peoples is materially inadequate;
- integration will improve their quality of life;
- interest in new technology on the part of indigenous peoples reflects a desire for integration;
- progress is inevitable.

Although all of these assumptions may be valid in certain respects, they are also very short sighted, and when they go unchallenged there is little basis for dramatically different policy alternatives. When considering the first assumption, it is apparent that indigenous peoples themselves do not find their way of life materially inadequate when they are still in control of their undepleted natural resources. Poverty is a product of the state, created by class systems that impose relative deprivation and cause resource depletion. These processes are endemic in modern states but conspicuously absent or minimal in tribal nations.

The assumption that integration into state society will improve the quality of life for indigenous peoples requires that they are indeed 'successfully' integrated somewhere above the lowest levels of society. In reality very few indigenous peoples can hope for such 'success'. Integration into the impoverished classes is likely to result in a significant decline in quality of life, and this is clearly reflected in health and nutrition.

The fact that indigenous peoples universally come to desire industrially manufactured goods such as knives, guns, and metal pots is invariably interpreted as both proof of the inadequacy of their way of life and a vote for integration into the dominant society. Although it is certainly true that new technologies are often seen as advantageous, when indigenous peoples are still in control of their lives and territories they select imported items that enhance their original adaptation. Some items, such as guns, may even be used to discourage outside intrusion. Of course the process of acquiring imports and actually using them may set in motion detrimental changes, but such results are rarely foreseen.

The argument that progress is inevitable actually represents the belief that there can be no place in today's world for independent tribal nations. It rests on all the preceding ethnocentric assumptions and implies a sort of Darwinian notion of survival of the fittest. This view certainly reflects recent historical trends, but it ignores the fact that indigenous peoples have only rarely surrendered their autonomy voluntarily. 'Progress' has been forced on them because expanding states required the resources they controlled. Such expansion clearly must have limits, and limits must be a matter of state policy. Ultimately, even states will be forced to acknowledge the value of careful resource management for genuine, sustained yield production. Ironically, the resources now controlled by indigenous communities are there only because they have been managed carefully by self-sufficient communal economies. Integration designed to exploit those resources for outside interests undermines local management and in the long run also threatens the resources. To argue that this is an inevitable process is both arrogant and shortsighted; it can also be a self-fulfilling prophecy that effectively blocks significant policy changes.

SELF-DETERMINATION

Perhaps the most significant new element confronting contemporary development planners is the directly expressed political demand of indigenous peoples themselves to retain their autonomy and to control their own future development. Tribal autonomy movements in Africa, the Philippines, India, North America, Central America, Indonesia, and Australia are discussed in parts VI and VII of *Tribal peoples and development issues* from which this chapter is taken. These are clearly a major force for change and a genuine expression of the will of tribal peoples. In response, some countries have already begun to modify the standard integration approach to reflect the possibility of self-determination by tribal peoples. This new approach represents a growing recognition that indigenous peoples have been systematically denied their basic human rights to occupy their traditional lands and to enjoy their own way of life. According to this view, tribal peoples have been politically oppressed and economically exploited as virtual internal colonies within the countries that claim control over them. Ideally, self-determination would put tribal peoples in charge of their internal affairs and territorial resources again. Even though this sort of self-determination has a clear basis in international law and human rights resolutions and is clearly in line with the wishes of indigenous people, states have difficulty relinquishing power of any sort to subgroups. Where self-determination has been adopted as official policy, it is granted within a very narrow framework defined by the state. Even so, this is clearly a significant new policy direction and raises a whole series of new issues.

Complete self-determination for indigenous people would involve full ownership of their traditional territory as well as political control within it. They would be free to manage their resources as they saw fit. They could bar entrance to undesirable outsiders and block the extraction of resources or the intrusion of dams and highways. They would also be free to maintain a self-sufficient, communal socio-economic system according to their own cultural traditions and social control mechanisms.

Achieving any degree of self-determination, particularly where control over valuable resources may be involved, is clearly an uphill struggle for indigenous peoples; it may also involve serious risks. Indigenous peoples may have little real political or economic power available to them, and they are likely to be at a distinct disadvantage in any struggle with the state for recognition of their rights. The need for political mobilization extending far beyond the traditional experience or capability of the society may encourage indigenous groups to form alliances with outside groups who may have their own agendas. Furthermore, political mobilization may in itself be incompatible with a decentralized, communal way of life. In spite of these obvious difficulties, the costs of integration policies and the potential rewards of self-determination are so high that where there is any possibility of success, indigenous peoples invariably choose it over integration.

The routes to self-determination for indigenous peoples are diverse. For example, those few groups who have not yet surrendered their autonomy, such as the Yanomami of Amazonia, might simply have their present independent status officially sanctioned, without ever going through a political mobilization self-defence process. For such special groups biosphere reserves, which are designed to preserve entire ecosystems, might be appropriate mechanisms. Other groups might require the restoration of lost territory and the creation of unique political structures, such as special provinces, to safeguard their interests.

Any suggestion that indigenous communities be allowed to either maintain or regain their distinctive self-sufficient, communal pattern may be met with a variety of misleading and irrelevant criticisms. For example, critics may misrepresent self-determination as a 'human zoo' approach, falsely assuming that it is based on naive assumptions such as the following:

- Indigenous peoples are noble savages.
- Culture is or should be static.
- Indigenous peoples must be isolated.
- Cultures are entities to be protected.

It can then easily be argued that there are no noble savages, no static cultures, and no isolated peoples and that culture itself is merely an abstraction. Furthermore, it might be argued, self-determination is politically unrealistic idealism. Of course, the real issue is whether communities will have the right to control their own lives within their own territories. It just happens that the way of life of indigenous people may be both very different and very successful. Whether indigenous peoples are noble savages, how much they change, how isolated they may be, and whether their way of life is protected are issues for them to decide in their own way.

Some official agencies have already adopted the language of self-determination but unfortunately not the substance. For example, the World Bank advocates official support for self-determination and cultural autonomy for indigenous peoples. However, it does not envision that indigenous peoples will actually be able to control their own resources; it sees them instead as sucessfully integrated ethnic minorities. Others argue piously that it is wrong to allow a few people to maintain 'inefficient' subsistence economies that lock up resources that might be used more profitably by the impoverished majority. This is of course the very heart of the issue, and it focuses attention on the real problem, which lies within the nation state itself. Inequitable and imbalanced growth in the nation state generates frontier expansion and converts the territory of indigenous peoples into internal colonies open for exploitation. Culture change is indeed required if indigenous peoples are to cease being victims of progress, but it is the state itself and its policies that must change.

In conclusion, it is the long-established policy of integration that fails to deal adequately with the real issues of indigenous peoples as victims of progress. Rather than integration, genuine self-determination should be promoted as the most just, humane, and intelligent policy. There is certainly a basis for hope, even though we know that historically indigenous peoples have been exterminated and/or absorbed by expanding states. This has been a global process, and it has been official state policy. However, neither the policies nor the victims have always been passive. The victims have often resisted, and in this century we have seen policies toward indigenous peoples move steadily, from conquest to the extension of humane treatment, to an emphasis on integration, to the preservation of ethnic identity, to a gradual recognition that self-determination may be a legitimate right of indigenous communities.

We have thus come almost full circle to a point at which some states may be prepared to permit the existence of independent indigenous nations. Such a self-determination approach must surely be the most ideal and humane policy. Although integration policies remain the dominant approach throughout the world, over the past decade the whole question of the status of indigenous people has been undergoing intensive reassessment. Full acceptance of self-determination policies is hampered by numerous conceptual and practical issues, but these issues are now being examined. As a contribution to this reassessment process, the collection of readings in *Tribal peoples and development issues* systematically reviews the historic pattern of interaction between tribal peoples and state societies and examines the dominant trends in state policy toward indigenous peoples worldwide during this century.

This chapter is taken from Bodley, J (1988) *Tribal peoples and development issues*. Mayfield Publishing Company, California.

Chapter 12

High Technology and Original Peoples: The Case of Deforestation in Papua New Guinea and Canada

Colin De'Ath and Gregory Michalenko[1]

These two case studies from Papua New Guinea and Canada highlight the convergence of interests between governments and multinational corporations to promote profitable development programmes that disrupt local subsistence economies while providing little compensation. In both cases development of forest resources has resulted in enormous environmental degradation, which has endangered previously viable local subsistence pursuits. Both governments clearly were in a position to profit from their support of the corporations in opposition to the demands of local people for a more reasonable share of the profits and for protection of their subsistence resources. The authors point out that in cases such as this, government policy must explicitly acknowledge that indigenous communities are engaged in 'dual economies'. If the government is concerned with citizen welfare, it may be just as necessary for it to support local subsistence economies as to facilitate penetration by the market economy. In Alaska government development planners have indeed calculated the cash-equivalent value of bush products to demonstrate the economic importance of hunting and fishing to native communities.

TECHNOLOGY, CORPORATIONS AND ORIGINAL PEOPLES

Technologies in industrialized countries can be characterized by the following features. First, they have become highly centralized in terms of bureaucratic control; second, they rely generally on a narrow spectrum of energy sources; third, organizations responsible for controlling technologies are large and complex, and relatively inflexible especially in terms of setting new goals; fourth, until quite recently such organizations assumed that resources were infinite and that growth and induced consumption by its clientele were unimpeachable goals.

The spread of Western technologies has been facilitated by a number of factors. More and more scientists spend their working lives making more and more sophisticated things, especially for the military. Through worldwide communications systems, the knowledge of the potential of new technologies becomes known rapidly by corporations. There is growing congruent thinking between transnational firms and national governments in terms of the supposed utility of various technologies. Emphasis in gauging the relevance of a given technology to a particular situation has usually been on whether there is a technological 'fit', rather than a social or ecological fit. There have not been dramatic advances in the social sciences equivalent to those in other sciences and their related technologies; this has led to centrally inspired technocratic fixes in lieu of sound social planning and follow-through. A fatalistic kind of technological determinism has been adopted, implying that the evolutionary trajectories of certain technologies once set in motion are irreversible – until, of course, there is a catastrophe. And, important for the purposes of this paper, the impact of technologies and their associated political institutions have not been adequately publicized.

In the heartlands of Western-type industrialization (whether in Europe, Japan, North America or what used to be the Soviet Union), there is some appreciation of what effects complex or 'high' technology has on ecological and social systems. Even there, however, technological change can be gradual enough for its participants not to realize what is happening to them and their natural environment. This anomaly is also attributable in part to a distortion in feedback. Because it is in the interests of governments and corporations to promote consumption, the illusion of improvement, growth and profits, it is very difficult for subject populations caught up in specialization, professionalism, machine-like routines, consumerism and 'commodification' to assess the direction in which change is moving.

The deterministic theories and assumptions of engineers, and physical and natural scientists generally, do not encourage ordinary citizens to participate in decision-making – the belief becomes widespread that the experts know what is right and appropriate in terms of resource exploitation and the efficient use of human beings.

Original peoples, such as those in northern Canada and in Papua and New Guinea, tend, in the face of the solvent power of Western money, science and technology, to be very vulnerable to corporate dynamism. In the following sections the results are examined of the playing out of this dynamism by a Japanese corporation, sanctioned by what was formerly the Australian colonial government in Papua New Guinea, and a British transnational assisted by Canadian governments.

THE TRANS-GOGOL

The Trans-Gogol is a tropical river valley about 60 kilometres from the town of Madang in Madang Province. This province is on the north coast of Papua New Guinea.

Although the authors' first study focused on the Gogol Timber Rights Purchase (TRP) area (52,265 hectares (ha)), the Jant (Japan and New Guinea Timbers) had interests in three other adjacent 'TRPs'.

The vegetation is tropical rainforest and somewhat different from that found in Indonesia and elsewhere in South East Asia. Compared with current knowledge of temperate forests, not a great deal is known about the ecological system. Foresters themselves, because of the role of natural disasters, cannot even agree on whether the forest has ever reached 'climax status'. Because of this knowledge deficit, and because there has never been massive clear-felling elsewhere in lowland forest areas in Papua New Guinea, it is not known what the impact on this complex life-system will be. Although the Man and the Biosphere Programme research in this area is well-intentioned, it has many gaps, particularly in the area of ethnobiological studies. Too, not enough is known about the role of the forest in stabilizing the valley which regularly floods but also goes through droughts, and its influence on the micro-climate. Another area which is of extreme importance is the effect on regeneration of extensive commercial tree cutting compared with that of the scattered cuttings of subsistence farmers who have been in the area for an unknown number of generations. Human settlements there may go back 50,000 years, although the current linguistic diversity could indicate that some groups are fairly recent arrivals.

The company and its position

Jant is a subsidiary of Honshu Paper. The latter began negotiations with the Australian Government in the late 1960s. At that time, the Australian Government was a United Nations trustee for the Territory of Papua and New Guinea (TPNG). After complex negotiations resulting in the incorporation of a TPNG/Australian/New Zealand saw log company, cutting operations began in 1973. (The subcontractor had been cutting saw logs for some time before this date.) Several agreements were involved in setting up the operation, namely between the local residents (who owned the land and the trees) and the government (which purchased the rights to cut them) and between the company and the government. The company agreed to pay royalties for cutting and entry rights. These agreements had a number of flaws.

It is doubtful whether the people understood what was about to happen to their land, their life-style, and their incorporation in (and exclusion from) an imposed cash economy. The agreements were vague about costs and benefits except for the initial purchase price for the trees.

The company's commitment to reforestation was not clearly spelled out; neither were rigid obligations to diversify agriculturally or industrially included in the agreements. Strict monitoring of company operations, particularly in the areas of staff localization, were not included in the agreement. Social, technological and ecological impact studies were not a prerequisite to setting up the operation.

Impact of the Operations

In Madang town itself, where a chip mill and the saw log facility are situated, probably only about half of the jobs promised materialized. All but two or three top executive positions are held by Japanese executives who live in comfortable housing areas. Most sub-contracting for goods and services is given to non-Papua New Guinea firms. The town, which is in economic decline, has not been revitalized by the foreign firms' presence. The area around

Madang – especially in Binnen harbour where Jant and Wewak Timber are located – is very beautiful and an ecologically sensitive area because of its estuarine and coral ecological systems. There was little consultation on the siting of these facilities, with the result that the town is saddled with an eyesore as well as an effluent-producing facility. Waste products include smoke, ash, sawdust and bark, and chemically polluted excess water.

The quarters for mill workers, not far from the mill and in the low-cost housing areas, leave much to be desired. A third of all workers come from outside of Madang Province. Employee turnover is very high. Employment for local women does not exist. There have been complaints about industrial safety and housing conditions, as well as a number of wild cat strikes. Jant has not paid any corporate tax since its operations commenced. It builds up considerable debts in Japan, and these must be serviced. It is doubtful whether the expertise exists in Papua New Guinea to monitor Jant's transfer costing and accounting practices and to ascertain whether these are ethical in terms of the welfare of a new nation. Jant suggests that its operations are not as profitable as envisaged, partly because of a depressed global pulp market. Much of its equipment, especially its bulldozers and log-hauling trucks, have become aged and unsafe. Because of overloading, the latter vehicles do much damage to the local road system (which is partially government maintained).

The Japanese, because of high staff turnover and their own ethnocentrism, have great difficulty in getting to know their own people. They are generally preoccupied with technical efficiency at the mill and in maintaining production targets. They consider that the welfare of their company is coterminous with the welfare of their employees and that the profitability of the company is always a prior consideration in dealing with workers about pay and working conditions. The philosophy of co-prosperity made very explicit before the Second World War, is alive and well within the firm, that is a profitable company automatically means an increase in the welfare of its employees and those involved in supplying resources to the corporation. When there are problems between the employers and local nationals, there is a strong tendency to leave the government the responsibility for their solutions. This has not been hard to do as there is still a strong paternalistic tradition within government; a heritage from the colonial Australians. Problems such as the size of royalty payments tend to take years to resolve.

Within Madang town, the size of the Office of Forestry's operations has grown considerably. A rough estimate for the Gogol TRP shows that between January 1974 and June 1977 the government received K1,398,968 in revenue and spent K1,962,000 in the operation (in March 1978, K1 = US$1.37). Nearly all of this could be classified as a subsidy for Jant's operations. Within the Office of Forestry however, there is some ambiguity about how the local foresters should function ie whether they should be:

1. revenue producers;
2. local royalty paymasters;

3. protectors of the environment;
4. company policemen;
5. facilitators of company activities (especially in times of crisis);
6. social planners acting on behalf of TRP residents.

From past operations, it becomes obvious that, despite the existence of an interdepartmental working group, the Office of Forestry does place emphasis on 1, 2 and 5. This is probably due to the training of the foresters and to the inevitably close contact which occurs between Jant and the Office of Forestry. Jant, with its entertainment and political contacts, can be a much more powerful lobby than can disorganized subsistence farmers who know little about the ways of transnational firms. Much infrastructural assistance by the government is given to the company through the provision of roads, electricity, harbour facilities, land and forest surveys, research and planning. Yet the company complains about its assessments for some of these items. (No comprehensive analysis has been done on the value of these services.) Local people receive some spin-off benefits but these, in the past, have tended to have an incidental effect rather than lead to a marked improvement in the standard of living in Madang town, whose population is 19,000.

The rural impact

At the commencement of the project there was much discussion about agricultural development – of the development of a small town and of employment prospects for Trans-Gogol inhabitants – and on both deforestation and reforestation. But little benefit has resulted from these proposals. Reforestation did not commence, and then only on a very limited scale, until 1978. Local people quickly became disillusioned with labouring for Jant at K25 per week (less than half the urban wage). Currently only about one-sixth of Jant's labour force is recruited in the timber areas. No comprehensive agricultural development has occurred, and the small town of Arar is still only a town planner's sketch.

The people's grievances include the following:

- **The size of royalty payments.** They get less than 50t ie half a K per cubic metre. Dressed timber in Madang town costs them K198 per cubic metre. The export value of 1 cubic metre of wood chips is K16.7, and Jant and Wewak Timber sell logs to one another for K11 per cubic metre.
- **Payment for road-building gravel.** The people receive nothing for the gravel, but elsewhere companies must pay a royalty of 5t per cubic metre.
- **Indifference of government, researchers and others to the needs of villagers.** Villagers say that visits by government officers may have become perfunctory and some social services are worse than they were previously.
- **Gardening and game.** Cut-over areas, because of bad logging practices, are unsuitable for gardening. Game, despite token reserve areas, has declined considerably. Many species of game used for food no longer exist.
- **The level of the water-table.** In many areas along levees and roads the

water-table has been heightened, leading to the death of vegetation and the existence of ponds where mosquito populations can increase. Malaria, consequently, has increased in the area. Some of the malaria is chloroquin-resistant, and some of the mosquito vectors are now resistant to DDT.

- **The style of Japanese operations.** Japanese supervisors tend to believe they are omniscient and, at the beginning of the project, made many mistakes in siting roads and bridges and in their actual construction. Few had any tropical forest experience and fewer still knew anything about local societies, tropical ecological sytems and reforestation. Their operations during the wet season were wasteful, and their initial use of labour profligate. Local folk complain that the Japanese never listen to them and that there are no real employment prospects. The company has been remiss in setting up meaningful training programmes which could lead to complete localization of their operations. In the five years of the company's operation, villagers have never risen higher than foreman.
- **The indifference of the company.** The company does not tell the villagers enough about its logging plans. Within the Trans-Gogol, the government embarked on a very expensive survey of clan boundaries primarily so that it could acquire land for reforestation and possibly pay royalties on a more equitable basis. But payments for leasing land that will be used for reforestation will be low, and there is a strong possibility that the landowners will again be disillusioned when they realize what valuation is put on their land and labour inputs.

There are a number of other negative effects.

A forest station has been established at Baku in the Trans-Gogol, not primarily to benefit the villagers but rather the company's operations. It is situated adjacent to Jant's base camp. The local police station also is sited there.

Many kilometres of roads have been constructed in the logging areas, but after one wet season many of these are washed out or have slumped. Poorly built logging bridges also collapse or may be washed out. Thus, after the company finishes its logging (except where there is reforestation), the roads rapidly become disfunctional, and the people are in as bad a shape for marketing their crops as they were before the logging. They do not have the means to maintain the roads, and the new provincial government has other priorities.

The company initially estimated that it would take 20 to 25 years to log its four TRPs. This has now been reduced to 11 years. The firm is seeking extensive concessions in areas nearer the Ramu River, which will mean very long, expensive hauls as well as expensive road maintenance. This will also increase the possibility of a road being linked to the highland roads where people are short of land. The Trans-Gogol and other Madang people are very apprehensive about the impact of such migrants. They have bad memories of land losses earlier to German and Australian occupants.

In many development scenarios it is assumed that technology and built features such as roads will lead to the well-being of all concerned. In the Trans-Gogol it is doubtful whether village people will benefit from Japanese controlled high technology. It is also doubtful whether the town of Madang and its regional hinterland has benefited. The entire nation probably also has been a loser, even if only traditional economic criteria are used. Certainly the ecological system will never recover its integrity. In this instance roads and high technology have led to a net outflow of resources. Because of the myths associated with business and government operations, and because of a colonial past in which these kinds of extractive operations were the norm, it is doubtful whether a great deal has been learned about how to manage – let alone enhance – the functioning of interdependent technological, social and ecological subsystems. The integrity of each of these subsystems, particularly the indigenous social subsystem which has taken thousands of years to refine to a point where it is self-sufficient in terms of food and material culture, has been breached.

Once their forest disappears and their land becomes unsuitable for food staples and game, the people in the Trans-Gogol, will be forced into a dependency relationship with the larger system. Even prior to the coming of the foreign timbering firm, the local people did not have great admiration for the exploitative behaviour of outsiders. The managed, at least during the period of 1871–1973, despite insistent demands for cash-crop surpluses and their labour, to keep their most valued resources (land and forest) intact. Now, however, the future is problematical.

THE OJIBWAY OF NORTHERN ONTARIO

The Ojibway live in the forested land lying between the Great Lakes and Hudson's Bay. Traditionally they have hunted moose, beaver, hare and fish, and lived in small local groups. Because of periodic or seasonal fluctuations in the resource base, there was always some movement of these small groups. 'Bush' resources were distributed on the basis of reciprocity or mutual sharing within the local group; this had the added effect of extending sharing to other groups of Ojibway in the area.

Contact with fur traders of European origin led to a profound shift and to a dual economy whereby a substantial supply of resources for direct consumption still came from the land, while furs (easily obtained through technological refinements of traditional trapping skills and existing ecological knowledge) were bartered and later sold for trade goods. Traditional seasonal movements of people were modified in accordance with this new dependence on trade goods.

Canadian colonial policy for indigenous peoples followed the pattern in the United States of extinguishing native rights to huge expanses of territory in return for tiny reservations. The Ojibway were one of the last peoples to sign away their land rights. Treaty No 9 in 1906 ceded their land to the federal state in return for small reserves, guarantees of the viability of and access to game stocks on the alienated territory, and tiny annual monetary grants. The Ojibway never had a concept of legal land entitlement, but rather one of land use. Thus surviving witnesses of the treaty signing maintain that they thought the treaty was merely a declaration of friendship, not a land transaction.

The dual economy functioned so long as fur prices remained high, but the prolonged depression in prices after the Second World War had drastic effects on original peoples across the country. Federal Government policy encouraged removal towards centralized settlements; pressed for compulsory schooling, even during the crucial seasons when endogenous education in subsistence pursuits such as trapping and hunting was most important; introduced complicated ad hoc systems of subsidiary payments like family allowances, old-age pensions and welfare; and favoured integration into the wage-labour market economy. Thus, by the 1960s, the nuclear family in relatively permanent settlements became the primary economic unit – and one increasingly dependent on the outside state. The contemporary native economy has not solved the problem of dependency on externally controlled agencies: direct government payments have had to replace productive labour as the main resource for obtaining trade goods, and self-destructive coping mechanisms (such as alcohol) have often marked the pace of cultural decline.

This discussion however ignores higher order political and economic shortcomings of the industrialized state – or even Western capitalism as a whole: foreign economic control by huge corporations, regional disparity, exploitation of the hinterland by the metropole, racism, the absence of long-range planning, and the growing severity of inflation linked with chronic unemployment.

The boreal forest

The boreal forest of the Ojibway lands is a mosaic of stands of poor species diversity growing in cycles largely initiated by fire. The most common species is the black spruce (*Picea marjana*) which has long fine fibres that are ideal for paper, can grow in both swampy land and on well-drained soils, but may attain a diameter of only 10 cm after 100 years because of the short growing season. Forest exploitation is controlled by the provincial government, which grants large timber licences to individual companies. Theoretically the nationally important pulp and paper industry has not reached maximal capacity, since the mean annual cut of combined operations does not reach the level of 'sustained yield'. This would be the level of cutting that could (ideally) be maintained in perpetuity, because the calculated increment of growth in all areas, no matter how distant from mills or difficult of access, is still greater than the amount taken in intensive operations in preferred, accessible stands.

Thus, official Ontario economic policy for the north is to stimulate quantitative additions to existing resource-extraction industries like forestry or mining, while ignoring apprehensions of foresters, silviculturists and conservationists that the supply of wood – or at least the wood that can be cut at reasonable cost – will run out by about 2020. What is so quietly ignored is that the Ontario pulp and paper industry cannot compete with the American in wages, productivity per worker, rate of re-growth, and woodcutting and hauling costs. This means that the effective circles of operation lie close to most mills, even if they are within large licensed tracts; necessary artificial regeneration and husbandry are neglected; and up to a third or more of cut-over land becomes unproductive in terms of preferred species.

Two situations result. In the first, old pulp factories are badly in need of major capital investment – they operate inefficiently, are the single greatest source of water pollution in the nation, and must utilize increasingly distant stretches of virgin forest lying beyond the belt of silvicultural islands that have resulted from past activities closer at hand. Some mills have closed, effectively killing the one-company towns that depend on them. Others have attempted to pass additional cutting costs to the worker, and one company faced a bitter strike with the deployment of imported 'scab' workers, elaborate security, and the ready co-operation of the police. In the other situation, new plants and new cutting operations are designed for maximal technocratic efficiency. Huge machines that cut, strip, prune and stack tree trunks are used over extensive clear-cut tracts. Capital costs are higher, labour is reduced, and the large denuded areas are much more difficult to regenerate than expensive strip cuts which leave intervening patches of seed stock. These practices are often likened to 'mining' the forests.

A paper company in north-western Ontario

In 1961, the Reed Paper Company, a subsidiary of the British multinational, Reed International, bought a pulp mill that had been built in Dryden, Ontario, in 1911. In 1970 it was discovered that the English-Wabigoon River system, downstream from the Dryden plant, was seriously polluted by mercury originating from a chlor-alkali unit installed in 1962 to produce bleach. Some 9000 kg were released into the river, while another 13,600 kg remain unaccounted for. Commercial fishing, especially important to the economy of the two downstream Indian reserves, Grassy Narrows and Whitedog, was closed; sports fishing was allowed to continue, although the number of tourists going on fishing trips declined and some local guides lost their jobs. Fish meat, now often containing thirty times the World Health Organization's limits for mercury, continued to be a major part of the reserve's diet. Government attempts to deal with the situation have been half-hearted, pointless and fumbling. For example, a freezer was given to one reserve to store uncontaminated fish flown in at some expense from distant lakes. One day the freezer was found warm and the fish spoiled. A ranking provincial cabinet minister accused the natives of politically motivated sabotage. It was soon discovered that a child at play had pulled out the electrical plug.

The two Ojibway communities face social destruction in the form of welfare dependency, unemployment, social apathy, and high rates of homicide, suicide, accidental death, crime, alcoholism, and gasoline-sniffing among the young. There have not been any striking cases of mercury poisoning as were seen in Minamata, Japan, but visiting Japanese specialists were able to diagnose a number of cases with definite early symptoms, such as narrowed visual field and muscular incoordination.

In 1974, the paper company announced plans for a new mill using the last unlicensed stands of commercial timber in a 50,000 km^2 block north of the town of Red Lake. Attempts by the newly formed indigenous 'umbrella' organization for northern Ontario, Grand Council Treaty Number 9, to obtain information about the scheme were fruitless, although the provincial government assured the original people that they would be consulted and their interests respected. Secret government documents that were leaked to the press later showed that this was a sham. Treaty 9 was forced to prepare its own information base for land use, treaty history, forest management, timber inventory, and the expected effects of large development projects. By 1976 the issue had become a public scandal, and a growing anti-paper company coalition of leftists, conservationists, organizations supporting native rights, church groups, labour, and independent research groups became increasingly effective.

A memorandum of understanding was signed by the provincial government and the Reed firm in October 1976 for the construction of a 1000 tonne/day mill that would use wood from the new forest tract. The tract includes several reserves and independent settlements, and also is hunted or trapped by other Indian communities to the north and east. Most of these communities can only be reached by hydroplane or canoe, although one, Osnaburgh House, is accessible by the only road into the district. Interestingly enough, the residents of Osnaburgh had been deported from their old scattered lakeside houses when a hydroelectric dam was built; the new centralized settlement on the road is suffering rapid social disintegration with chronic alcoholism as the main symptom. Yet even in Osnaburgh, 41 per cent of the Indians' livelihood is still derived from traditional bush pursuits.

Capricious economic and social planning

Ontario government principles for land use planning stress that 'public participation is an essential part of the planning process', 'fairness is required when dealing with the people concerned', 'plans should be made for the long term and should provide for future options', and 'the public good must take precedence over the private good'. These have proved to be hollow sentiments as far as Ontario Indians are concerned. The natives of the communities affected by the paper firm's mercury pollution continue to be ignored at the best of times and are the subject of arbitrary decisions at others.

Although commercial fishing is banned in contaminated waters, sport angling under the slogan 'fishing for fun' (ie angling is allowed, but eating the poisonous catch is discouraged), is still promoted. Indian spokesmen stressed that this made it difficult to dissuade their people, many of whom eat fish daily, from eating the deceptively healthy looking fish. There is general agreement that the English-Wabigoon waters will remain badly polluted for many years.

In true bread-line style, healthy fish are provided by the government for immediate consumption yet no systematic programmes have been proposed to replace the economic losses suffered by native commercial fishermen.

The indigenous people were eventually forced to resort to legal manoeuvres against Reed; the lawsuits seem destined to drag on until they falter in the complexities of trying to prove a direct link between metallic mercury discharges and social disintegration – suicide, gasoline-sniffing, anomie – against the skills of sophisticated and well-paid company lawyers. What is most sadly lacking in the government is the simple, honest will to admit that a tragedy has occurred and that redress is a moral necessity. The only recognition that either the government or industry has given to the situation is a curious attentiveness to the possibility of legal liability when negotiating the sale of the Dryden mill to another company in 1979.

An investigation of the possibility of Minamata disease occurring in Grassy Narrows and Whitedog was delegated by the government to a specialist, who eventually prepared an inconclusive preliminary report that skirted the main questions. Yet it earned the specialist a prestigious position in occupational health at the University of Toronto and a position on a provincial government board with important powers to review environmental problems. And, in a most macabre misunderstanding of the point of the whole matter, the government licensed fish that were too poisonous for Canadian consumption to be exported to countries that had no mercury-contamination standards for foodstuffs.

Some third-order consequences

Planning for the new mill is completely contradictory. A new plant was advocated for somewhere in the area, even though it is admitted that too much reliance is put on extractive primary industry, especially forestry with its spectacularly erratic production cycles. It is still not publicly known whether the eventual guidelines for regional development advocating a mill were developed to fit a purely political decision by powerful elected allies of the paper firm or whether Reed's decision was a genuine response to an objective development decision. Negotiations were secret; leaked government memoranda showed that natives were being misled to believe that there would be consultation before any decisions were made.

The government and company both operated from deficient information bases. A 1967 timber inventory that showed a shortage of wood in the proposed area lay hidden until an independent researcher discovered it in 1979. Neither Reed nor the government undertook forest inventories before the agreement was signed in 1976 and Reed's subsequent, much publicized, environmental assessment (1976) was only a site-selection study that ignored the forest the mill would depend on. The government's reaction to information leaks from its disgruntled civil service was to launch an internal witchhunt, take peevish punitive actions against at least one sincere civil servant, and subject Treaty 9 workers and independent researchers to police interrogation.

Alleged police surveillance of outspoken Indian political leaders across Canada became a major topic of conversation among those involved in fighting for native issues. Even the company's timber inventory, conducted as a condition of the agreement, may not be accurate. One of the authors spent an evening in an isolated northern pub in August 1977 with exhausted members of the survey crew who complained that the exigencies of time, weather, biting flies, and protracted isolation forced them to make quick, objective guesses of parameters that required objective measurement.

The marginality of native concerns

Although several thousand Ojibway live in or near the new lease, they were intentionally not consulted. When the natives, through Treaty 9, insisted on making themselves an issue, the company offered not to cut in small 130-km^2 blocks around the three reserves in the lease area. Reed was not aware that not all natives lived on reserves, and a slim volume available on native people, in its environmental assessment, merely displayed maps of official reservations. A query to Treaty 9 would have established that one important, independent community also existed in the lease.

The final lease boundaries also excluded a small strip in the north and a block in the east that were included in an earlier map. These changes conveniently removed a number of reservations from being within the *geographical* boundaries of the lease, while ignoring the existence of the use areas of these same communities within the lease.

In addition, no studies were made of possible direct effects of cutting operations on wild rice (*Zizania aquatica*) harvesting, fishing, trapping or hunting. Logging roads would also open lakes to outside anglers and forests to big-game hunters, yet the effects of such competition on the traditional economy was not considered.

The bogey man of wage labour and development

The government's northern economic policies are based on a 1977 document, 'Design for Development: Initiatives and Achievements'. Key programmes so far have been communications (roads, airstrips, electrification), administration, law enforcement (courts, gaols, police), and schools. All these provide the infrastructure for the development of the national market economy and European culture as well as the infiltration of state control 'from the south'. There has been no job programme to suit endogenous talents, skills or needs.

The place of natives in provincial forestry policy was first admitted in 1972: 'Many kinds of forestry work are well suited to the employment of relatively unskilled labour of the type found in rural areas and amongst the native population – a labour force that is otherwise wasted'. A study by Reed echoes this view and discusses the residents of Osnaburgh not as a community but rather as data of available man-hours of unemployed, able-bodied males between 15 and 64 years of age.

Thus, in the north, natives remain hewers of wood and crushers of rock. At Osnaburgh, they worked as flagmen, blasters, or operators of jackhammers in road construction, or held hot, dusty jobs as rock sorters in the mill at the nearby Union Minière copper mine.

Racism is also important. Natives are disliked by many employers who claim that they are unreliable, leave without notice, have poor stamina, or just cannot manage in 'white' occupations. It may be true that the impoverished diet of some acculturated natives affects their health. But there is evidence that the indigenous can adapt to a variety of jobs, will take to work unrelated to traditional pursuits (and not just 'analogous' activities such as tour guides and fire wardens), and do not show a high turnover rate. Natives have shown that they prefer jobs that are outside, keep them near their families, and are reasonably steady. Thus they are actually a highly selective labour force with an unorthodox emphasis on non-material employment values.

Statistics show that natives usually are given jobs that last only a few months, that are distant from home and friends where they must live in bunkhouse-style social isolation, or are dirty and noisy. Credentialism – pre-employment prejudice based on inflated or artificial educational requirements or restricted professional licences – also bars natives from jobs. It has been found that, for natives, there is no correlation between job quality and years of schooling, but only between having a high school diploma (secondary school certificate) or not having one. The most important positive variables in one study of native employment are the work supervisor and identification with the goals of plans and jobs. This suggests that community control of, or involvement in, job-generating activities is essential, and that natives have not yet been acculturated to the condition of classically alienated labour found in industrial societies.

SOME CONCLUSIONS

Wage labour in the market economy poses a subtly invidious threat to the traditional economy and the social structures around it. It could undermine the traditional value of economic equality and create classes of rich and poor, because it is most often young unmarried men who get jobs. This would concentrate wealth in the hands of those with fewest economic responsibilities who may be least willing or able to use it in socially useful ways. Wage labour also helps to undermine the value of, or esteem for, traditional labour that does not produce disposable income. Thus, industry presentations to administrative tribunals considering major projects in the north (such as oil pipelines) give wild meat a low monetary rating.

If valuation were done properly in comparing the two sides of the dual economy, then it would have to be based on use value – or at least on replacement costs of purchased non-indigenous goods or services. If such comparisons were made, it would be realized that natives in viable traditional communities are not impoverished or unemployed because, having some bush income, they do not need as much employment. Their low labour-participation rates, employment levels, and per capita earned incomes cannot thus be fairly used as additional arguments by the corporations for establishing major projects without first examining the prevailing economic characteristics of the communities.

Where the bush economy is stable and significant, wage labour itself may then become the problem, rather than unemployment. In the long run, it is also possible that economic flexibility will be lost if the dual economy is superseded by the narrowed skills of, and dependence on, jobs. In an area solely dependent on an industry with as wildly fluctuating fortunes as forestry, long-run economic survival may actually depend on the ability to switch back to an earlier vocational skill.

This chapter is taken from Bodley, J (1988) *Tribal peoples and development issues*, Mayfield Publishing Company, California.

NOTES

1. Professor De'Ath, now at the University of Waterloo, Waterloo, Ontario, Canada, is a native of Turua, New Zealand. He holds degrees from universities in his country and in the United States as well as the Certificate of the Australian School of Pacific Administration, Sydney. Dr Michalenko, his co-author, studied biology at the University of Saskatchewan and now teaches environmental studies at the University of Waterloo. He has been involved in studies of pulp mills in both Saskatchewan and Ontario.

REFERENCES

Australian Unesco Committee for Man and the Biosphere. *Ecological Effects of Increasing Human Activities on Tropical and Sub-tropical Forest Ecosystems*. Canberra, Government of Australia, 1976.

Bernard, H (ed). *Technology and Social Change*. New York. Macmillan, 1972.

Bishop, C. *The Northern Ojibway and the Fur Trade: An Historical and Ecological Study*. Toronto, Holt, Reinhart and Winston, 1974.

Canadian Association in Support of Native Peoples. For Generations Yet Unborn: Ontario Resources North of 50. *Bulletin*, Vol 18, No 2, entire issue, 1977.

De'Ath, C. *The Throwaway People: Social Impact of the Gogol Timber Project, Madang Province*. Monograph 13. Boroko, Papua New Guinea, Institute of Applied Social and Economic Research, 1980.

Hartt, E. *Interim Report and Recommendations*. Toronto, Royal Commission on the Northern Environment, Government of Ontario, 1978.

Hutchinson, G and Wallace, D. *Grassy Narrows*. Scarborough, Ont, Van Nostrand, Reinhold, 1977.

Jalee, P. *The Pillage of the Third World*. New York, Modern Reader Paperbacks, 1968.

Michalenko, G and Suffling, R. Social Impact Assessment in Northern Ontario: The Reed Paper Controversy. In *India SIA: The Social Impact Assessment of Rapid Resource Development on Native Peoples*, ed Geisler, G C, Usner, D, Green, R and P West, Monograph 3, 274–89. Ann Arbor, University of Michigan Natural Resources Sociology Research Lab, 1982.

Ontario Public Interest Research Group. *Quick Silver and Slow Death: A Study of Mercury Pollution in Northwestern Ontario*. Waterloo, Ont, University of Waterloo, 1976.

Ontario Public Interest Research Group. *Reed International: Profile of a Transnational Corporation*. Waterloo, Ont, University of Waterloo, 1977.

Paijmans, K (ed). *New Guinea Vegetation*. Canberra. Commonwealth Scientific and Industrial Research Organization, Australian National University, 1976.

Papua New Guinea Office of Environment and Conservation. *Ecological Considerations and Safeguards in the Modern Use of Tropical Pulpwood: Example, The Madang Area, PNG*. Port Moresby, Department of Natural Resources, Government of Papua New Guinea, 1976.

Shkilnyk, A M. *A Poison Stronger Than Love: The Destruction of an Ojibwa Community*. New Haven, Yale University Press, 1985.

Suffling, R and Michalenko, G. The Reed Affair: A Canadian Pulp and Paper Controversy. *J Biol conservation*, Vol 17, No 5, 1980.

Troyer, W. *No Safe Place*. Toronto, Clarke, Irwin, 1977.

Vayda, A. *Peoples and Cultures of the Pacific*. New York, Natural History Press, 1968.

Winslow, J (ed). *The Melanesian Environment*. Canberra, Australian National University, 1977.

PART VI
BIOLOGICAL DIVERSITY CONSERVATION

Chapter 13

An Introduction to the Technical Aspects of Biodiversity and its Conservation

Nels Johnson

INTRODUCTION

The extinction of species has been a predictable feature of evolution since life first appeared on earth some 3.6 billion years ago. At times, mass extinctions have taken place over a relatively short period of geological time. These extinctions are thought to have involved a substantial percentage of the species present during that period. Mass extinctions at the end of the Cretaceous era (the Age of the Dinosaurs – about 100 million years ago) are the best example of such phenomena (Raven, 1985). This period saw a sudden decline of large reptilian life forms which, in turn, set the stage for the rapid speciation of modern mammalian vertebrates and angiospermous plants. Yet, many ecologists believe the current worldwide deterioration of natural environments is causing species to become extinct at a rate unprecedented in earth's history. Some estimates place the current extinction rate at 400 times that recorded through recent geological time and all available evidence suggests it is accelerating rapidly, and far outpacing speciation. This has created what E O Wilson (1985) calls the 'biological diversity crisis'.

The magnitude of the decline is not precisely known. This imprecision is a reflection of the fact that the 1.7 million species formally identified and classified may represent only 5 per cent of the total number of species now present on earth. Estimates of species numbers vary widely. Over the last two decades these estimates have risen steadily as previously unexplored habitats have been sampled. Until very recently the most commonly cited figures ranged from five to ten million species (eg Myers, 1979). However since researchers have started sampling the canopies of tropical moist forests, estimates of the number of insect species alone have increased to as many as 30 million (Erwin, 1983). It is quite conceivable that total species counts will increase as tropical moist forest canopies are surveyed for other flora and fauna. In addition, comprehensive surveys of tropical reef and estuarine habitats and warm deserts are likely to identify many new species. The distribution of species is unequal in both geographical and taxonomic terms: approximately two-thirds of all species are found in the tropics (Wilson, 1985) and insects make up 80 per cent of all species identified to date (Erwin, 1983).

Biological diversity encompasses the number of species of microbes, plants and animals, but species number is only one dimension of biological diversity. Other dimensions include the genetic variation within each species (eg the obvious diversity among members of the Homo sapiens species) and the variety and complexity of habitats and ecosystems that support and are supported by these species.

Habitat destruction

Worldwide, the destruction of natural environments is reducing biological diversity at all levels. Tropical rainforests – the most biologically diverse of all terrestrial habitats – are being cleared at an annual rate that some estimate to be as high as 1 per cent. Estimates of tropical deforestation range from 5.6 million hectares to 20 million hectares per year (Caufield, 1982).The causes of deforestation are varied and include timber harvesting, slash and burn agriculture, pasture conversion, tree plantations, and pest control (OTA, 1984). Threats to tropical forest habitats are well documented and can be found in virtually any discussion of the loss of biological diversity (eg Aiken and Leigh, 1985; Caufield, 1982; Guppy, 1984; Myers, 1979; Prescott-Allen, 1983; Raven, 1985). The fact that tropical rainforests are at the centre of debates about biological diversity is quite understandable given the high visibility which tropical deforestation has gained in the media, but while policy makers and the general public commonly view biological diversity as synonymous with tropical rainforests, the issue is a much more wide-ranging one. Biological diversity exists nearly everywhere. Placing such emphasis on a single place or biome may focus attention too narrowly, at the expense of diversity elsewhere. The level of scientific study and public concern regarding threats to tropical reefs and estuaries, for instance, is relatively low. Reefs and estuaries in the tropics represent the most biologically diverse marine environments, and while their destruction is extensive and accelerating, the number of species in these habitats has not yet been accurately studied. The causes and consequences of reef destruction are, if anything, more complex than that of tropical forests (Bryceson, 1981; Carr, 1982). Unfortunately, reefs and associated near-shore ecosystems such as mangroves and seagrass beds are in the zone of maximum development for urban growth, coastal transport, fisheries, recreation, military and other uses. Damage resulting from these activities is taking place before there has been a chance to assess the significance of even a minute fraction of this diversity.

It is not only natural ecosystems and species that are threatened by a loss of diversity. Biological diversity losses, particularly at the genetic level, may have tremendous implications for the practice of agriculture worldwide (Prescott-Allen and Prescott-Allen, 1983). In South and South East Asia, the 'green revolution' has led to dramatic gains in agricultural productivity, but these have often come at the expense of genetic variation. In India, where there were once several thousand rice varieties, most farmers now grow only a handful of high-yielding varieties (Reichert, 1982). Throughout much of the world, localized polycultures have been replaced by more productive monocultures. The narrowing of the gene pool for these crops has limited the options available for selecting disease resistant traits and other improvements in the varieties. Decisions have been made to increase productivity at the expense of stability and resilience. The loss of these genetic resources is irreversible as many of the natural species which gave rise to common agricultural crops have become very restricted in both a genetic and geographical sense, and the technology does not yet exist to re-create a species without a natural template to copy.

It is only in the last few decades that biological diversity research efforts have become more extensive, and the attention of policy makers and the general public is of an even more recent date. We are just beginning to grasp the scale and magnitude of biological diversity, some of its determinants, some roles of diversity in ecosystem functioning and the importance of biological diversity to our own species. This paper will begin by reviewing the major scientific findings on biological diversity to show something of the complexity and remaining uncertainties surrounding the subject. Then follows a summary of the main technical issues which conservation planners face in designing suitable protected areas. Finally biological diversity conservation will be considered in relation to different social, cultural, political and economic contexts, since these often determine the success or failure of conservation efforts.

SCIENTIFIC RESEARCH ON BIOLOGICAL DIVERSITY

The diversity of life has been of interest to mankind since pre-historic times. Cro-Magnon cave paintings near Lascaux, France, feature a variety of mammals, and pictographs and hieroglyphs of subsequent cultures around the world relied heavily on symbols of various life forms. There has also been a long history of attempts to classify the various components of the natural world. Aristotle invented one of western civilization's first taxonomic classifications, of animal, vegetable and mineral. Yet it wasn't until the eighteenth century that the Swedish botanist Linneaus developed a classification system that was based on the biological characteristics of life forms. Since then nearly two million species have been formally identified and classified. Such diversity was unimaginable in Linnaeus' time, yet it represents only a fraction of the biological diversity on earth, with much still left to be discovered. Following the development of taxonomy, other biological disciplines have emerged to facilitate our understanding of biological diversity. Biogeography, genetics, evolutionary biology and ecology have helped address the questions of where, when, why and how many, that accompany the investigation of biological diversity.

The challenge of taxonomy

The identification, description and classification of species continues to challenge taxonomists. The 1.7 million species identified so far represent only about 5 per cent of the latest estimate of a worldwide total species count of 30 million (Wilson, 1985). As Wilson (1985) and Tangley (1985) point out, this comes at a time when the field of taxonomy is in a state of decline. According to Wilson (1985), the number of professional taxonomists is decreasing as research funding is reduced, professional opportunities decrease and other disciplines are accorded higher priority. Even in the USA, where a vast amount of information exists on native flora and fauna, there are still very major gaps in knowledge – indeed less than one-third of all species have been fully documented (Tangley, 1985). Worldwide the situation is considerably worse. There is a substantial price to pay for this lack of information: biological research is hampered, and the development of applications such as biological pest control and biotechnology is slowed.

Why is there so much biological diversity?

In a classical paper published in 1959, G E Hutchinson posed the question 'why are there so many kinds of organisms?' In the 35 years since, scientists from field naturalists to mathematicians have attempted to answer this question, and have formulated numerous hypotheses (see reviews by MacArthur, 1964; Whittaker, 1977; Gentry, 1982). In a review of these efforts, Brown (1981) concludes 'Despite two decades of intensive investigation, Hutchinson's question remains largely unanswered', and suggests that energetics may hold the key to generating a universal theory of diversity.[1]

The problem with many of the proposed hypotheses is that they do not allow for the role of other possible mechanisms.[2] Pianka (1966, 1979) suggests that several mechanisms may function together, with the relative importance of each dependent on the particular area in question. Recent research by Schmida and Wilson (1985) provides a clear example of how several mechanisms working together can account for the presence and distribution of species diversity in an ecosystem. They identify four categories of biological determinants which cause and maintain species diversity: niche relations, habitat diversity, mass effects, and ecological equivalency. Niche relations and habitat diversity can be seen as expressions of how species divide up the limited resources of the ecosystem to make a living. They may divide up the available space (eg by selecting different habitats) or energy resources (eg by adopting different diets). Mass effect determinants reflect the fact that natural communities are never closed. Organisms disperse and in times of rapid population increase some individuals of a species may become established in a habitat where it was absent previously. Thus, the diversity within that habitat is increased, at least temporarily. Finally, ecological equivalence refers to different species having effectively identical niche and habitat requirements. Using data from Israel and California, the authors demonstrated how the relative importance of these factors changed according to the scale of the area considered. At the smallest scale, niche relations were clearly the most dominant. Diversity at a slightly larger scale was generated primarily by habitat diversity but with significant contributions from niche relations and mass effect. At a still larger scale, mass effects become the most important determinants while contributions from niche relations and habitat diversity declined and ecological equivalence enters the equation for the first time. At a macro scale ecological equivalence explains additional diversity with a declining contribution from habitat diversity. At a worldwide level, additional diversity is entirely explained by ecological equivalence.

An added complexity derives from the fact that the patterns and determinants of diversity vary greatly between regions. The differences observed in the correlation between species diversity and rainfall levels are a case in point. In the Pacific Northwest of the USA the diversity of forest communities increases along a gradient of decreasing maritime influence, with highest diversity in the drier areas (del Moral and Fleming, 1979). Similarly, Mediterranean woodland and shrubland communities in Israel exhibit greater diversity as rainfall decreases to relatively low levels (Naveh and Whittaker, 1979). These findings are contrary to strong correlations in the tropics between high rainfall and high levels of diversity (Gentry, 1982). Another factor whose influence on species diversity varies from region to region is that of grazing. Whereas diversity decreases steadily with imposed grazing rotations (ie the introduction of cattle and sheep) in North America, diversity was found to increase to relatively high levels under similar regimes in Israel (Naveh and Whittaker, 1979). Regional differences in species diversity patterns is also evident for animal species. In Panama, bird species diversity is greatest in mature, primary forest (Schemske and Brokaw, 1981) whereas diversity levels in Indonesia for some animals, particularly primates, were highest in moderately disturbed forest areas rather than in undisturbed primary forest (Johns, 1983; Wilson and Johns, 1982).

Biological diversity and ecosystem functioning

The question of what causes and maintains biological diversity has often been reversed to ask: what role does diversity play in ecosystem functioning? The search for a paradigm to explain this role has been a preoccupation of ecologists for some time. The concept of a relationship between biological diversity and ecosystem stability has received most attention. The catalyst for much of the theorizing involved was the publication of a short paper by MacArthur (1955). He postulated that an ecosystem with numerous energy pathways, ie a foodweb with many species, would change less upon the removal or addition of a species than would an ecosystem with few energy pathways. For nearly two decades MacArthur's argument went essentially unchallenged until May (1973) demonstrated the weaknesses of the premise.[3] It now seems unlikely that a positive relationship between biological diversity and ecosystem stability can be established. While in some cases diversity and stability of an ecosystem are linked (eg Margalef, 1969; Pielou, 1975) there are just as many cases where there is no such link (see eg Kimmerer, 1984; Smedes and Hurd, 1981).

Predicting the distribution of biological diversity

The science of biogeography attempts to predict the distribution of biological diversity, without explicitly addressing the difficult issue of causal relationships between biological diversity and ecosystem functioning. Following theoretical advances in the subject, particularly the 1967 publication of MacArthur and Wilson's *The theory of island biogeography*, the field has become a dominant topic in the conservation literature. Simply stated, the theory holds that the number of species and the species composition of an island is dynamic (ie constantly changing), and is determined by the equilibrium between the immigration of new species and the extinction of those already present.

One of the central tenets of biogeography is that dispersal leads to creation. The dispersal of a species from its area of origin occasionally leads to a segment of the spreading population becoming geographically isolated. As the isolated population stops exchanging genes with other populations of the same species, it may evolve into a diffeent species as a result of its restricted gene pool and, perhaps, in response to natural selection pressures that are unique to their isolated location. While these processes occur all around the world, it is difficult to measure species migration and define isolated populations unless areas are clearly isolated, such as is the case with oceanic islands, peninsulas and mountain tops (eg MacArthur and Wilson, 1967; Diamond and May, 1977; Brown, 1978).[4] A number of factors affect the ability of isolated areas to support species. Diversity is limited by, among other factors, distance from sources of colonizers, size of the isolated area, and the amount of resources available (MacArthur and Wilson, 1967). Assuming there is no change in the size of the island, the distance to the nearest source of species or the resources available, the species diversity found on an island is thought to represent an equilibrium between opposing rates of colonization and extinction (Stenseth, 1979).

As forest habitats have shrunk because of deforestation, many of them have become isolated and can be thought of as islands. Whereas oceanic islands acumulate species, habitat islands created by rising water levels or by human disturbance begin with a full complement of species at the time of their isolation (Karr, 1982). Over time, these islands lose species, for a number of reasons. First, smaller islands have smaller populations which are more likely to become extinct due to random events (Terborgh and Winter, 1980). Second, the smaller the island, the lower the probability that it will intercept dispersing organisms. Third, habitat diversity decreases as island size decreases (Wilcox, 1980). And fourth, local extinction may occur as populations fluctuate in response to environmental change (Karr, 1982). One of the most ambitious and long-term studies of forest fragmentation is being carried out in the Brazilian rainforests of the Amazon Basin (Lovejoy *et al*, 1983). Isolated fragments of tropical forest ranging in size from 1 hectare to 10,000 hectares are being studied to determine the minimum critical size needed to conserve the forest's original diversity. Preliminary results show that in the smaller patches, many species of primates and other large mammals and birds disappeared within two years. Island biogeography predicts the loss of tree species in small fragments next. While the study will be of most value for the Amazon basin, it may yield some insights that can be applied to increasingly isolated forest areas elsewhere in the tropics.

A number of researchers argue for caution in the application of the island biogeography theory. Connell (1978) argues that tropical forest ecosystems are rarely, if ever, in a state of equilibrium, and that therefore one of the most important assumptions underlying island biogeography theory is not applicable to tropical forests. Shaffer and Samson (1985) point out that the minimum population and area needed for a species to survive cannot be determined solely on the basis of island biogeography, and that minimum area and population size are likely to increase when demographic factors, such as the sex ratio and age of populations, and environmental fluctuations are taken into account. Another important factor is biotic interaction: different ecosystems will exhibit different responses to the extinction of a species depending on how much 'control' that species exerted on the ecosystem (Ehrlich and Mooney, 1983). So-called 'keystone' species, which provide essential ecosystem services such as pollination, predation, and scavenging of dead organic material, exert more control than others. Another critic of island biogeography, Karr (1982) argues that the mechanisms underlying species colonization and extinction are often better explained by random events and physical features of the landscape than by island biogeography theory.

The effect of disturbance on biological divesity

Perhaps some of our best understanding of conservation biology derives from the considerable amount of research that has been carried out on the impact of disturbance on the biological diversity of ecosystems (eg Leidy and Fiedler, 1985; Dritschillo and Erwin, 1982; Schemske and Brokaw, 1981; Wilson and Johns, 1982). Modeling exercises suggest that small disturbances lead to highest species diversity, whereas large disturbances are likely to favour the persistence of competitive, opportunistic species of both plants and animals, thereby decreasing natural species diversity (Miller, 1982). These so-called 'sub-climax' species generally are able to reproduce fast and disperse widely, and are of smaller size and shorter lifespan. The 'climax' organisms of undisturbed ecosystems, on the other hand, tend to be highly adapted to stable equilibrium conditions, but are less flexible and more vulnerable to change. They are generally longer lived, and do not disperse as well.

Descriptive studies tend to confirm the results of these modelling exercises. Schemske and Brokaw (1981) provide empirical evidence showing that a certain degree of natural disturbance favours diversity. They found that moderate disturbances in tropical moist forests such as the ones caused by natural treefalls resulted in the greatest diversity of bird species. Areas where the canopy was unbroken had considerably fewer bird species, presumably because of the fact they contained fewer niches. Wilson and Johns (1982) found the diversity of large animals to be greatest in areas that had been selectively logged 3–5 years previously. This diversity, however, depends on a ready supply of colonizers from nearby undisturbed forest areas. Leidy and Fiedler

(1985) provide one of the best documented examples of the effect of human disturbance on native species diversity levels. They sampled 125 streams in the San Francisco Bay drainage basin ranging from pristine habitats to concrete channels, and defined five disturbance categories. Species diversity levels were highest in streams with moderate levels of disturbance. While diversity levels were equally low in undisturbed and highly disturbed habitats, there was a great qualitative difference in the species present. In the undisturbed areas only native species were found, while only a few exotic species were present in the highly disturbed streams. Areas of intermediate disturbance supported a diverse assemblage of native and exotic species. Similar findings have been reported by Clout and Gaze (1984) for New Zealand. They found that most of the bird species diversity in tree plantations was accounted for by exotic species while undisturbed mature forests were populated predominantly by native bird species.

One of the implications of the fact that moderate disturbance often gives rise to higher biological diversity, is that it is not always desirable to exclude human activity from nature conservation areas. In areas where man has long been an intimate part of the landscape and had much to do with its recent evolution, discontinuing human interference may disturb an equilibrium resulting from thousands of years of moderate grazing, coppicing, cropping and induced burning. Diverse plant communities in the Mediterranean, for instance, become overgrown and are choked out by relatively few, shrubby plants once they are left to themselves.

SCIENCE APPLIED TO BIOLOGICAL DIVERSITY CONSERVATION

The scientific understanding of biological diversity is, without a doubt, far from comprehensive. And while the urgency of the conservation agenda dictates intervention, it is difficult to manage poorly understood phenomena. Waiting for a more complete information base before acting would mean losing much biological diversity in the meantime. Conservation biology is therefore a crisis discipline, attempting to conserve dwindling resources with restricted funds, imminent deadlines and limited information.

Just as medicine draws on the basic sciences of cell and molecular biology, conservation biology is based on the science of ecology. Applying the findings and methods of a pure science to everyday reality inevitably leads to some tension between practitioners on either side of the fence. Academic researchers are concerned that the theories they helped develop are being applied inadequately or out of context, and that the resulting failure will reflect poorly on their work. Managers of nature reserves, on the other hand, may feel that much research does not address the practical problems with which they are grappling. On the whole, however, interactions are positive. The contributions of science to practical conservation activity are substantial and increasing. Many scientists are actively involved in designing protected areas, prioritizing species and ecosystems conservation programmes, assessing habitat requirements and ecosystem processes, conserving germplasm and a host of other activities. In addition, conservation practice often provides new insights on issues of concern to theoretical biologists.

Areas and species to be protected: setting priorities

Given the rate of ecosystem degradation and the limited resources available for conservation activities, it is clear that some species and ecosystems will vanish because of human activity. It is not possible to conserve all biological diversity, so priorities must be set, and difficult questions answered. Which species are most likely to persist because of conservation efforts? What areas contain the greatest relative species diversity or highest percentage of endemic species? What types of organisms might have the greatest economic potential? What species are essential to ecosystem functioning? Which threatened ecosystems and species are of most interest to science? How does the relative importance of these questions vary from place to place, or over time?

Species may be conserved either *in situ*, ie in their natural habitat, or *ex situ*, eg in zoos, botanical gardens or germplasm collections. Protected areas such as wildlife sanctuaries offer species protection in an environment similar to the one they would inhabit in the wild. Habitats themselves may be the targets of conservation activities and nature reserves can be established to protect threatened habitats which in turn provide shelter for particular species of plants and animals. Where habitat fragmentation has occurred to the extent that *in situ* conservation of species is jeopardized – ie when the reduced habitat can no longer support certain species – biodiversity conservation may require restoration of the habitat in areas surrounding the surviving fragments. Biodiversity conservation can also be achieved in ecosystems which are managed first and foremost for productive purposes. In such instances the productive activities (eg selective logging, or controlled grazing or cropping) can help maintain favourable habitats for the target species as well as providing valuable income for the conservation effort.

In the past, protected areas have often been selected for their scenic value, presence of large populations of popular animals (eg elephants), and their accessibility to visitors, particularly those who spend money.[5] They have rarely been chosen for their potential to protect the greatest variety of rare and endangered species or ecosystems. As a result, existing protected areas rarely coincide with areas having the most diverse biotic communities. More recently, the development of better methods for establishing conservation priorities has become a major preoccupation of conservation biologists. A number of methods have been developed to help set conservation priorities, some based on a single criterion such as overall species number or the presence of rare species, others based on a mix of qualitative and quantitative factors.

Many tropical countries contain large numbers of species with small geographical ranges, called endemic species or endemics.[6] Because of their restricted range, endemics are especially vulnerable to extinction by habitat loss. Therefore, endemism is often cited as a useful criterion for selecting areas in need of protection. Because endemics are typically small organisms with modest spatial requirements, the critical issue for their conservation is the location, not the size of reserves. With just a few well-placed reserves, most of the endemic species in a region could be protected. Using Ecuador and Colombia as examples, Terborgh and Winter (1983) developed a method for siting parks based on endemism. They transformed distributional data for 156 bird species from major museum collections into range maps which were then superimposed to determine which areas had the highest concentration of endemics. Unfortunately, there was little correspondence between existing protected areas and points of maximal endemism.

Recent advances in theoretical, behavioural and mathematical ecology have given rise to more and more complex methods for determining conservation priorities. Adamus and Clough (1978) developed a numerical method to rank species according to conservation priority, based on a weighted summation of scores on 13 qualitative and quantitative characteristics (such as area requirements, spatial distribution, scarcity, endemism, and habitat specialization). Kirkpatrick (1983) developed a similar method for the ranking of areas. The method is based on an iterative process of elimination: once an area is selected for protection, the species or ecosystems represented in this area have less influence on priorities established in the next round. The process can then be repeated until all pre-determined conservation goals are met. The problem with these complex methods is that many of the data needed are not known for many species and ecosystems, and that weighting of different characteristics is highly subjective. Margules and Usher (1981) argue that advances in multivariate statistical techniques will produce more sophisticated ways to aggregate heterogeneous data in the next decade.

Apart from sheer numbers of species, qualitative considerations are also important. As was discussed in the previous section, moderately disturbed ecosystems often have the highest diversity, but this is at the expense of native species. The creation in a protected area of additional edge habitat may favour diversity at that level, especially of opportunistic exotic species, but at the same time may lead to the loss of local species which require large homogenous habitats. So while total species numbers in any one protected area may increase with the creation of edge habitats, the species would be largely exotic and perhaps the same exotics as present in other reserves where edge habitat has also been maximized. To avoid such duplication and repeated sacrifices of native species, biological diversity conservation needs to be examined at a regional landscape level (eg Noss, 1983; McCoy, 1982; Samson and Knopf, 1982).

Key considerations in the design of protected areas

Scientists have yet to reach a consensus on the optimal size and shape for protected areas. How much is enough? How large an area is needed and what populations sizes are required to ensure that reserves maintain current biodiversity levels? There is increasing evidence, however, that large reserves are best for most terrestrial ecosystems (eg Lovejoy, 1980; Lovejoy *et al*, 1983; Nilsson, 1978; Shaffer, 1981). Ng (1983) argues large reserves are needed to conserve tropical lowland rainforests because species are present at low population densities and many species have separate male and female plants. Other arguments for large reserves include seasonal movements of fauna, often associated with seasonal changes in habitat (eg floodplains). These are, nevertheless, strong advocates for putting an equal amount of land into greater numbers of small reserves. One of the most prominent and consistent critics of the 'biggest is best' approach is Daniel Simberloff (eg Simberloff and Abele, 1982; Simberloff, 1982; Simberloff and Gotelli, 1984). In remnant prairies and forests in Minnesota, he found that 'archipelagos' of small sites contained more species than did single large sites of the same total area.[7]

Therefore, both large and small reserves may be needed to maximize species diversity. To complicate matters further, plant and animal species may need different conservation strategies. Nilsson (1978) documented decreases in bird species diversity following habitat isolation, but observed no change in plant species diversity in the same habitat fragments. Even within groups of similar animals, conservation requirements may vary. Lynch and Whigham (1984) found that highly migratory bird species were most abundant in extensive, mature, floristically diverse forests, whereas short distance migrants and permanent residents were less affected by habitat size and shape.

Boecklen and Gotelli (1984) sound a warning note about overly sophisticated approaches to determining conservation priorities and designing protected areas: 'Most conservation decisions will continue to be essentially pragmatic and small-scale, concentrating on single species or communities.' In other words, in the real world we often have to take what we can get. Design options are rarely available.[8] An additional complication in the choice of sites for protected areas is the dynamic nature of species and ecosystems. Species and ecosystem ranges may change in response to any number of environmental and biotic factors (Pianka, 1979). Indications that atmospheric accumulations of carbon dioxide will soon result in substantial alterations in climate need to be taken into account in the planning of new protected areas (Peters and Darling, 1985). Anticipating warmer and drier conditions, and a resulting northward shift of species and ecosystem ranges in the USA, they suggest protected areas be located at the northern margins of present ranges. Similarly, protected areas which include large differences in altitude would allow for shifts in species ranges in response to changing climatic conditions.

Buffer zone management and restoration

Worldwide, many habitats are becoming isolated fragmented segments of what were once extensive ecosystems. Immigration of new species into these fragments will be impossible if all sources of recolonization have disappeared (Pickett and Thompson, 1978). One cannot assume that the protection of fragments of natural habitats alone will safeguard the survival of many highly specialized and sporadically occurring species (Adamus and Clough, 1978). So the key to maintaining diversity in these habitat fragments is to ensure surrounding landscapes are able to accomodate the movement of a maximum number of species contained in the fragment. This can be accomplished by managing areas surrounding reserves to maintain a variety of habitats (Harris, 1984). If activities such as timber harvesting (Roche, 1979) and grazing around protected areas are carefully coordinated, they may contribute to their conservation value, rather than detract from it.[9] This is especially relevant for tropical forest biodiversity conservation for two reasons. First, only a small percentage of tropical forests are protected, and substantial increases in protected areas are not expected in most countries. And second, many tropical forest reserves contain migratory species that spend considerable time outside these reserves, such as elephants and tigers.

Unfortunately, there has been little applied research on how natural resources can be used without threatening the diversity and integrity of the ecosystems concerned. More work will need to be done on modes of natural resource utilization that do not fundamentally disturb key ecosystem processes. A good example is the work conducted by wildlife biologists and managers on game harvesting. They have looked at how controlled harvesting of game animals can mimic the regulative functions of the large predators that man has displaced. Another example is the research of Skorupa and Kasenere (1984) on the impact of logging on biological diversity in tropical forests. They compared the impact of selective timber harvesting with that of natural treefalls in the highland forests of Uganda. The latter create gaps of various sizes in the forest canopy, in which tree species regenerate, allowing the forest to maintain itself. They found that if timber felling and extraction resulted in the destruction of more than 50 per cent of the trees in logged-over areas, large gaps were formed which exposed surrounding trees to damage from wind, direct sunlight and reduced soil moisture levels. As a result forest succession was considerably altered and biological diversity lost. Timber harvesting operations which damaged less than 35 per cent of standing trees on the other hand did not seem to threaten forest ecosystem integrity and diversity.

Over half of the world's tropical moist forests have been cut, cleared, converted to pasture or have in some other way been transformed so that normal ecological succession processes no longer operate (Lovejoy, 1985). This may be justified on socio-economic grounds where productive and stable agricultural systems are established in place of the forest, but often this is not the case. Extensive areas of former rainforest in the Amazon basin are now dominated by herbaceous perennials, contain few if any woody species, and have productivity levels that are only a fraction of what they were previously (Buschbacher, 1986). Attempts to reconstruct biotic communities and ecosystems, called 'restoration ecology' by scientists such as Diamond (1985) and Abers and Jordan (1985), are likely to play an increasingly important role in conserving biological diversity. In some cases, restoring biotic communities is the only way to prevent the loss of biological diversity, as in the case of the natural tallgrass prairies in the American Midwest, which have been reduced to small widely scattered communities. They have been restored with considerable success (Abers and Jordan, 1985; Liegel, 1981). The most advanced restoration programme in tropical forests is in the severely depleted dry forests on the Pacific slopes of Costa Rica. The intention is to reforest nearly one thousand square kilometres in the Guanacaste National Park, by encouraging regeneration of native tree species in grasslands around remaining islands of trees. Techniques used include controlled burning and grazing, and planting pasture with trees and hedgerows of native species to attract seed-dispersing birds (Janzen, 1986). Ecological restoration of tropical forests may in some cases be achieved by planting exotic species which improve the environment for native tree species to re-establish themselves, as illustrated by native broadleaves re-establishing themselves in pine plantations in the Himalaya foothills. Restoration ecology may also be used to help restore productivity of degraded land, which can help to ease development pressure on remaining natural areas (Lovejoy, 1985; Abers and Jordan, 1985; Diamond, 1985).

BIOLOGICAL DIVERSITY CONSERVATION: SOCIAL, CULTURAL, POLITICAL AND ECONOMIC DIMENSIONS

While scientific knowledge of biological diversity is important for its conservation, there are other factors which often have more influence on the success or failure of conservation efforts. The social, cultural, political and economic contexts in which conservation takes place often determine whether the science of biological diversity conservation will provide results in practice. Conservation rarely succeeds on its own merits. When it does it is usually because it coincides, or is perceived to coincide, with the interests of those people whose support is crucial to the conservation effort. The constituency may be a single autocrat, a bureaucratic few or the public at large. In short, the key to success is to frame the benefits of conservation in terms of the self-interest of the relevant constituency.

Traditional use and conservation of biological resources

Some of the best examples of the coincidence of conservation and self-interest are provided by the traditional ways in which indigenous peoples control the use of natural resources (eg Adisewojo *et al*, 1984; Alcorn, 1984; Connelly, 1984; Johannes, 1984; Kartawinata *et al*, 1984; Kwapena, 1984; May, 1984; Ovington, 1984; Weinstock, 1985). Depending on the circumstances, their conservation practices may be intentional and based on an intuitive understanding of the environment, while at other times it may be incidental or part of a negative feed-back loop (eg over-utilization of a resource which is depleted to the point that it affects the indigenous population who then modify their use of the resource). Very often it may manifest itself in spiritual beliefs and practices (Boyd, 1984). Indigenous people are often more knowledgeable about species characteristics and ecology than western researchers who are unaccustomed to the area. Indeed scientists would do well to observe and document traditional uses of various species and where traditional customs, uses and conservation practices are still in effect, it would be ill-advised to impose laws and regulations based upon western paradigms. Maintaining or reinforcing local traditional customs may be a much more effective conservation tool.

An illustration of traditional conservation and utilization of wildlife is provided by Kwapena (1984) for the Maopa people of Papua New Guinea. The Maopa actively practise conservation and wildlife continues to play a key role in the social structure of their societies. At present, wildlife management areas are run by committees of people with traditional land and resource use rights to these areas. Their rules are enacted only with unanimous consent and have the authority of law. A typical example of their rules is a seasonal ban on hunting every three to four years, and a ban on the harvest of crabs until after the breeding season. Infractions by local people are very rare. Johannes (1984) describes a similar case of traditional conservation practice in the tropical pacific. Because of the narrowness of the continental shelf off the islands in this region, fisheries have long been considered by native peoples as valuable and limited resources. Traditional customs have dictated the sustainable use of these resources. Conservation laws based on American or European models that ignore or undermine traditional fishing rights have proven much less effective than fear of punishment by traditional custom. Not all such traditional conservation practices are exclusively subsistence-oriented. In Borneo, growing rattan, a liana fibre used to make wicker, in a secondary forest fallow of a swidden food production system promotes ecologically balanced swidden agriculture (Connelly, 1984; Weinstock, 1985). The most commercially valuable rattan comes from secondary forests that are relatively well-developed after regenerating for 20 years or more. Areas left fallow for this period regain fertility levels that permit successful agriculture for several years before they are again left fallow. This indigenous strategy for producing a cash crop (rattan) as well as food in a tropical rainforest is sustainable and minimizes development pressure on primary forests. Although such a system is not immune to disruption by market forces, the investigation of other perennial cash crops that can co-exist with the fallow vegetation of tropical rainforest swidden could prove useful elsewhere in the conservation of biological diversity.

One way of furthering biological diversity conservation is to ensure the survival of traditional cultures such as those described above. In turn, the conservation of biological diversity is likely to help maintain traditional cultures. The relationship between conservation and culture is illustrated by an example from Australia (Ovington, 1984). In the northern part of the country the landscape contains an unusually rich diversity of vegetation, wildlife and geological features, some of which are found nowhere else in the world. The area also includes a rich collection of Aboriginal art sites. Recently, Aboriginal people from the area leased over 6000 square kilometres to the Australian national Park and Wildlife Service as a national park. Since then, the

Aborigines have been very actively involved in the planning and management of the park. They continue to live in the park and practice traditional customs but are given every opportunity to learn about more conventional management techniques. Many do. They believe themselves to be guardians of the park and its resources as well as of their own heritage.

Balancing conservation and development

Notwithstanding the above-mentioned examples of traditional societies maintaining a balance between resource use and conservation, it has to be recognized that in most of the world this balance no longer exists. And where it still does, it is often a delicate one that is easily threatened when new technologies are introduced, populations grow rapidly, other groups migrate into the area, or land and other resources are privatized. Conservation approaches based on traditional customs will be of little use where the authority of such customs has been eroded. Kartawinata *et al* (1984) describe how the existing balance between resource use and conservation in one area of East Kalimantan was upset by government intervention. In this area, shifting cultivation continued as it had for centuries, rotating on secondary forests (primary forest was not used) and respecting long fallow periods, until the government started implementing a programme to resettle the shifting cultivators. The land deforested was unsuited for the irrigated rice production and other sedentary agriculture advocated by the programme, and new primary forests had to be cleared as crops failed and the deforested land was degraded.

In many countries, conservation is seen as an unaffordable luxury, a drain on limited national resources. Economic arguments are needed to convince governments to invest in conservation efforts in these countries. The World Conservation Strategy (IUCN, 1980) was the first international planning document to put economics firmly on the conservation agenda, and to demonstrate that conservation can be achieved by a combination of the wise use of biological resources and the protection of wild species and ecosystems: sustainable utilization and preservation. In this way, sustainable agricultural practices which replace unsustainable ones may be considered conservation measures (eg Nicholaides *et al*, 1985). And conservation measures such as the protection of wetlands can be seen as an efficient use of resources, since protected wetlands provide such ecosystem services as water recharge, water filtration and flood control – services which are very costly and difficult to replace with man-made measures (Ehrlich, 1982; Pearshall, 1984). Apart from playing a role at the national level, economic arguments may also be necessary to encourage participation of people living near protected areas. Lusigi (1981) suggests a concept of a 'conservation unit', economic returns of which would accrue to the local population. Controlled grazing and browsing of livestock and some agricultural practices would be allowed in the reserve and local people could be given the opportunity to gain employment from tourism in the reserve and receive a share of the returns from the harvest of surplus game populations. Western and Henry (1979) go further, saying that national parks in developing countries should have a clearly stated economic justification to ensure their successful establishment. This may not always be possible, however, as conservation benefits may be very difficult to quantify, even if protected areas yield services which are highly valued by all peoples. These services include water recharge and purification, flood control, conservation of genetic resources, and contributions to spiritual well-being.

An important lesson can be drawn from the observation of both traditional and newly emerging conservation efforts: local people need to be involved from the beginning in the planning of conservation projects which will affect them. No other factor is likely to be as important in determining the success of the project. Almost without exception successful examples of conservation projects have paid close attention to the needs and attitudes of local people (eg Halffter, 1981; Johannes, 1984; Kwapena, 1984; Ovington, 1984). These initiatives have built on local priorities compatible with conservation rather than imposing an outside paradigm. This approach often entails a shift away from conservation through pure preservation towards conservation through sustainable use. Ultimately, the decisions and responsibility for conservation rests with national governments. Educating decision makers on the value of their countries' biological resources and the cost of not conserving them must therefore be accorded high priority.

This chapter originally appeared in 1987 as an unpublished report to the International Institute for Environment and Development, Washington, DC.

NOTES

1. Brown (1981) proposes looking first at how the energy resources available in any given physical environment determine its capacity to support life, and then at how these energy resources are subdivided among species. Once these general calculations have been made, it would be possible to predict the number of species an environment could support.
2. None of the hypotheses purporting to explain why some areas are more biologically diverse than others fit well to all bodies of data (Schmida and Wilson, 1985).
3. See Kimmerer (1984) for a formal expression of MacArthur's hypothesis and an explanation of its weaknesses.
4. It is probably no coincidence that Darwin's thinking on evolution was shaped to some extent by his observations on the Galapagos Islands.
5. In many cases, protected areas were sited on land that was deemed unsuitable for any productive use at the time of establishment.
6. Endemism is especially common in the neotropics. For example, nearly one-quarter of the resident bird species of South America are endemics with ranges of less than 50,000 square kilometres (Terborgh and Winter, 1983). Areas that served as Pleistocene forest refugia in the Amazon are rich both in terms of endemics and overall biodiversity (Prance, 1982), but high rates of endemism do not always coincide with high floral and faunal diversity. Isolated habitats on mountain peaks or in valleys are often rich in endemics but poor in overall species diversity.
7. Two qualifications of these findings should be noted. First, as the authors admitted, the remnants they examined had only been isolated for a short time, so the effects of local extinction (if any) had not yet become evident. Second, it is not clear whether or not the individual islands in the 'archipelagos' were so close to each other that dispersal between them was possible.

8. A major exception to this rule are marine parks for near-shore reefs, which, in the absence of difficult tenure issues prevalent in terrestrial habitats, can often be designed in a more logical fashion. This is extremely important as reefs derive their populations of fish and coral largely from currents sweeping upstream sources. The latter need to be protected to guarantee ecosystem integrity.
9. Areas surrounding reserves which are managed to increase their potential for conserving biological diversity are often referred to as buffer zones (see eg Oldfield, 1987).

REFERENCES

Abers, J D and Jordan III, W R (1985) 'Restoration ecology: environmental middle ground' *BioScience* 35:399.

Adamus, P R and Clough, G C (1978) 'Evaluating species for protection in natural areas' *Biol Cons* 13: 165–179.

Adisewojo, S, Tjokronegro, S and Tjokronegro, R (1984) 'Natural biological compounds traditionally used as pesticides and medicine' *Environmentalist* 4 (Supplement 7): 11–14.

Aiken, S R and Leigh, C H (1985) 'On the declining fauna of Peninsular Malaysia in the post-colonial period' *Ambio* 14: 15–22.

Alcorn, J B (1984) 'Development policy, forests and peasant farms: reflection on Huastec-managed forests' contributions to commercial production and resource conservation' *Econ Bot* 38: 389–406.

Boecklen, W J and Gotelli, N J (1984) 'Island biogeographic theory and conservation practice: species-area or specious-area relationships? *Biol Cons* 29: 63–80.

Boyd, J M (1984) 'The role of religion in conservation' *Environmentalist* 4 (supplement 7): 40–44.

Brown, J H (1981) 'Two decades of homage to Santa Rosalia: towards a general theory of diversity' *Am Zool* 21: 877–888.

Brown, J H (1978) 'The theory of insular biogeography and the distribution of boreal birds and mammals' *Great Basin Nat Mem* 2: 209–227.

Bryceson, I (1981) 'A review of some problems of tropical marine conservation with particular reference to the Tanzanian coast' *Biol Cons* 20: 163–171.

Buschbacher, R J (1986) 'Tropical deforestation and pasture development' *BioScience* 36: 22–28.

Carr, A (1982) 'Tropical forest conservation and estuarine ecology' *Biol Cons* 23: 247–259.

Caufield, C (1982) *Tropical Moist Forests: The Resources, The People, The Threat.* Earthscan Paperbacks, London. 67pp.

Clout, M N and Gaze P D (1984) 'Effects of plantation forestry on birds in New Zealand'. *J Appl Ecol* 21: 795–815.

Connell, J (1978) 'Diversity in tropical rainforests and coral reefs'. *Science* 199: 1302–1310.

Connelly, W T (1984) 'Copal and rattan collecting in the Philippines'. *Econ Bot* 39: 39–46.

Diamond, J M (1985) 'How and why eroded ecosystems should be restored'. *Nature* 313: 629–630.

Diamond, J M and May, R M (1977) 'Species turnover rates on islands: dependence on census interval'. *Science* 197: 266–270.

Dritschillo, W and Erwin, T L (1982) 'Responses in abundance and diversity of cornfield carabid communities to differences in farm practices'. *Ecology* 63: 900–904.

Ehrlich, P R (1982) Human carrying capacity, extinctions and nature reserves. *BioScience* 32: 331–333.

Ehrlich, P R and Mooney, H A (1983) 'Extinction, substitution and ecosystem services'. *Bioscience* 33: 248–254.

Erwin, T L (1983) 'Beetles and other insects of tropical forest canopies at Manaus, Brazil sampled by insecticidal fogging'. In Sutton, S L, Whitmore T C and A C Chadwick, eds. *Tropical Rain Forest: Ecology and Management*, Blackwell, Edinburgh. pp 55–79.

Gentry, A H (1982) Patterns of neotropical plant species diversity. *Evol Biol* 15: 1–84.

Guppy, N (1984) 'Tropical deforestation: a global view'. *Foreign Affairs* 62: 928–965.

Halffter, G (1981) 'Local participation in conservation and development'. *Ambio* 10: 93–96.

Harris, L D (1984) *The Fragmented Forest*. University of Chicago Press. 212 pp.

Hutchinson, G E (1959) 'Homage to Santa Rosalia, or why are there so many kinds of animals?' *Am Nat* 93: 145–159.

IUCN (1980) *The World Conservation Strategy*. International Union for the Conservation of Nature, Gland, Switzerland.

Janzen, D H (1986) 'Guanacaste National Park: tropical, cultural and ecological restoration'. San Jose: Editorial Universidad Estatal Distancia.

Johannes, R E (1984) 'Marine conservation in relation to traditional lifestyles of artisanal fishermen'. *Environmentalist* 4 (supplement 7): 30–35.

Johns, A D (1983) 'Tropical forest primates and logging – can they co-exist?' *Oryx* 17: 114–118.

Karr, J R (1982) 'Population variability and extinction in the avifauna of a tropical land bridge island'. *Ecology* 63: 1975–1978.

Kartawinata, K, H Soedjito, Jessup, T, Vayda, A P and Colfer, C P (1984) 'The impact of development on interaction between forests and people in East Kalimantan: a comparison of two areas of Kenyah Dayak settlement'. *Environmentalist* 4 (supplement #7): 87–95.

Kimmerer, W J (1984) 'Diversity/stability: a criticism'. *Ecology* 65: 1936–1938.

Kirkpatrick, J B (1983) 'An iterative method for establishing priorities for the selection of nature reserves: an example from Tasmania'. *Biol cons* 25: 127–134.

Kwapena, N (1984) 'Traditional conservation and utilization of wildlife in Papua New Guinea'. *Environmentalist* 4 (supplement 7): 22–26.

Leidy, R A and Fiedler, P L (1985) 'Human disturbance and patterns of fish species diversity in the San Francisco Bay Drainage, California'. *Biol Cons* 33: 247–267.

Liegel, K (1981) 'Restoring American prairie'. *Oryx* 16: 169–175.

Lovejoy, T E (1985) 'Rehabilitation of degraded tropical forest lands'. *Environmentalist* 5: 13–20.

Lovejoy, T E (1980) 'Discontinuous wilderness: minimum areas for conservation'. *Parks* 5: 13–15.

Lovejoy, T E and Bierregaard, R O, Rankin, J M and Schubert, H O R (1983) 'Ecological dynamics of tropical forest fragments'. In Sutton, S L, Whitmore, T C and A C Chadwick, eds *Tropical Rain Forests: Ecology and Management* Blackwell, Edinburgh. pp 377–384.

Lusigi, W J (1981) 'New approaches to wildlife conservation in Kenya'. *Ambio* 10: 87–92.

Lynch, J F and Whigham, D F (1984) 'Effects of forest fragmentation on breeding bird communities in Maryland, USA'. *Biol Cons* 28: 287–324.

MacArthur, R H (1964) 'Patterns of species diversity'. *Biol Rev* 40: 510–533.

MacArthur, R H (1955) 'Fluctuations of animal populations and a measure of community stability'. *Ecology* 36: 533–536.

MacArthur, R H and Wilson, E O (1967) *The Theory of Island Biogeography*. Princeton University Press, Princeton. 203 pp.

Margalef, R (1969) 'Diversity and stability: a practical proposal and a model of interdependence'. *Brookhaven Symp Biol* 22: 25–27.

Margules, C and Usher, M B (1981) 'Criteria used in assessing wildlife conservation potential: a review'. *Biol Cons* 21: 79–110.

May, R M (1984) 'Prehistory of Amazonian Indians'. *Nature* 312: 19–20.

May, R M (1973) *Stability and Complexity in Model Ecosystems*. Princeton University Press, Princeton. 162 pp.

McCoy, E D (1982) 'The application of island biogeographic theory to forest tracts: problems in the determination of turnover rates'. *Biol Cons* 22: 217–227.

Miller, T E (1982) 'Community diversity and interaction between the size and frequency of disturbance'. *Am Nat* 120: 533–542.

del Moral, R and Fleming, R S (1979) 'Structure of coniferous forest communities in western Washington: diversity and ecotope properties'. *Vegetatio* 43: 143–154.

Myers, N (1979) *The Sinking Ark*. Pergamon Press, Oxford. 307 pp.

Naveh, Z and Whittaker, R H (1979) 'Structural and floristic diversity of shrublands and woodlands of northern Israel and other Mediterranean areas'. *Vegetatio* 41: 171–190.

Nicholaides, J J *et al* (1985) 'Agricultural alternatives for the Amazon Basin'. *BioScience* 35: 279–285.

Ng, F S (1983) 'Ecological principles of tropical lowland rain forest conservation'. In Sutton, S L, Whitmore, T C and A C Chadwick, eds, *Tropical Rain Forest: Ecology and Management*. Blackwell, Edinburgh. pp 358–376.

Nilsson, S G (1978) 'Fragmented habitats, species richness and conservation practice'. *Biol Cons* 23: 86–95.

Noss, R F (1983) 'A regional landscape approach to maintain diversity'. *BioScience* 33: 700–706.

Oldfield, S (1987) *Buffer zone management in tropical moist forests: case studies and guidelines*. IUCN Tropical Forest Programme Paper No 5, Cambridge.

OTA (1984) *Technologies to Sustain Forest Resources*. Office of Technology Assessment, Washington, DC. 343 pp.

Ovington, J D (1984) 'Aboriginal People – guardians of a heritage'. *Environmentalist* 4 (supplement 7): 36–39.

Pearshall III, S H (1984) 'In absentia benefits of nature preserves: a review'. *Environ Cons* 11: 3–10.

Peters, R L and Darling, J D (1985) 'The greenhouse effect and nature reserves'. *BioScience* 35: 707–717.

Pianka, E R (1979) *Evolutionary Ecology* (Second Edition) Harper and Row, New York. 356 pp.

Pianka, E R (1966) 'Latitudinal gradients in species diversity: a review of the concepts'. *Am Nat* 100: 33–46.

Pickett, S T and Thompson, J N (1978) 'Patch dynamics and the design of nature reserves'. *Biol Cons* 13: 27–38.

Pielou, E C (1975) *Ecological Diversity*. Wiley, New York. 162 pp.

Prance, G T (ed) (1982) *Biological Diversification in the Tropics: Proceedings of the Fifth International Symposium of the Association for Tropical Biology*. Columbia University Press, New York. 714 pp.

Prescott-Allen, R and Prescott Allen, C (1982) *Genes from the Wild*. Earthscan, London. 102 pp.

Raven, P H (1985) 'Disappearing species: a global tragedy'. *Futurist* 19: 8–14.

Reichert, W (1982) 'Agriculture's diminishing diversity: increasing yields and vulnerability'. *Environment* 24: 6–11.

Roche, L (1979) 'Forestry and the conservation of plants and animals in the tropics'. *For Ecol Man* 2: 103–122.

Samson, F B and Knopf, F L (1982) 'In search of a diversity ethic for wildlife management'. *Trans 47th N Amer Wildl and Nat Res Conf:* 421–431.

Schemske, D W and Brokaw, N (1981) 'Treefalls and the distribution of understorey birds in a tropical forest'. *Ecology* 62: 938–945.

Shaffer, M L (1981) 'Minimum population sizes for species conservation'. *BioScience* 31: 131–134.

Shaffer, M L and Samson, F B (1985) 'Population size and extinction: a note on determining critical population sizes'. *Am Nat* 125: 144–152.

Schmida, A and Wilson, M V (1985) 'Biological determinants of species diversity'. *J Biogeog* 12: 1–20.

Simberloff, D (1982) 'Big advantages to small refuges'. *Nat Hist* 91 (4) 6–15.

Simberloff, D and Abele, L G (1982) 'Refuge design and island biogeographic theory: effects of fragmentation'. *Am Nat* 120: 41–50.

Simberloff, D and Gotelli, N (1984) 'Effects of insularization on plant species richness in the prairie-forest ecotone'. *Biol Cons* 29: 27–46.

Skorupa, J P and Kasenere, J M (1984) 'Tropical forest management: can rates of treefalls help guide us?' *Oryx* 18: 96–101.

Smedes, G W and Hurd, L E (1981) 'An empirical test of community stability: resistance of a fouling community to biological patch-forming disturbance'. *Ecology* 62: 1561–1572.

Stenseth, N C (1979) 'Where have all the species gone? On the nature of extinction and the Red Queen Hypothesis'. *Oikos* 33: 196–227.

Tangley, L (1985) 'A national biological survey'. *BioScience* 35: 686–690.

Terborgh, J and Winter, B (1983) 'A method for siting parks and reserves with special reference to Colombia and Ecuador'. *Bio Cons* 27: 45–58.

Terborgh, J and Winter, B (1980) 'Some causes of extinction'. In Soule M E and B A Wilcox, eds, *Conservation Biology*. Sinauer, Sunderland, Mass. pp 95–117.

Weinstock, J A (1985) 'Rattan: A complement to swidden agriculture'. *Unasylva* 36: 16–22.

Western, D and Henry, W (1979) 'Economics and conservation in Third World national parks'. *BioScience* 29: 414–418.

Whittaker, R H (1977) 'Evolution of species diversity in land communities'. *Evol Biol* 10: 1–87.

Wilcox, B A (1980) 'Insular ecology and conservation'. In Soule, M E and B A Wilcox, eds, *Conservation Biology*. Sinauer, Sunderland, Mass. pp 95–117.

Wilson, E O (1985) 'The biological diversity crisis: a challenge to science'. *Iss in Sci and Tech* 2: 20–29.

Wilson, W L and Johns, A D (1982) 'Diversity and abundance of selected animal species in undisturbed forest, selectively logged forest and plantations in East Kalimantan, Indonesia'. *Biol Cons* 24: 205–218.

Index

Index compiled by John Tooke